Petroleum Migration

Geological Society Special Publications
Series Editor J. BROOKS

GEOLOGICAL SOCIETY SPECIAL PUBLICATION NO 59

Petroleum Migration

EDITED BY

WILLIAM A. ENGLAND & ANDREW J. FLEET

Exploration and Production Division,
BP Research,
Sunbury-on-Thames, UK

1991

Published by

The Geological Society

London

THE GEOLOGICAL SOCIETY

The Geological Society of London was founded in 1807 for the purposes of 'investigating the mineral structures of the earth'. It received its Royal Charter in 1825. The Society promotes all aspects of geological science by means of meetings, special lectures and courses, discussions, specialist groups, publications and library services.

It is expected that candidates for Fellowship will be graduates in geology or another earth science, or have equivalent qualifications or experience. All Fellows are entitled to receive for their subscription one of the Society's three journals: *The Quarterly Journal of Engineering Geology*, *Journal of the Geological Society* or *Marine and Petroleum Geology*. On payment of an additional sum on the annual subscription, members may obtain copies of another journal.

Membership of the specialist groups is open to all Fellows without additional charge. Enquiries concerning Fellowship of the Society and membership of the specialist groups should be directed to the Membership Manager, The Geological Society, Burlington House, Piccadilly, London W1V 0JU.

Published by the Geological Society from:
The Geological Society Publishing House
Unit 7
Brassmill Enterprise Centre
Brassmill Lane
Bath
Avon BA1 3JN
UK
(*Orders*: Tel. 0225 445046)

First published 1991

A catalogue record for this book is available from the British Library

ISBN 0-903317-66-4

Distributor USA:
AAPG Bookstore
PO Box 979
Tulsa
Oklahoma 74101–0979
USA
Tel. (918) 584-2555

Distributor Australia:
Australian Mineral Foundation
63 Conyngham Street
Glenside
South Australia 5065
Tel. (08) 379-0444

Printed and bound by
The Cromwell Press Limited,
Broughton Gifford, Melksham, Wiltshire

Contents

Introduction

W. A. ENGLAND & A. J. FLEET

*Exploration and Production Division, BP Research, Chertsey Road, Sunbury-on-Thames,
Middlesex TW16 7LN, UK*

'Petroleum Migration' follows petroleum from its generation in source rocks through migration to the reservoir or the surface. The volume is divided into four parts. Part I deals with both the generation of petroleum by the thermal breakdown of kerogen and the expulsion (primary migration) of the petroleum from the source rock. Emphasis, however, is on understanding and quantifying expulsion driving forces and processes so that the composition and phase of petroleum from a given source rock can be predicted. Predicting the direction of expulsion is also critical. Upward or downward expulsion determines what migration route the petroleum enters, which traps it may reach and so the operative plays in the basin.

Part II considers secondary migration: the processes which control petroleum behaviour during its movement through relatively permeable carrier beds from the mudrock sequences, which contain source intervals, to the reservoir in the structural culmination of the carrier bed or other trap. Secondary migration may be negligible where expulsion is direct into the trap or operate over distances of several hundred kilometres (e.g. in the West Canada Basin). Typically migration distances are of the order of tens of kilometres. Processes occurring during secondary migration may modify the phase and composition of the expelled petroleum and so determine the nature of the petroleum charge reaching the trap. They may also lead to 'loss' (retention) of petroleum along the migration pathway and so govern how much petroleum arrives at the trap.

The case studies of Part III shows how understanding of generation, expulsion and secondary migration can be used to explain the distributions of oil and gas in a basin, and so, more importantly, be used to predict the nature of the petroleum in an undrilled prospect. Such predictions must take into account not only generation and migration but also post-accumulation alteration in the trap by, for instance, biodegradation or oil-to-gas cracking.

Last, but by no means least, the papers of Part IV highlight leakage from accumulations. Understanding leakage is important for two reasons. Firstly it must be understood if the significance of surface occurrences of petroleum, such as oil seeps, and near-surface phenomena, like shallow subsurface gas clouds, are to be understood and used in exploration. Secondly its influence must be appreciated quantitatively if volumes of petroleum involved in migration and entrapment are to be estimated.

Previous reviews: the identified problems

Previous reviews of petroleum migration have included two key collections of papers: the American Association of Petroleum Geologists' (AAPG) 'Problems of petroleum migration' (Roberts & Cordell 1980), and the Institut Français de Pétrole's (IFP) 'Migration of hydrocarbons in sedimentary basisn' (Doligez 1987). These, in turn, developed or challenged earlier individual contributions (e.g. primary migration: Chapman 1972; secondary migration: Schowalter 1979; phase behaviour: Price *et al.* 1983; hydrodynamics: Hubbert 1953; see Tissot 1987 for comprehensive reference list).

In introducing the AAPG papers, the editors, Roberts & Cordell (1980), suggested that the title of the volume 'might be taken as a confession of ignorance'. They went on to stress what they concluded was the conceptual/conjectural nature of understanding on petroleum migration; an understanding 'based on interpretation of factual observations, sponsored and enhanced by imagination'. No consensus emerged from the AAPG volume. Roberts & Cordell found 'honest agnosticism' among the authors. The topics covered, like those in this 'Petroleum migration' volume, range from primary migration through secondary migration to surface seepage. The papers on primary migration focus on the mechanisms. Primary migration in aqueous solution is considered improbable (McAuliffe) with migration occurring either as a molecular film/interaction over kerogen surfaces or between rock grains and pore fluids (McAuliffe; Hinch) or in a pore-centre oil filament (Barker). The forces proposed for driving primary migration arise from compaction, thermal expansion of water, clay-mineral conversion and osmosis (Magara;

From England, W. A. & Fleet, A. J. (eds), *Petroleum Migration*
Geological Society, Special Publication No. 59, pp. 1–6.

1

Barker). Ideas presented on secondary migration in the AAPG volume centre very much on the role of water, either carrying petroleum in solution or driving it along hydraulically. McAuliffe and Jones each, however, argue that transport in water solution is unlikely and that oil and gas migrate in liquid or gas petroleum phases.

The thirty or so papers in IFP's 'Migration of hydrocarbons in sedimentary basins' (Doligez 1987) reflect the burgeoning of interest in migration which occurred through the mid-1980s. In particular they reflect the desire not just to understand migration but to attempt to model it quantitatively. In his introductory historical perspective Tissot argues firmly for petroleum-phase expulsion from source rocks as a result of pressure-drive. Secondary migration is seen as occurring in a separate phase from the water which saturates the carrier beds and as being buoyancy-driven (e.g. Lehner et al.). Various authors (e.g. Welte; Ungerer et al.; Bethke & Corbet; Nakayama & Lerche) advance their ideas of how to build on these basic concepts so as to be able to model fluid flow in sedimentary basins. Such modelling, though, will only be of use for determining the petroleum charge to a trap, if our quantitative, rather than qualitative, understanding of migration is enhanced. In their reviews for the IFP meeting, Tissot and Demaison identified major problems as follows.

(1) Quantitative estimation of expulsion: what controls do sedimentary fabric (intra- and inter-bed) and mineral matrix exert? What factors determine if expulsion is in the liquid or gas phase?

(2) Quantitative estimation of migration efficiency: how are losses from the expelled petroleum during secondary migration, through residual saturation or formation of sub-economic accumulations, accounted for? What are the implications of two-phase flow (liquid or gas and water) turning into three-phase flow (liquid and gas and water) during secondary migration? How may other (probably gas) phases affect migration?

To these Durand (1988), in a separate review, added the need for quantitative understanding of petroleum phase behaviour under subsurface conditions. In addition to the need for these technical requirements, Demaison highlighted the need for better communication between researchers and specialists and exploration personnel. He noted that advances in understanding of migration made over the last decade had often not been communicated to explorers and, in a number of instances, had not been taken up by petroleum geology text books. Unless these gaps are closed, he concluded, 'the more advanced computer-aided methods, which aim at quantification of migration processes, may encounter skepticism or, worse, indifference at industry management levels'.

It was against this background of the need for communication and of the key problems listed above that the Geological Society's 'Petroleum Migration' meeting was arranged. We summarize below how the papers which arose from the meeting approach those key problems and attempt to highlight the main conclusions of the specialist workers for the general geoscientist.

Part I: Generation and expulsion (primary migration)

If an understanding of petroleum migration is to be used to predict the nature of the petroleum charge reaching a trap from a source-rock 'kitchen', the first step is to assess how petroleum generation and expulsion proceeds in a given source rock with increasing maturity (i.e. increasing time and temperature due to burial). The timing of generation depends on the nature of the source rock and the rate at which it is heated. Various approaches have been taken to model generation ranging from those which attempt to simplify the source-rock system as far as possible (e.g. Quigley & Mackenzie 1988) to those who tend to treat each source rock individually (e.g. Ungerer 1984; Burnham et al. 1988). The former can lead to oversimplification, for instance lack of distinction of early generation from carbonate source rocks, while the latter may not be directly applicable to 'frontier' areas where no source rocks samples are available. In such 'frontier' areas any modelling of generation has to assume the type of source rock present based on geological interpretation. The type of petroleum generated (liquid versus gas) and its composition (gas : oil ratio; condensate : gas ratio; sulphur and wax contents etc.) depends on the type of kerogen within the source rock and the stage of generation (maturity) of the source rock.

The nature and amount of the generated petroleum which leaves the source rock depends on the expulsion behaviour of the source rock. This must in turn depend on the mechanism, or pathway, followed by the petroleum and the behaviour of the individual compounds which make up the petroleum. The mechanism, and especially the efficiency of the expulsion, can have a profound influence on the hydrocarbons released. This is because unexpelled oil (retained in the source rock) will be thermally broken down to more easily expelled gas if the source rock is subsequently exposed to higher temperatures.

The first three papers in Part I of the volume present approaches to modelling expulsion behaviour. **Pepper** and **Duppenbecker et al.** each base their models on different expulsion processes but find good agreement between their results and observations. **Pepper** argues that the kerogen itself is probably the prime control on expulsion behaviour. He suggests that unreactive kerogen retains, by adsorption, petroleum generated from the reactive portion of the kerogen, so that adsorption tends to increase with decreasing hydrogen index (S_2/TOC). In taking this approach **Pepper** argues against significant expulsion of petroleum through the pore volume of the source mudrock once some critical level of saturation of the pores has been reached. In contrast, **Düppenbecker et al.** base their model on expulsion through this pore network and through microfractures which they claim develop in a predominantly bed-parallel orientation during overpressuring of the source rock as a result of compaction disequilibrium and volume expansion due to petroleum generation. At first sight, the difference in expulsion processes favoured by **Pepper** and **Düppenbecker et al.** does nothing to resolve current uncertainties about how petroleum gets out of source rocks. The two papers may, however, point to a number of controls operating on expulsion, the dominance of any particular one depending on the source rock. For instance, adsorption onto kerogen, as in **Pepper's** model, may only become critical in source rocks with relatively moderate or low hydrogen indices (e.g. of <300–400). Only further careful calibration and testing of competing models is likely to resolve how different approaches can be applied effectively.

Leythaeuser & Poelchau consider the expulsion of different hydrocarbons rather than overall expulsion of petroleum. They focus on Type III kerogen which yields petroleum products in the gas phase. They show how temperature, pressure and supply of generated gas determine the nature of the expelled petroleum, explaining how Type III kerogen in overpressured source rocks can yield light oils or gas-condensates.

Overpressure, as discussed above, is widely accepted as the driving mechanism of petroleum expulsion. Estimation of overpressure is, therefore, a critical step in modelling petroleum migration (e.g. see Burrus et al. this volume; England et al. 1987; Nakayama & Lerche 1987; Bethke et al. 1988; Mann & MacKenzie 1990). **Mudford et al.** examine the development of fluid pressure in a compacting basin using an 1D model. They identify the problematical question of shale permeability as critical to any model, and, using the Scotian Basin as an example, highlight the need to identify causes of overpressure, other than compaction, which may be significant in individual basins.

Part II: Secondary Migration

Secondary migration of petroleum through a carrier bed from a source rock, or source-interval containing mudrock, to a reservoir is driven mainly by the buoyancy which results from the density contrast between the petroleum and the water which saturates the carrier bed (e.g. Schowalter 1979). If there is hydrodynamic flow, it may act with buoyancy or against it. The main resistive force opposing buoyancy is capillary pressure which is a function of the pore-throat radii of the carrier bed, petroleum–water interfacial tension and wettability (e.g. Schowalter 1979). Quantitative estimates of buoyancy, hydrodynamic flow and capillary pressure, therefore, must be key inputs to numerical basin-scale fluid flow models. Production of such models, as mentioned above, was a major development of the 1980s. One group of papers in Part II of 'Petroleum Migration' look beyond the models themselves at the feasibility of applying the models in real situations and at constraining the model input parameters. Others develop ideas for constraining phase behaviour during migration.

Burrus et al. go straight to the heart of the matter and ask the question 'are numerical models useful in reconstructing the migration of hydrocarbons?' They consider a case study on the Northern Viking Graben using IFP's THEMIS (TEMISPACK) program in coming to their answer. They find that integrated modelling, constrained by available data (e.g. maturity, pressure), can reconstruct well the 'dynamic geological framework' of basin (i.e. burial history, petroleum generation, overpressure). It is less able, however, to handle definitively migration phenomena which require knowledge of capillary pressures, relative permeabilities and petroleum fluid properties; parameters which are poorly constrained, particularly at the basin level. **Burrus et al.** conclude that modelling alone cannot at present provide quantitative estimates of the volumes of petroleum in structures. What it can do is allow assessment of various migration scenarios and their sensitivities to the 'flow' parameters which are poorly constrained.

Verbal contributions to the 'Petroleum migration' meeting, which were not offered as papers to the volume, echo the conclusions of **Burrus** *et al*. Schowalter, in his abstract for the meeting, pointed out that the physical principles of secondary migration are well known but that quantitative estimates of the controlling phenomena on a basin scale are very difficult. As a consequence he suggested that quantitative geologic modelling of migration losses and trap efficiency is 'complex to the point of being insolvable. Qualitative estimation of migration charge and trap efficiency should be adequate in many risk assessment situations'. Lehner, in his abstract for the meeting, also highlighted the dependence of models on input data such as permeability of the source and carrier beds. He concluded that 'an important use of migration models lies in testing the quantitative consequences of different assumptions or alternative hypotheses about poorly constrained data'.

Sylta takes modelling of secondary migration a step further by including modelling of phase behaviour using an equation of state of a multi-component hydrocarbon mixture. He overcomes the problems discussed above relating to flow parameters by considering a synthetic basin. He presents his results for the accumulations in his hypothetical prospects in the light of how sensitive they are to factors such as expulsion efficiency and loss during secondary migration. His approach is best applied to areas where the composition of petroleum in some traps is known. This information can then be used to constrain the modelling by iteratively inputing the best values for 'flow' parameters and, thus, using the model to predict the nature of petroleum in undrilled prospects.

Another way of constraining a basin-scale parameter is suggested by **Chapman** *et al*. They analyse perturbations in heat flow across basins which they ascribe to advection resulting from fluid flow. Their results allow them to infer basin-scale permeabilities.

The role which hydrodynamics can play in secondary migration and modelling is highlighted by **Davis**. He emphasizes that, although hydrodynamic analysis has largely been applied to individual traps, it can also be applied at the basin-scale. As with petroleum migration modelling, hydrodynamics modelling is sensitive to the geological inputs which describe the subsurface flow paths. However, if a hydrodynamic regime can be well described from measurements, its characteristics can themselves be used to constrain geological parameters for input into modelling programs.

Understanding the processes which occur during secondary migration is not only important for predicting the probable nature of petroleum accumulations in prospect but also critical for interpreting the composition of petroleum samples recovered by drilling or from seeps. One aspect is to understand how individual components fractionate between petroleum phases when gas and liquid phases separate. **Larter & Mills** report some initial results on the behaviour of certain compounds which are used in making geochemical interpretations of the source and maturity of petroleums. Their results relate to just one fractionation step and show small effects, but they emphasise that multiple phase separations during migration could lead to more significant fractionations. In particular they question the interpretation of gas condensates using 'traditional' geochemical methods. **Kroos** *et al*. consider other processes, collectively referred to as geochromatography, which may affect the compositons of petroleums during migration. They define it as including the gas–liquid fractionation effects discussed by **Larter & Mills** but also comprehensively include liquid–solid and gas–solid fractionations.

Part III: Case studies

Many of the papers in Parts I and II of 'Petroleum migration' include case studies against which ideas have been tested or calibrated. The case studies in Part III relate to geographically discrete petroleum systems which the authors discuss and interpret in terms of petroleum migration.

Barnard & Bastow review the generation, migration and alteration processes which they believe account for the petroleum accumulations of the Central and Northern North Sea. Their approach is to assess the timing and extent of petroleum generation and expulsion throughout the area, and to interpret the resulting secondary migration in the light of the distribution of known accumulations. They favour faults as major conduits for petroleum moving from the Jurassic to Tertiary reservoirs. Their explanation of altered accumulations involves degradation during migration.

Thompson's case study of petroleums from the Gulf of Mexico focuses on the different processes which have affected oils in Tertiary clastic and Mesozoic carbonate reservoirs. Thompson presents detailed arguments for biodegradation and fractionation between the oil and a latter gas charge determining the compositon of the Tertiary oils, and for water washing affecting the Mesozoic oils.

Piggott & Lines review the vast petroleum system of the West Canada Basin. They describe three discrete 'hydrocarbon cells' made up of Devonian, Mississippian–Neocomian and mid-Upper Cretaceous strata each of which were sourced from intervals within the cell. Regionally lateral migration over hundreds of kilometres dominated petroleum accumulation. Locally, however, vertical migration has played a part and **Piggott & Lines** present an example where phase behaviour was critical in determining the variation in petroleum type across one area.

The final two case studies focus on the role of tectonism in migration. **Butler** presents a study through the Arnecy district of the French Alps which examines petroleum generation and migration in relation to thrusting. **Roberts** focuses down from this regional view to consider fluid migration along a thrust during faulting.

Part IV: Trap Leakage and subsequent migration.

Any assessment of petroleum migration in a basin must attempt to take account of leakage to the surface and loss of petroleum from traps. This is particularly true of any attempt to balance quantitatively expelled volumes against accumulated volumes and losses. Similarly any interpretation of petroleum seeps at the surface in terms of subsurface accumulations, migration pathways etc. must appreciate how the seeps relate to petroleum migration through the basin.

Clarke & Cleverly discuss the variety of forms which seeps take on the basis of a review of about 6000 seeps. They consider the near-surface processes which lead to the dispersal and alteration of seeps particularly in sub-aerial environments. Finally they emphasize the uncertainty which underlies our knowlege of seep occurrence and the need for better methods of seep detection.

Vik et al. look at gas leakage from deep reservoirs using the Haltenbanken area, mid Norwegian shelf, as an example. This leakage has resulted from severe overpressuring in Jurassic reservoirs and consequent seal failure of the cap rocks. Subsequent lateral migration in the Cretaceous and Tertiary has displaced the gas laterally so that no direct link can be expected to be recognized between shallow gas occurrences and the deep reservoir.

The outstanding problems

Above we listed the problems which keynote papers in previous reviews of petroleum migration had identified. These centred on:

> quantitative estimation of the generation and expulsion

> quantitative estimation of secondary migration efficiency (including the ability to handle loss of petroleum by leakage to surface)

> quantitative understanding of subsurface petroleum phase behaviour

> better communication of migration understanding to explorers.

The studies presented in 'Petroleum Migration' suggest that all of the first three topics are areas of active concern but the problems are still very much current. Enhanced understanding of processes and model development will help further progress but little is likely to be achieved without rigorous calibration and testing. A refreshing aspect of the 'Petroleum Migration' meeting was the very practical approach taken by all to modelling particularly of secondary migration. Modelling was presented as a tool to help weigh up choices and test the sensitivities of different hypotheses to different input parameters, not as a panacea. Such an approach will hopefully help to foster communication between specialists and multi-disciplinary explorers and engender progress on understanding petroleum migration across the total spectrum of industry-academia.

We would like to thank all who participated in the 'Petroleum Migration' meeting and the staff of the Geological Society who ensured its smooth running. We would also thank especially the contributors to this volume, including those who reviewed the papers, for their timeliness and patience or perseverance. Last, but definitely not least, we thank Angharad Hills at the Geological Society Publishing House, for her guidance and assistance in producing this volume.

References

BURNHAM, A. K., BRAUN, R. L. & SAMOUN, A. M. 1988. Further comparison of methods for measuring kerogen pyrolysis rates and fitting kinetic parameters. *In*: MATTAVELLI, L. & NOVELLI, L. (eds) *Advances in Organic Geochemistry 1987, Organic Geochemistry*, **13**, (4–6), Pergamon Press, 839–845.

BETHKE, C. M., HARRISON, W. J., UPSON, C. &

ALTANER, S. P., 1988. Supercomputer analysis of sedimentary basins. *Science*, **239**, 261–267.

CHAPMAN, R. E. 1972. Primary migration of petroleum from clay source rocks. *Bulletin of the American Association of Petroleum Geologists*, **56**, 2185–2191.

DOLIGEZ, B. (ed.) 1987. *Migration of hydrocarbons in sedimentary basins*. Editions Technip, Paris.

DURAND, B. 1988. Understanding of HC migration in sedimentary basins (present state of knowledge). *In*: MATTAVELLI, L. & NOVELLI, L. (eds) *Advances in Organic Chemistry 1987, Organic Chemistry*, **13** (4–6), Pergamon Press, 445–459.

ENGLAND, W. A., MACKENZIE, A. S., MANN, D. M. & QUIGLEY, T. M. 1987. The movement and entrapment of petroleum fluids in the subsurface. *Journal of the Geological Society, London*, **144**, 327–347.

HUBBERT, M. K. 1953. Entrapment of petroleum under hydrodynamic conditions. *Bulletin of the American Association of Petroleum Geologists*, **37**, 1954–2026.

MANN, D. M. & MACKENZIE, A. S. 1990. Prediction of pore fluid pressures in sedimentary basins. *Marine and Petroleum Geology*, **7**, 55–65.

NAKAYAMA, K. & LERCHE, I. 1987. Two-dimensional basin analysis. *In*: DOLIGEZ, B. (ed.) *Migration of hydrocarbons in sedimentary basins*. Editions Technip, Paris, 597–611.

PRICE, L. C., WEGNER, L. M., GING, T. & Blout, C. W. 1983. Solubility of crude oil in methane as a function of pressure and temperature. *Organic Geochemistry*, **4**, 201–221.

QUIGLEY, T. M. & MACKENZIE, A. S., 1988. The temperatures of oil and gas formation in the subsurface. *Nature*, **333**, 549–552.

ROBERTS, W. H. III & CORDELL, R. J., 1980. *Problems of petroleum migration*. The American Association of Petroleum Geologists Studies in Geology, **10**.

SCHOWALTER, T. T. 1979. Mechanics of secondary hydrocarbon migration and entrapment. *Bulletin of the American Association of Petroleum Geologists*, **63**, 723–760.

TISSOT, B. 1987. Migration of hydrocarbons in sedimentary basins: a geological, geochemical and historical perspective. *In*: DOLIGEZ, B. (ed.) *Migration of hydrocarbons in sedimentary basins*. Editions Technip, Paris, 1–19.

UNGERER, P., 1984. Models of petroleum formation: how to take into account geology and chemical kinetics. *In*: DURAND, B. (ed.) *Thermal phenomena in sedimentary basins*. Editions Technip, Paris, 235–246.

PART I:

GENERATION AND EXPULSION (PRIMARY MIGRATION)

Estimating the petroleum expulsion behaviour of source rocks: a novel quantitative approach

ANDREW S. PEPPER

BP Research, Chertsey Road, Sunbury-on-Thames, Middlesex, UK
Present address: BP Exploration Operating Company Ltd, PO Box 2749, Jakarta 16001, Indonesia

Abstract: Estimates of potential volume, composition and phase of petroleum charges to traps are an integral part of petroleum prospect evaluation. Such estimates are directly sensitive to the timing and efficiency of petroleum expulsion from the source rock.

To date, geochemists have adopted two approaches to prediction of expulsion behaviour. The first is an heuristic approach based on a limited number of case studies. The second, a theoretical approach based on extrapolation of permeability concepts used in reservoir engineering, has popular appeal but may be of limited application to the pores of fine-grained petroleum source rocks, where the behaviour of interstitial water is rather different. Consequently the processes controlling expulsion are still a subject for debate.

It remains possible to challenge established relative permeability models, and to argue that the concepts of petroleum saturation and discrete-phase flow are either unimportant in the expulsion process, or else involve such low concentrations of petroleum within the pore volume that they can be considered negligible. Instead, kerogen itself may exert the prime control on expulsion behaviour by retaining (adsorbing) generated petroleum. If so, a simple quantitative model can be devised to predict expulsion efficiency of both oil and gas as a function of the initial kerogen composition (proportions of petroleum-generative versus petroleum-retentive kerogen).

Expulsion behaviour modelled in this way compares favourably with previously observed behaviour of a wide variety of source rocks. In particular, a case history presented demonstrates the capability of the model in predicting the differences in the expulsion behaviour of different coals, and the consequences in terms of composition of the expelled products.

Petroleum expulsion from source rocks is one of the most important, yet at the same time poorly understood, subsurface processes which must be quantified in the geochemical evaluation of petroleum prospects (e.g. Mackenzie & Quigley 1988). The most obvious way in which expulsion processes impact on exploration is in controlling the bulk mass of petroleum expelled (M_e) from a source rock, which can be written:

$$M_e = P^0 \times \text{PGI} \times \text{PEE} \qquad (1)$$

where PGI is the fraction of petroleum potential generated and PEE is the fraction of the generated petroleum expelled. Both are fractional and dimensionless so that M_e has the same units used to quantify the initial petroleum potential (P^0), e.g. kg kg^{-1}, kg tonne^{-1}, kg m^{-3}, as appropriate to the exercise in hand.

However, the strong influence that expulsion behaviour exerts on the overall composition of the expelled petroleum and hence ultimately its subsurface phase (England *et al.* 1987) is less frequently appreciated. (In this paper, composition refers to the mass proportions of gas (C_{1-5}) and oil (C_{6+}) molecules in the petro-

leum). Figure 1 considers the case of a source rock containing wholly oil-generative kerogen, which has been matured by burial first through the 'oil window' and then through the 'gas window', i.e. a source rock which would conventionally be ranked as now 'overmature'. If a trap of sufficient size formed before the onset of expulsion (i.e. a trap which has been able to collect all of the petroleum expelled) the ultimate petroleum composition in that trap will be determined by the relative proportions of:

the mass of oil-rich (low GOR) petroleum expelled during passage through the 'oil window';

the mass left behind as feedstock for cracking and expulsion as gas-rich (high GOR) petroleum during passage through the gas window.

Since oil expulsion from some source rocks is recognized to be highly efficient (Cooles *et al.* 1986; Larter 1988; Rullkotter *et al.* 1989) it follows that the overall mass balance of expulsion products from some 'overmature' source rocks may be dominated by the oil generated

From England, W. A. & Fleet, A. J. (eds), *Petroleum Migration*
Geological Society, Special Publication No. 59, pp. 9–31.

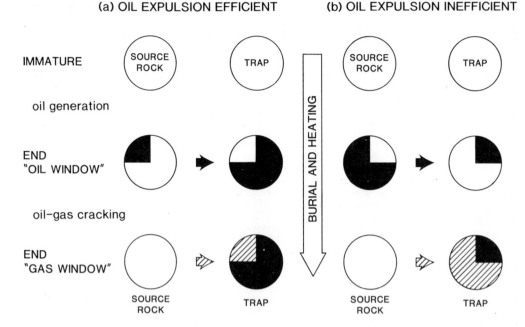

Fig. 1. Cartoon showing conceptual behaviour of two oil-generative source rocks which differ only in their oil expulsion efficiencies. Source rock (**a**) expels much of its oil at low thermal stress; little gas is available for expulsion following oil–gas cracking (an example of a Class 1 source rock *sensu* Mackenzie & Quigley 1988). In contrast, the inefficient expeller (**b**) holds back much of the oil it generates, so that a large feedstock is available for cracking and expulsion as gas-rich petroleum (an example of a Class 2 source rock *sensu* Mackenzie & Quigley 1988). At equally high levels of thermal stress, the cumulative mass balance favours an oily composition expelled from rock (**a**), compared to the gassy composition expelled from rock (**b**).

during the 'oil window' (Fig. 1a). The consequence for subsurface predictions is that low GOR 'black oil' accumulations may be the end-product of expulsion from one source kitchen (Fig. 1a), whilst high GOR 'gas-condensate' pools can be charged from another at *the same 'overmature' level of thermal stress* (Fig. 1b). Thus it is largely meaningless to evaluate the likely petroleum composition and phase to be encountered in a prospect based simply on whether the source rock in the 'oil or gas window' (conventionally these windows are often determined using vitrinite reflectance thresholds). Similarly, elaborate geothermal–geochemical modelling studies are of little practical value in exploration if they fail to account for expulsion (e.g. Yalcin & Welte 1988; Yalcin *et al.* 1987).

Whilst it is recognized almost universally that petroleum expulsion is influenced by a wide range of geological and geochemical factors, satisfactory treatment of this seemingly intractable problem clearly lies in reducing the problem to a manageable level of complexity. All too often geochemists construct elaborate determin-

istic models for which the input data are then assumed or inferred by analogy with other systems. In contrast, the pragmatic approach adopted below shares a common philosophy with that behind previous work published by BP Research (Cooles *et al.* 1986; Quigley *et al.* 1987; England *et al.* 1987; Quigley & Mackenzie 1988; Mackenzie & Quigley 1988).

After a brief review of the current state of knowledge, this paper goes on to develop a novel quantitative scheme directed at better quantifying the expulsion process. The model has been designed around these previously published approaches; familiarity with the methodologies of Cooles *et al.* (1986) and Mackenzie & Quigley (1988) is assumed in the quantitative aspects of this paper.

The final case study shows how this new approach predicts expulsion efficiencies within a laterally variable suite of coal facies, allowing improved understanding of the variation in both reservoired petroleum compositions and subsurface phases within an Australasian petroliferous basin.

Current wisdom

There are currently two basic types of approach to predicting expulsion efficiencies of source rocks in the subsurface. The first is heuristic: rules of thumb are derived from the results of a number of case studies where mass balance calculations are used to compare mature source rocks with their immature equivalents. The second is deterministic: such approaches are forced to assume that the pore system of the source rock behaves in a similar fashion to that of a petroleum reservoir, allowing the flow of petroleum fluids through the pore network to be described using Darcy's law. Two-phase flow is modelled using the concept of relative permeability, again as applied in petroleum reservoir engineering.

Heuristic approach

The carbon mass-balance approach of Cooles *et al.* (1986) adopted elsewhere in various modified forms (e.g. Larter 1988; Rullkotter *et al.* 1989), has shown convincingly that petroleum expulsion from some source rocks can be very efficient. Petroleum expulsion efficiency (PEE) can reach 80–90% during oil generation. The mass balance calculations of Cooles *et al.* (1986) involved comparison of thick sections of source rock, thus avoiding the effect of local capillary forces which are known to promote petroleum expulsion within a few metres of source rock–carrier bed boundaries (e.g. Mackenzie *et al.* 1987; Leythaeuser *et al.* 1988a, b).

One of the most significant conclusions drawn from the results of this study was based on an apparent dependency between PEE and the initial petroleum potential, P^0 (Fig. 2, taken from Mackenzie & Quigley 1988). In terms of conventional geochemical screening measurements, P^0 is the initial (S1 plus S2) yield from Rock-Eval type pyrolysis of immature organic-matter (Espitalie *et al.* 1977). Their 'rules of thumb' derived from Fig. 2 can be summarized as follows.

> Petroleum expulsion (during oil generation) from rocks whose P^0 exceeds 0.005 kg kg^{-1} (5 kg tonne^{-1}) is efficient. PEE is commonly in the region 0.6–0.8 (60–80%).

> Expulsion from rocks with P^0 less than 0.005 kg^{-1} (5 kg tonne^{-1}) is inefficient. PEE is less than 0.4 (40%).

Mackenzie & Quigley (1988) used these observations to support a causal link between the

expulsion efficiency and P^0: the richer the source rock, the larger will be the excess of petroleum available for expulsion following saturation of primary migration pathways in the rock (Durand 1983).

As will be seen later in this paper, the empirical basis of such an heuristic approach is its major limitation: empirical trends may not be extrapolated with confidence outside the range of data on which they were calibrated.

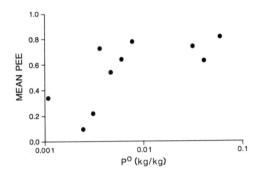

Fig. 2. Mean Petroleum Expulsion Efficiency (PEE), calculated between Petroleum Generation Index (PGI) limits 0.48 and 0.8, as a function of initial petroleum potential (P^0) for source rocks. From Mackenzie & Quigley (1988).

Deterministic approach

Two-phase flow (petroleum and water) in productive petroleum reservoirs is simulated using Darcy's law and the concept of relative permeability. Stated simply this means that, in a two-phase flow regime, the subordinate phase will not flow unless it occupies some critical volume of the pore space. Figure 3 shows some typical relative permeability curves for water-wet pore systems in different reservoir lithologies. Typically, flow of the petroleum (in this case, oil) phase and the pore water phase is equal when the petroleum phase occupies about a third of the pore volume.

These ideas, developed for reservoir rocks, have become increasingly popular in modelling fluid expulsion from source rocks (Durand 1983; Durand & Paratte, 1983; Durand *et al.* 1987), and they now lie at the heart of many of the so-called 'basin modelling' packages such as THEMIS (Ungerer *et al.* 1984; Doligez *et al.* 1986). However, the 'Achilles heel' of such an approach is the lack of relative permeability data for fine grained (source) rocks, since relative permeability is an empirically derived relationship having no strict theoretical foundation

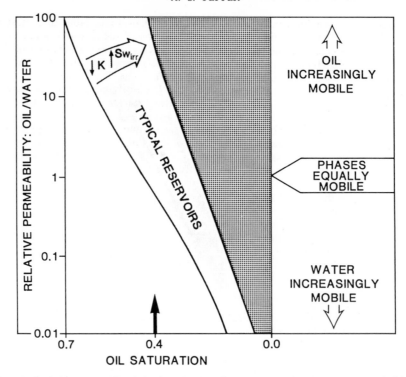

Fig. 3. Range of relative permeability behaviour for the oil–water system in water-wet reservoir lithologies (light toned area). Note the wide range in behaviour: curves are shifted clockwise with decreasing absolute permeability (K) and increasing irreducible water saturation (Sw$_{irr}$). At oil saturations of 40% (suggested, for example, by Mackenzie & Quigley (1988) as the critical threshold saturation for expulsion from potential source rocks) the relative permeability to oil varies by two orders of magnitude, depending on the petrophysical properties of the reservoir! No relative permeability data are available for source rock lithologies; however, due to their low permeability and high irreducible water saturation (bound water), it may be speculated that their relative permeability curves should be situated far to the right of the reservoir curves (dark toned area), i.e. oil will flow preferentially to water at low oil saturations, relative to the total pore space.

(Ungerer *et al.* 1984). This problem is readily appreciated when reviewing the literature, where estimates of the volume of pore space that might be occupied in discrete phase petroleum flow range from as little as 1% (Dickey 1975) to effectively 100% (Leythaeuser *et al.* 1987), i.e. over two orders of magnitude! Furthermore, the much more complex situation of three-phase flow (oily, gaseous, aqueous fluids) remains to be simulated.

Void space and permeability in fine-grained rocks

In order to reconcile this diversity of opinion in the literature, it is necessary to think carefully about the type of void system present in fine grained rocks, and whether the petroleum reservoir represents a suitable analogue.

Pore system. The pore system of an unfractured source rock differs from that of a reservoir rock in two important respects.

(1) Pores, pore-throats and absolute permeabilities are many, many orders of magnitude smaller

(2) The mineral matrix is rich in clay minerals which present a much larger surface area to volume ratio compared to detrital quartzo-feldspathic or carbonate grains. A much greater proportion of the pore fluid is in contact with the pore surfaces, compared to even the most clay-rich of reservoir lithologies. This is true also of most 'carbonate' source rocks, which commonly occur in low energy calcareous facies containing significant detrital clays (e.g. Brosse *et al.* 1988).

The immense surface area to volume ratio of clay pore systems provides sites for adsorbed water, often referred to as 'bound' or 'structured' water. The precise definition and quantification of bound water volumes is again a source of much discussion and debate in the literature (Low 1976; Barker 1980; Hinch 1980; Magara 1980; Tissot & Welte 1984). Clay minerals on deposition probably have many layers, but with progressive compaction and burial these are removed as more and more of the pore water is expelled. Honda & Magara (1982), demonstrated this in a quantitative experimental study of Japanese Tertiary mudstones, showing that all the water is structured in rocks with porosities lower than 15%. Note that most mature potential source rocks would satisfy this criterion. This is consistent with the view of Talukdar et al. (1988), who suggested that the pore network of La Luna source rocks in the Maricaibo Basin is effectively filled by bound water during petroleum generation. The consensus on bound water appears to be that two molecular layers are present, and that these occupy most, if not all, of the pore space in source rocks at the depth/pressure/temperature range over which petroleum is generated.

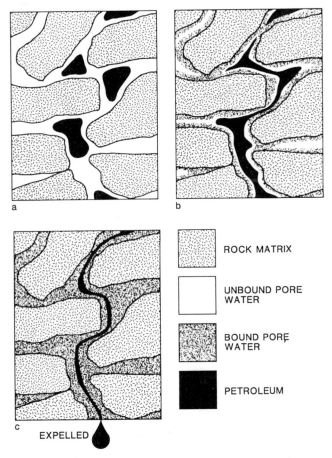

Fig. 4. Importance of bound water in petroleum expulsion. Cartoons a, b and c show potential source rock pore systems with identical pore volume, identical volumes of petroleum generated, but with different proportions of bound water in each case. (a) None of the pore water is bound. The petroleum concentration is insufficient for continuous filaments to form, resulting in globules of petroleum in the pores, unable to move through the interconnecting pore throats due to opposing capillary forces. Petroleum expulsion can not occur: relative permeability to the petroleum phase is zero. Only water is expelled. (b) Some of the pore water is bound. The same petroleum concentration as in (a) now occupies a large proportion of the 'free' pore space: the globules coalesce to form continuous filaments and, although only pore water has been expelled so far, expulsion of the petroleum phase may now begin. (c) Effectively all the pore water is bound. The petroleum volume (still the same as in (a) and (b) readily forms continuous filaments, and relative permeability to the petroleum phase is high. A large proportion of the petroleum available is actually able to be expelled (expulsion efficiency is high).

The implications for discrete-phase expulsion of petroleum was first discussed by Dickey (1975), although these appear to have been largely neglected in the 15 years since: the more pore volume occupied by bound water, the smaller the volume of petroleum required to attain saturation sufficient for discrete-phase flow, since only the remaining 'free' pore volume is experienced by the petroleum fluid phase. Figure 4 illustrates this in cartoon form. The reason for the widely differing views on critical oil saturations (1–100%) discussed above, becomes apparent. In fact our level of understanding of pore fluid behaviour is sufficiently low that it is still possible to question whether flow through the pore network is the dominant pathway in petroleum expulsion (Stainforth 1988; Stainforth and Reinders, 1989; Wang & Barker 1989).

Fracture system. The potential role of fractures deserves brief comment. Is it possible that fracture networks provide the avenues which must be saturated by petroleum attempting to migrate out of the source rock? The closest analogues to the source rock in terms of pore size and permeability are probably the fractured chalk reservoirs such as those in the North Sea Basin. Where matrix permeabilities are low, natural fractures enhance significantly the rates of petroleum flow and drainage of the reservoir. However, the actual volume of the reservoir occupied by fractures is almost negligible. For example, the fracture network in the Skjold reservoir, Danish North Sea, represents a fractional porosity of 0.001 or 0.1% (Oen et al. 1986). Saturation of such a small portion of rock volume is unlikely to present a significant impediment to petroleum expulsion. Thus, in the final analysis, it is rather unhelpful to draw a distinction between pores and fractures: it can be argued that neither present a significant volumetric impediment to petroleum flow through the source rock.

Experimental data versus extrapolation. Unfortunately, unless a method is devised to measure accurately the relative permeabilities of source rocks to two or three fluid phases, and this seems unlikely at present, present expulsion models will continue to rely on data borrowed from reservoirs, or at least *extrapolations* of such data. Extrapolating the observed trends in reservoir rock behaviour (Fig. 3), it seems conceivable that minimal concentrations of petroleum (compared to the *total* pore space) might be required

to initiate bulk petroleum flow in such fine-grained rocks.

Perhaps data from environmental studies into soil behaviour will provide a closer analogue to compacted clay-rich source rocks. Fernandez & Quigley (1985) performed sequential permeation experiments on a water-wet clay soil samples with porosities of around 50% and permeabilities of around 10^{-5} Darcy. They observed that simple hydrocarbons occupied only about 10% of the pore space in passing through the soils. This is much lower than the critical saturations typical of oil reservoirs (as suggested in Fig. 3). Over the petroleum generation window source rocks typically have porosities an order of magnitude lower, and permeabilities several orders of magnitude lower still, than the studied soil: it follows that critical saturations at the *lower end* of the 1–10% range suggested by Dickey (1975) are actually quite realistic.

Geophysical log data. There are further data which need to be included in this discussion. Wireline geophysical logs are a source of evidence much cited in favour of the pore saturation concept. Since the early observations of Murray (1968) and Meissner (1976, 1978), it has been fashionable to believe that the high resistivities observed in mature problem source rocks result from replacement of conductive water by significant quantities of resistive petroleum *in the pore spaces*. Again we are required to believe that the reservoir, for which such tools were developed and on which their calibration and intepretation is based, provides the analogue for the source rock. In fact, the interpretation of 'shaley sand' reservoirs (Schlumberger 1987) is one of the most difficult problems in formation evaluation, so that any attempt to interpret saturations in source rock pores must be regarded as speculative, at best.

Figure 5 shows a typical gamma-resistivity-sonic log suite through a mature, expelling oil source rock: the upper part of the Kimmeridge Clay Formation (KCF) at 3.5 km depth in the North Sea. The gamma and sonic logs, confirmed by geochemical analyses not presented here, show that the organic richness of the KCF increases upwards towards the Jurassic–Cretaceous boundary, providing a characteristic 'funnel'-shaped log motif. However, the resistivity log motif on Fig. 5 exhibits exactly the same form: the concentration of the petroleum in the KCF is proportional to the organic matter content. This has been confirmed by regional geochemical analysis of mature KCF sections, showing a remarkably constant proportionality between the pyrolysis P1 yield (similar to the Rock-Eval S1 yield; Espitalie *et al.* 1977) and the TOC, of 0.108 kg kgC^{-1} or 108 mg gC^{-1} (R. G. Miller, BP Research, pers. comm.).

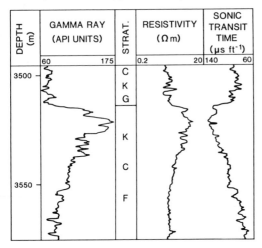

Fig. 5. Wireline log response of the Kimmeridge Clay Formation (KCF) source rock, overlain by organically barren mudstones of the Cromer Knoll Group (CKG) in the North Sea Basin. Note that the resistivity log anomaly (recording oil concentration in the rock) follows closely the gamma ray and density log anomalies (recording organic matter concentration in the rock).

However, such observations do not rely on access to modern proprietary analytical data. Pertinent data has been in the public domain for some 20 years, since Tissot *et al.* (1971) showed plots of extract yield versus TOC from three wells in the Paris basin. All three lower Toarcian well sections display a remarkably linear relationship between petroleum yield and organic carbon content. The two shallow (<1500 m), marginally mature wells display a gradient of around 0.1 kg kgC^{-1} (100 mg gC^{-1}), while the Montmirail well in the basin centre (fully mature and proven to have expelled oil into superjacent Dogger reservoirs) displays a gradient of 0.2 kg kgC^{-1} (200 mg gC^{-1}). Interestingly, wireline geophysical log patterns similar to those shown in Fig. 5 can be seen in lower Toarcian well sections in the centre of the Paris Basin, except that there the organic richness increases downwards so that a 'bell'- rather than 'funnel'-shaped log motif results.

It is difficult to reconcile these independent geophysical and geochemical observations, showing that the retained petroleum and kerogen contents are related, with pore saturation theory, according to which no demonstrable

relationship between petroleum concentration and kerogen concentration would be expected. Returning to Fig. 5, note that there is no resistivity anomaly associated with the underlying organically lean and overlying organically barren mudstones, through which oil must have passed on its supposed pressure-driven vertical route through the mudrock pores to the carrier bed (England *et al.* 1987). The inescapable conclusion is that petroleum migration through the pore network of such rocks is extremely efficient, with little residual petroleum being left *en route*. For example, detailed proprietary analytical results in a mature KCF core from the Norwegian North Sea (K. J. Myers BP Research, pers. comm.) show that primary migration avenues through the lower, organically lean sections contain little migrated petroleum. Even if the measured P1 yields are assumed to occupy only the pore network, they are found to represent an oil saturation of no more than 8% of total porosity.

It is worth noting that many of the petroleum saturations quoted in recent studies (e.g. Leythaeuser *et al.* 1987; Mackenzie & Quigley 1988; Mann *et al.* 1989; Hvoslev *et al.* 1989) are derived by making this same flawed assumption, i.e. inferring that the petroleum measured via solvent extraction or pyrolysis is sitting only in the pore system. No wonder that the estimates of saturation vary considerably!

Why then, if movement through the pore system can be argued to be so efficient, can high concentrations of oil be found in source rocks? S1 pyrolysis yields in coals can be many tens of kg tonne^{-1} (e.g. even reaching 44 kg tonne^{-1} in *immature* coals studied by Khorasani 1987). To my mind, the answer has been staring us in the face since the pioneering studies of Tissot *et al.* (1971), some 20 years ago: the site of origin of the petroleum, the host kerogen itself, exerts the supreme control on petroleum release from the source rock. The concept of an oleophilic kerogen network (Young & McIver 1977; Jones 1980; McAuliffe 1980) lost popularity through the 1980s but is now making something of a comeback based on various independent areas of research (Stainforth 1988; Behar & Vandenbroucke 1989; Pepper 1989; Stainforth & Reinders 1989; Wang & Barker 1989).

Before expanding on this idea, the potential role of clay mineral surfaces requires discussion. There is some experimental evidence showing that some *dry*, pure clay minerals and clay-rich rock samples are able to adsorb significant amounts of petroleum (e.g. Espitalie *et al.* 1980). However, other more realistic physical experiments involving sequential permeation of already water-wet clay-rich soils (Fernandez & Quigley 1985) have shown that simple hydrocarbons (with low dielectric constants) can not compete with water molecules (with high dielectric constant) for adsorptive sites on clays. Because water is always associated with the surfaces of clay minerals from the time of submergence beneath the water table it is reasonable to assume, as in the model presented below, that clay minerals represent the hydrophilic/oleophobic part of the source rock system.

Figure 5 and the accompanying discussion showed that geophysical log data are consistent with the view that the amount of petroleum in the source rock is determined by the organic matter concentration. There is much independent supporting evidence. Because petroleum geochemists have traditionally studied mainly rich, marine source rocks for which the empirical rules of thumb governing expulsion appear to work satisfactorily (e.g. Mackenzie & Quigley 1988), much of this support comes from coal science.

Coals have long been appreciated as having a structure akin to molecular 'cages' within which mobile phases (including petroleum) can be trapped (van Krevelen 1961; Hanbaba et al. 1968; Peters & Juntgen 1968; Given et al. 1986). This petroleum is not readily quantified by conventional geochemical extraction measurements using mild solvents, but there is a large body of evidence derived from n.m.r. (e.g. Given et al. 1986), progressive extraction (e.g. Marzec et al. 1979), repeated extraction after acidification (Behar & Vandenbroucke 1989) and even repeated extraction with the same mild solvent (e.g. Durand et al. 1987), that the organic network in coals can trap substantial amounts of petroleum. This casts doubt on the argument which invokes coals as efficient expellers of oil, based on the observation that S1 and conventional extract yields sometimes do not appear to increase significantly in generating coals (e.g. Huc et al. 1986). Even precursor humic acids are known to be strong adsorbers of hydrophobic organic compounds, including alkanes (Khan & Schnitzer 1972).

Petrographic evidence is also consistent with retention of petroleum by the organic phase. A common property of the exinite and perhydrous vitrinite maceral groups is their fluorescence under U. V. light (Teichmuller 1974; Stach et al. 1975; Radke et al. 1980). Fluorescence of individual, readily identified, macerals in both conventional source rocks and coals is clear and unambiguous evidence that they are sites for retention of generated oil! Furthermore, extracts from different macerals yield oils of a similar composition, indicating that generated oil is mobile throughout the organic network (Allen & Larter 1984). Indeed, Durand & Paratte (1983) showed a coal micrograph in which oil generated by an exinite maceral appears to be permeating the surrounding vitrinite. More specifically, for example, Horsfield et al. (1989) have suggested that the role of aromatic oils generated by fluorescing vitrinite and resinite macerals is to deactivate adsorptive sites in the organic framework of Talang Akar coals, thereby enhancing the expulsion efficiency of the saturate fraction of the oils. A further source of evidence comes from studies of coal-bed methane, a major resource in the USA (Rightmire 1984). Coals can store large volumes of low molecular weight petroleum (gas) via adsorption.

Analytical data have shown that light hydrocarbons can be retained by sedimentary organic matter in marine rocks (C_1–C_8; Whelan et al. 1984) as well as coals (C_1–C_2; Friedrich & Juntgen 1972).

A novel approach

Viewed overall, the evidence for retention of petroleum (both oil and gas) in coals appears convincing; but the relevance of these observations to 'conventional' source has largely been overlooked. This is a short-sighted view. For example, the kerogen in a boghead coal differs from the Green River Shale only in its concentration per unit mass of rock; its fundamental composition (freshwater algal/bacterial biomass) is similar. Likewise, kerogens in humic coals and some clastic source rocks (e.g. the Upper Cretaceous of the Douala Basin; Huc et al. 1986) differ only in the concentration of (vascular plant-derived) organic matter. Interestingly,

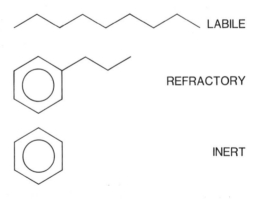

Fig. 6. Simplistic compositional elements of kerogen (scheme of Cooles et al. 1986).

SCHEMATIC REPRESENTATION OF EXPULSION

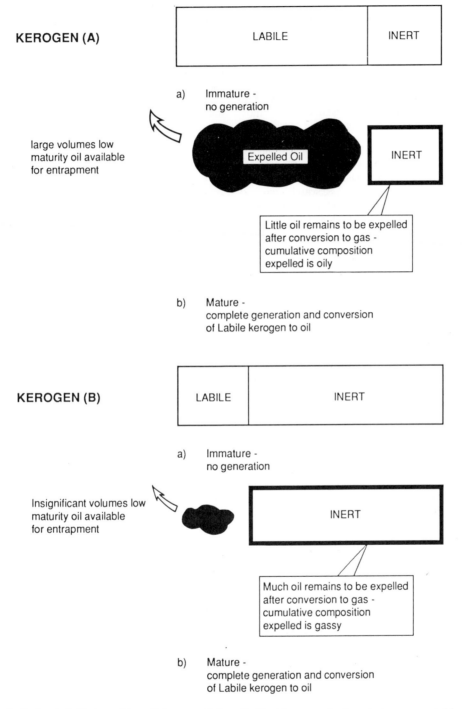

Fig. 7. Rationale behind expulsion efficiency prediction. Residual kerogen adsorbs petroleum generated by breakdown of the labile kerogen portion; differing labile: inert kerogen ratios result in differing ultimate oil expulsion efficiencies (cf. Fig. 1).

Durand & Paratte (1983), while arguing in the opposite sense that coals should not be necessarily worse expellers than conventional source rocks, demonstrated that interactions at the molecular level should be equally important in both. If, as argued above, the movement of petroleum through a source rock can be considered effectively independent of any control exerted by the pore network, could the expulsion behaviour of *all* source rocks be modelled as dependent simply on the kerogen composition, and independent of its concentration in the rock?

In order to appreciate how kerogen composition might be used to predict expulsion behaviour, it is first necessary to classify the types of kerogen present. Again, following the simple scheme of previous BP Research workers (Cooles *et al.* 1986; Quigley *et al.* 1987; Quigley & Mackenzie 1988; Mackenzie & Quigley 1988), three basic kerogen components are allowed: labile, refractory and inert (Fig. 6).

(1) Labile kerogen is derived principally from algal/bacterial/higher land plant lipid precursors and generates high molecular weight petroleum (oil with subordinate gas) at relatively low temperatures (typically 100–150°C).

(2) Refractory kerogen is derived principally from a higher land plant lignin precursor and expels low molecular weight petroleum (gas) over a temperature range where oil-gas cracking proceeds at significant rates (typically >150°C).

(3) Inert kerogen is sufficiently hydrogen-poor that it is unable to generate petroleum.

As the name suggests, the Inert portion plays no part in the generation process, although its highly aromatic structure provides potential adsorptive sites which, if capable of retaining petroleum in sufficient quantity, could influence expulsion of the petroleum generated by the reactive components. The simple example in Fig. 7 shows how different proportions of oil-generative labile and oil-retentive inert kerogen in a source rock could lead to different expulsion efficiencies. The novel aspect of the approach below is in its *quantitative* application to predicting the expulsion efficiencies of mature source rocks. Is there any quantitative data to support this idea?

Figure 8 shows the results of extending the Cooles *et al.* (1986) mass balance, in slightly modified form, to include a larger dataset of 20 source rocks, including some coals. The graph (similar to Fig. 2) shows mean expulsion efficiency attained during labile kerogen breakdown (PEE$_L$) plotted as a function of the initial petroleum potential (P^0). The results are, to a degree, consistent with those of Cooles *et al.* (1986) in that a relationship between PEE and

P^0 is seen, where P^0 is less than about 0.05 kg kg^{-1} (50 kg tonne^{-1}). This is encouraging since it shows that minor modifications to their approach result in little change in the results, i.e. the approach is robust. However, by extending the dataset to include very organic-rich samples (coals), this apparent relationship breaks down. Based on Fig. 2 and the pore saturation concept, these source rocks should be the most efficient expellers; yet, in agreement with experience in the field, many coals are relatively poor oil source rocks (Durand & Paratte 1983).

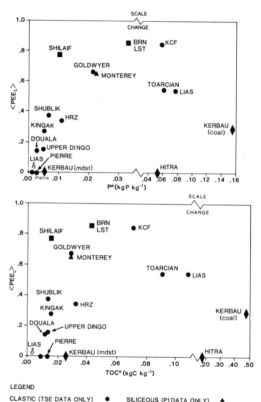

Fig. 8. Average Petroleum Expulsion Efficiency attained during labile kerogen degradation (PEE$_L$) versus initial petroleum potential (P^0) for 20 source rocks (including coals, and the 10 solely marine examples previously presented in Fig. 2). Calculation of PEE$_L$ involves a correction to account for subordinate concentrations of gas present in the petroleum generated by labile kerogen. The previously noted simple apparent dependency of PEE on P^0 (Fig. 2) does not hold true when coals are included in the data set.

Figure 8 indicates that there is some other important factor at play in the expulsion process. Rather than building a complex model which considers both source rock richness and composition, I have assumed, as argued earlier in the paper, that movement through the pore network is so efficient that the losses of petroleum there can be neglected; I consider instead that the release of petroleum from kerogen is the only process determining expulsion efficiency.

To see whether this is a reasonable assumption, PEE_L can be replotted as a function of the initial Inert kerogen proportion on Fig. 9 (for the same sample set as in Fig. 8). The portion of initial kerogen which is inert ($= C_{KI}^0/\{C_{KL}^0 + C_{KR}^0 + C_{KI}^0\}$) is derived simply from the initial hydrogen index (HI^0) using:

$$C_{KI}^0/\{C_{KL}^0 + C_{KR}^0 + C_{KI}^0\} = 1 - W.HI^0, (2)$$

where the factor W is the average mass fraction of carbon in petroleum produced during pyrolysis of the reactive (labile plus refractory) portion of the kerogen. W is taken at $0.85\,kgC\,kg^{-1}$ (Cooles et al. 1986); it varies between 0.75 and 0.86 over the complete range of homologous n-alkanes seen in kerogen pyrolysates.

Note that, in contrast to Fig. 8, Fig. 9 shows a good correlation for all the source rocks: coals as well as lean and rich clastic/carbonate/siliceous source rocks lie on a single trend. In other words, there is a systematic increase in expulsion efficiency with increasing kerogen quality, irrespective of absolute petroleum potential of the source rock. This of course does not *prove* a causal link (any more than did Fig. 2, interpreted in terms of pore saturation theory), but it does provide encouragement to formulate and test an algebraic scheme to predict expulsion behaviour as a function of kerogen type.

Algebraic scheme

Following Cooles et al. (1986), the petroleum expulsion efficiency (PEE) can be written:

$$PEE = 1 - (\text{petroleum remaining/petroleum generated}) \quad (3)$$

thus if the term 'petroleum remaining' can be predicted, it becomes straightforward to predict the expulsion efficiency as a function of the amount of petroleum generated i.e. to construct

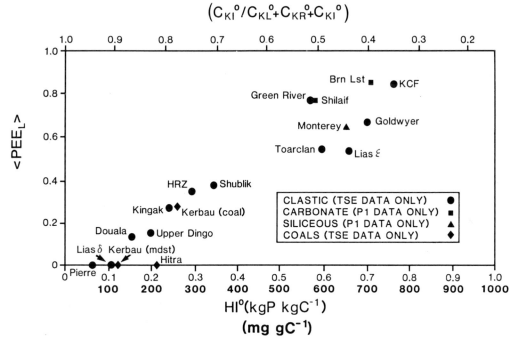

Fig. 9. Average PEE_L, as defined in Fig. 8, now plotted versus initial Hydrogen Index (HI^0) for the same 20 source rocks in Fig. 8. Note the now excellent correlation between expulsion efficiency and initial organic matter quality, irrespective of the absolute initial petroleum potential or TOC of the rocks; i.e. rocks with widely differing organic matter contents (e.g. coals and mudstones) will behave the same so long as they share a common organic matter type.

an expulsion-generation relationship for any source rock of interest.

Cooles et al. (1986) showed how equation (3) can be written in terms of the various carbon quantities in petroleum (oil, C_O; gas, C_G), and kerogen (labile, C_{KL}; refractory, C_{KR}; and inert, C_{KI}). This study is concerned with predicting expulsion efficiencies during two different maturation stages: during firstly labile and secondly refractory kerogen breakdown. The bulk petroleum generation index can be considered as the sum of two maturity increments, acting in series: PGI_L and PGI_R, which are the components due to labile and refractory kerogen degradation, respectively:

$$PGI = \frac{\{PGI_L.(C_O{}^0 + C_{KL}{}^0)\} + \{PGI_R.C_{KR}{}^0\}}{(C_O{}^0 + C_{KL}{}^0 + C_{KR}{}^0)} \quad (4)$$

where

$$PGI_L = \frac{(C_O{}^0 + C_{KL}{}^0) - C_{KL}}{(C_O{}^0 + C_{KL}{}^0)} \quad (5)$$

and

$$PGI_R = \frac{C_{KR}{}^0 - C_{KR}}{C_{KR}{}^0} \quad (6)$$

Note that PGI_L includes an initial petroleum term which is never zero; hence PGI_L is non-zero for immature potential source rocks.

PEE_L represents that fraction of the total petroleum generated, at a particular stage of labile kerogen degradation, which has been actually expelled. It can be written:

$$PEE_L = \frac{(C_O{}^0 + C_{KL}{}^0 - C_{KL}) - (C_O + C_G)}{(C_O{}^0 + C_{KL}{}^0 - C_{KL})} \quad (7)$$

Similarly, PEE_R represents the fraction of the total petroleum generated, at a particular stage of refractory kerogen degradation, which has actually been expelled. It is written:

$$PEE_R =$$
$$\frac{(C_O{}^0 + C_{KL}{}^0 + C_{KR}{}^0 - C_{KL} - C_{KR}) - (C_O + C_G)}{(C_O{}^0 + C_{KL}{}^0 + C_{KR}{}^0 - C_{KL} - C_{KR})} \quad (8)$$

In equation (7), the term $(C_O + C_G)$ will be predicted assuming that it represents the high molecular weight petroleum adsorbed onto the residual kerogen mass during labile kerogen breakdown. Thus:

$$(C_O + C_G) = (a_L/W).(C_{KL} + C_{KR}{}^0 + C_{KI}{}^0), \quad (9)$$

where a_L is the adsorptivity coefficient for high molecular weight (oil-rich) petroleum onto kerogen (kg kgC^{-1}).

In equation (8), the term $(C_O + C_G)$ will be predicted assuming that it represents the low molecular weight petroleum adsorbed onto the residual kerogen mass. Thus:

$$(C_O + C_G) = (a_R/W).(C_{KR} + C_{KI}{}^0), \quad (10)$$

where a_R is the adsorptivity coefficient for low molecular weight (gas-rich) petroleum onto kerogen (kg kgC^{-1}).

By substitution of equations (9) and (10) into equations (7) and (8), respectively, we arrive at:

$$PEE_L =$$
$$\frac{(C_O{}^0 + C_{KL}{}^0 - C_{KL}) - (a_L/W).(C_{KL} + C_{KR}{}^0 + C_{KI}{}^0)}{(C_O{}^0 + C_{KL}{}^0 - C_{KL})}$$
$$\quad (11)$$

and

$$PEE_R =$$
$$\frac{(C_O{}^0 + C_{KL}{}^0 + C_{KR}{}^0 - C_{KR}) - (a_R/W).(C_{KR} + C_{KI}{}^0)}{(C_O{}^0 + C_{KL}{}^0 + C_{KR}{}^0 - C_{KR})}$$

Having established the mathematical formalism to predict PEE_L and PEE_R as a function of the residual carbon concentration, values are required for the adsorptivity coefficients governing retention of high molecular weight petroleum during labile kerogen breakdown (a_L), and low molecular weight petroleum during refractory kerogen breakdown (a_R).

Values for a_R are the more readily estimated from studies of gas storage in coals. Using data from such studies (e.g. Ancell et al. 1980; Meissner 1987; Satriana 1980), I have assumed that, over the window of temperatures and pressures at which refractory kerogen degrades, a_R is a constant which can be expressed in terms of a mass of petroleum per unit mass of carbon, equivalent to 0.02 kg kgC^{-1}.

Values of a_L are more difficult to estimate, since data on adsorption of high molecular weight petroleum at high temperatures and pressures are lacking. However, it is interesting to note that a global review of typical maximum extract yields attained in mature marine source rocks suggests a value of around 200 mg gC^{-1} is typical (cf. Tissot et al. 1971). Assuming that this represents the petroleum that is adsorbed onto kerogen surfaces (i.e. taking the opposite but equally defensible extreme view to the proponents of the pore saturation concept), and that over the window of temperatures and pressures at which labile kerogen degrades a_L is a constant, it too can be expressed in terms of a mass of petroleum per unit mass of carbon, as 0.2 kg kgC^{-1}, i.e. an order of magnitude greater than the value of a_G. Note that, to some extent, the value of a_L depends on the geochemical measurement used to quantify petroleum in the rock. Extracts and thermal volatilates measure different types of 'mobile' organic matter, for example Tarafa *et al.* (1988) showed that some of the high molecular weight material measured by solvent extraction may actually appear in the S2 yield of a sample which undergoes conventional S1–S2 pyrolysis (Espitalie *et al.* 1977). Because extract yields were used to characterise 'petroleum' in the mass-balance approach of Cooles *et al.* (1986) I have followed the same convention here. However, it is important to stress that, so long as a_L is derived using the same type of measurement as is used to define 'petroleum' in the mass balance calculations, the results will be compatible. No doubt a_L will actually depend on oil type and hence a lower constant will be appropriate if 'petroleum' is defined using thermal volatilate yields, which contain a lower molecular weight range than solvent extraction yields. Since different kerogen types generate different oil types, some natural variation in a_L might also be expected. However, such possibilities are beyond the scope of this initial, simple model.

Testing of results using natural datasets

Expulsion profiles constructed using this quantitative approach are compatible with, and can be compared with, 'observed' expulsion efficiencies in naturally matured sample suites. When comparing predicted versus observed expulsion efficiencies, it is important to remember that the 'observed' values are not raw data, but are themselves the results of algebraic manipulations with inherent assumptions and data limitations. All 'observed' values are prone to confidence ranges, which are especially large at low values of PEE.

Figure 10 shows a comparison of predicted versus observed expulsion efficiency for two contrasting source rocks from the 20 studied. Example (a), the Monterey Formation of California, is a Class 1 source rock *sensu* Mackenzie & Quigley (1988), containing high quality organic matter (HI^0 = 657 mg gC^{-1}). The observed expulsion profile shows a short lag

between onset of expulsion and generation; thereafter expulsion efficiencies rise rapidly to very high values (PEE about 0.9), typical of good oil source rocks in general. There is little discernible additional increase in expulsion efficiency at very high maturities (PGI > 0.9), when the small refractory component is breaking down. The modelled expulsion profile predicts an early onset (at PGI of about 0.2), and a subsequent rapid rise in PEE with further maturation, ultimately reaching efficiencies around 0.9 at the end of labile kerogen degradation. A further small increase in expulsion efficiency is predicted and observed during refractory kerogen degradation.

SOURCE ROCK (A): HIGH HYDROGEN INDEX

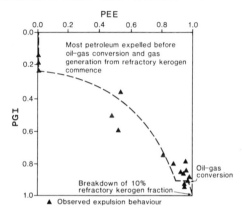

SOURCE ROCK (B): LOW HYDROGEN INDEX

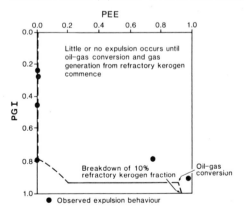

Fig. 10. Comparison of observed and predicted generation-expulsion profiles for two source rocks of contrasting quality (cf. Figs. 1 and 7): (a) High quality (marine, siliceous) source rock (HI^0 = 657 mg gC^{-1}); (b) Poor quality (terrestrial, clastic) source rock (HI^0 = 155 mg gC^{-1}).

Example (b), the Cretaceous of the Douala Basin, is a poor quality source rock as reflected by a low initial hydrogen index: $HI^0 = 155$ mg gC^{-1} (0.155 kg kgC^{-1}), whose expulsion behaviour is very different to (a). Petroleum explusion does not commence until almost all of the labile kerogen is degraded (PGI > 0.8); thereafter, PEE rises sharply during refractory kerogen breakdown to high values (PEE > 0.7).

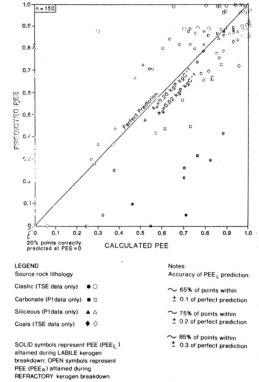

Fig. 11. Cross-plot of predicted PEE_L versus 'observed' PEE_L for individual data points (cf. Fig. 10), compared at the appropriate level of maturity (value of PGI_L). Same for PEE_R. Note the good correspondence at high values of PEE, where the uncertainty in 'observed' PEE is least. Scatter at low values of 'observed' PEE_L is the unavoidable consequence of scatter in raw geochemical measurements which is carried through the Cooles *et al.* (1986) mass balance methodology.

This is an example of Class 2 source rock behaviour *sensu* Mackenzie & Quigley (1988). The modelled profile predicts that no expulsion should take place until high maturity is reached (PGI > 0.8), and that PEE should not exceed about 0.2 even at the end of labile kerogen degradation. The predicted high expulsion efficiency (PEE > 0.9) during refractory kerogen

degradation is a consequence of the change in molecular weight range, which results in a reduction in adsorptivity coefficient from 0.2 to 0.02 kg kgC^{-1}. The predicted, sudden jump at the onset of refractory kerogen degradation is merely a function of the order of magnitude step function in adsorptivity in this simple model; in reality, a more gradual change in adsorptivity coefficient might be envisaged.

As far as can be determined from attempts to match such 'observed' expulsion profiles (e.g Fig. 10), model predictions are consistent with the results of the carbon mass-balance field 'data'. Figure 11 represents an attempt to compare predictions against observations for the whole global dataset. This sort of plot suffers from a great deal of scatter, primarily because of the previously discussed scatter in observed PEE, at low values. However, where confidence in observed PEE is highest, it is encouraging to note a good agreement between predicted and observed expulsion efficiency.

Rules of thumb governing expulsion behaviour

The algebra outlined above, although simple, is cumbersome in everyday use; these calculations are best performed in a computer programme or spreadsheet. Alternatively, nomograms can be constructed to allow rapid evaluation of the oil expulsion potential of a source rock, just as they are commonly used in quick-look formation evaluation (Schlumberger 1987).

Figure 12 shows a nomogram designed to predict PEE_L as a function of PGI_L (which can in turn be related to total bulk PGI using equation (4)), given a full suite containing all the required geochemical data. However, our experience is that pyrolysis-GC is not universally used as a routine screening technique; therefore Fig. 13 shows further nomograms devised for rapid screening of source rocks when only initial Hydrogen Index data are available.

Global implications

Predictions of expulsion efficiency, and the attendant implications for the composition of expelled petroleum, should always be consistent with observations on the gross basin scale, as well as on the local scale. Below is an investigation of some of the extremes in source rock behaviour predicted by the nomograms (Figs 12 and 13), examining their implications for the type and composition of expelled petroleum products, and commenting on whether these are consistent with observation.

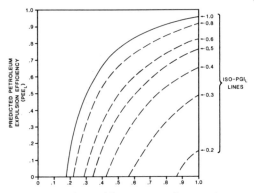

PARAMETER 'i' = $\{C_O^*+C_{KL}^*\}/\{C_O^*+C_{KL}^*+C_{KR}^*+C_{KI}^*\}$ (kgP kgC^{-1})

Fig. 12. Nomogram to predict variation of PEE_L with PGI_L and organic matter composition parameter 'i' = $[C_O^{0'} + C_{KL}^{0'}]/[C_O^{0'} + C_{KL}^{0'} + C_{KR}^{0'} + C_{KI}^{0'}]$. Starting on the x-axis at the parameter value 'i' calculated for the immature source rock of interest, interpolate between the iso-PGI_L curves to derive the fraction of labile kerogen which must be degraded before expulsion may commence ($PGI_{L(c)}$). E.g. if 'i' = 0.4 kgP/kgC, $PGI_{L(c)}$ = 0.44. Then read vertically upwards, reading off on the y-axis the value of PEE_L reached at any required value of PGI_L. Cross-plot PEE_L versus PGI_L to derive expulsion-generation profiles similar to those shown in Fig. 10.

Lacustrine 'Type I' source rocks

At one extreme of the hydrogen index spectrum are 'Type I' source rocks: an extremely high value of 898 mg gC^{-1} is typically quoted for the lacustrine Green River oil-shales (e.g. Tissot *et al.* 1987a). At a glance, Fig. 13 shows that such source rocks (HI^0 approximately 0.9 kg kgC^{-1}) should be highly efficient expellers of oil (attaining PEE_L > 0.9). Oil expulsion should begin after a minimal fraction of the labile kerogen has degraded ($PGI_{L(c)}$ <0.2). The field evidence is that so-called "immature" oils, asphalts and bitumens have indeed long been recognized as a characteristic feature of the Uinta Basin (Hunt *et al.* 1954).

Coaly 'Type III' source rocks

At the other end of the hydrogen index spectrum lies coals, which have been a source of difficulty in organic geochemistry for some time (Durand & Paratte 1983): although the ability of almost all coals to generate oil is not disputed, they are more often associated with gas deposits than

with oil (e.g. UK Southern North Sea Carboniferous). Part of the confusion in understanding the behaviour of coals arises from their blanket classification as "Type III' kerogens (e.g. Tissot *et al.* 1987b). In fact, coals are a highly diverse family of source rocks whose initial hydrogen indices are determined by the type of organic input and diagenesis; 150–350 mg gC^{-1} (0.15–0.35 kg gC^{-1}) might be regarded as typical of 'conventional' humic coals (the variation arising from various mixtures of liptinite and vitrinite), although it is not difficult to find examples outside this range (e.g. Khorasani 1987).

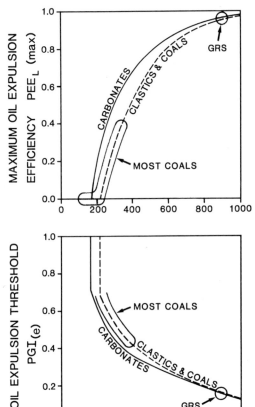

Fig. 13. Nomograms to estimate: (a) the expulsion threshold in terms of the amount of petroleum potential which must be realised before oil expulsion begins $PGI_{(c)}$ and (b) the maximum value of expulsion efficiency attained during labile kerogen breakdown (PEE_L), if only hydrogen index data are available for the immature source rock. These curves rely on global data to define default values for the initial oil concentration and the initial refractory kerogen proportion for three source rock facies: carbonate, clastic, and coaly sediments.

Using Fig. 13 it is easy to appreciate that even a relatively limited range in HI^0 between 0.15–0.35 kg kgC^{-1} implies a huge range in expulsion behaviour. Those at the lower end of the range (HI < 200 mg gC^{-1}) will be unable to expel until oil–gas cracking and refractory kerogen degradation have commenced: these coals are chiefly sources of gas, with variable amounts of associated highly mature, low molecular weight oil (usually appearing as associated condensate). As initial hydrogen index increases above 200 mg gC^{-1}, coals will be able to expel increasing amounts of the oil they generate, commencing at successively low levels of maturity. Better quality coals such as the Miocene of the Mahakam Delta (average HI^0 about 330 mg gC^{-1}) should be able to start expelling oil after about half the labile kerogen is degraded, reaching ultimate expulsion efficiencies (before oil–gas cracking commences) of about 50%. This is not as favourable as most typical 'Type II' marine source rocks (HI^0 500–700 mg gC^{-1}), and accounts for the fact that many of the oil accumulations derived from coals often have significant associated gas. For example, in the Mahakam Delta (the often-quoted classic example of a coal-sourced oil province) the overall gas : oil ratio of the province is a very high 10 000 scf bbl^{-1}, equivalent to a gas mass fraction of approximately three quarters (van de Weerd et al. 1987); even the Handil 'oilfield' is a complex stack of individual oil and gas reservoirs with an overall gas mass fraction of about one quarter (Total 1985).

Application of the model in a basin case study

The case study illustrated is an Australasian petroliferous basin (Fig. 14). This choice was dictated by the following criteria:

the basin contains a variety of pools containing petroleums with widely differing compositions (and subsurface phases);

the only source rocks in the basin are coals which vary laterally in their geochemical characteristics, thus providing a test of the model's success in predicting the gross expulsion behaviour of such 'problematic' source rocks;

the sourcing system is well-understood (oil-source correlation studies have related the oils to the coals; absence of alternative source rocks is proven), secondary mi-

gration is relatively simple, and biodegradation has not altered the petroleum composition in the traps.

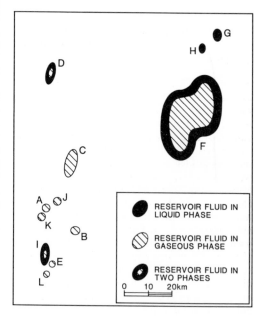

Fig. 14. Map showing relative locations and subsurface petroleum phases of pools in the basin case study.

Using the prospect evaluation approach of Mackenzie & Quigley (1988), but modified to account for the expulsion model described above, it is possible to explain retrospectively how the composition and phase of petroleum in different parts of the basin is related to the local coal chemistry, and hence expulsion behaviour. The following data were used:

average maturity level reached in the kitchen for the prospect;

characterization (pyrolysis, pyrolysis-GC) of the source rock in the kitchen, when immature (taken from the nearest well control).

The results can be compared with the actual outcomes based on drilling results in order to establish the capability of the simple model. The following data were used:

composition(s) and phase(s) of the reservoired petroleum(s);

biomarker data to establish maturity of the reservoired petroleum;

present reservoir depth (z);

average regional pressure gradient (hydrostatic);

average regional temperature function (surface temperature $= 10°C$ and temperature gradient $= 30°C\,km^{-1}$).

The first check was to test whether the predicted expulsion thresholds were consistent with the maturity of the reservoired petroleums (the higher the predicted threshold, the more advanced should be the thermally-sensitive molecular reactions recorded in the certain biomarker ratios). All the examined petroleums were sufficiently mature that the commonly used hopane and sterane isomerization reactions (Mackenzie 1984) were complete. However, the Methyl Phenanthrene Index of Radke (1987) and related aromatic hydrocarbon ratios are useful at higher levels of maturity. These allow vitrinite reflectance equivalent maturity to be estimated using empirical correlations with vitrinite reflectance (e.g. Radke 1987), and ultimately, temperature. Such maturity levels are difficult to interpret in *reservoired* oils and condensates since these fluids may represent the cumulative products expelled from the source kitchen over a broad range of thermal stresses. However, such data do give a relative feel for petroleum maturity levels.

Figure 15 confirms that the modelled expulsion threshold (PGI_{IL}), predicted from the kerogen quality of the locally developed coal facies, is consistent with the relative maturity level (%R_o equivalent) of the expelled oil or condensate. Note in the case of pool I the two different oils from different parts of the pool have significantly different maturities. This is a common feature of many petroleum pools which inherit compositional inhomogeneities caused by spatial and temporal variation in the original petroleum charge composition (England & Mackenzie 1989).

Table 1 shows a comparison of predicted versus observed in-place composition for 12 petroleum pools in the basin. Composition varies continuously with maturity, but in this study the composition is reported at three maturity stages (Table 1): 160, 180 and 220°C, temperatures which represent approximately the onset of oil–gas cracking, end of oil–gas cracking, and end of refractory kerogen breakdown, respectively. Where the maximum average kitchen temperature was able to be determined with confidence, a precise value of predicted composition was determined by simple interpolation.

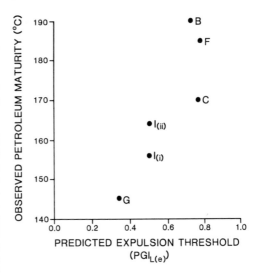

$I_{(i)}$ and $I_{(ii)}$ are from different segments of the same pool complex

Fig. 15. Comparison of observed petroleum maturity (as recorded by aromatic hydrocarbon ratios in oils and condensates) with predicted expulsion threshold of local coals using the expulsion model. The most mature petroleum originates from coals which are predicted to expel at the highest expulsion thresholds.

The comparison with predicted composition is best made on a cross-plot such as Fig. 16, which compares the predicted composition at the three stages with the observed petroleum in-place. The latter varies widely in composition from 0.05 to 0.92 kg gas kg^{-1}. In general the simple model yields reasonably accurate predictions of petroleum composition encompassing this wide range. For example, compositions of five out of the six oiliest pools were predicted in the correct order, a valuable result since this basin, like many of the world's basins, is prospective mainly for oil, with gas accumulations having minor or negligible economic worth at the current time.

The usefulness of these compositional predictions can be further demonstrated by using them, in turn, to understand petroleum phase in the twelve pools. The necessary tool is a phase envelope for the basin, using the pressure/temperature gradients stated above and the correlations of England et al. (1987) to construct bubble- and dew-point composition curves. It is then possible to estimate whether the predicted petroleum composition should exist in a single liquid phase ('black oil'), two phases ('oil leg plus gas cap'), or a single gaseous phase ('gas-condensate') at the reservoired depth of the accumulation (Fig. 17). The results are shown in

Table 1. *Comparison of predicted versus observed in-place composition for 12 petroleum pools in the studied Australasian basin*

	Observed petroleum composition in trap (kg gas kg^{-1})	Mean initial HI (HI^0) of coal (mg gC^{-1})	Predicted composition in trap at each maturity stage		
			Stage A at 160°C (kg gas kg^{-1})	Stage B at 180°C (kg gas kg^{-1})	Stage C at 220°C (kg gas kg^{-1})
A	0.88	192		0.88	0.67
B	0.84	284	0.24	0.56	(0.61) 0.80
C	0.79	363	0.26	0.61	(0.70) 0.99
D	0.51	135		0.79	0.97
E	0.84	210		0.71	0.75
F	0.78	281	0.20	0.62	(0.73) 0.63
G	0.19	529	0.15	(0.15) 0.26	0.37
H	0.05	529	0.10	(0.10) 0.23	0.35
I	0.32	401	0.14	0.40	(0.43) 0.55
J	0.82	228		0.72	0.98
K	0.86	192		0.87	0.96
L	0.92	218		0.83	0.93

Compositions in brackets are derived by interpolation, where the average maturity of the source kitchen is known.

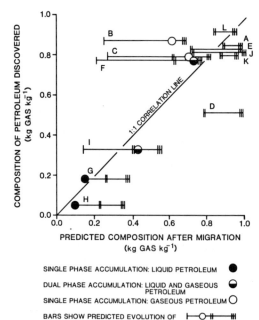

SINGLE PHASE ACCUMULATION: LIQUID PETROLEUM ●

DUAL PHASE ACCUMULATION: LIQUID AND GASEOUS PETROLEUM ◒

SINGLE PHASE ACCUMULATION: GASEOUS PETROLEUM ○

BARS SHOW PREDICTED EVOLUTION OF CUMULATIVE COMPOSITION AFTER PETROLEUM CONTRIBUTIONS A.B.C.   A ▲ B C   INTERPOLATED COMPOSITION WHERE MATURITY KNOWN

Fig. 16. Comparison of petroleum composition (mass fraction gas in total petroleum) predicted by the model to be expelled from the locally developed coal facies, with that observed in twelve pools. All relevant data and results are given in Table 1.

Table 2. *Comparison of predicted versus observed sub-surface phase for twelve petroleum pools in the studied Australasian basin*

Pool	Observed phase(s)	Predicted phase(s)
A	G	G
B	G	G + L
C	G	G
D	G + L	G
E	G	G
F	G + L	G + L
G	L	L
H	L	L
I	L + G	L + G
J	G	G
K	G	G
L	G	G

Predicted phase(s) are derived by locating the predicted compositions after migration from Table 1, on a phase envelope (e.g. Fig. 17) at the reservoir depth of interest.

L, occupies single liquid phase at reservoired depth; L + G, occupies two (liquid plus gaseous) phases at reservoired depth (volumetrically most significant phase given first); G, occupies single gaseous phase at reservoired depth.

Table 2, which shows that ten out of the twelve subsurface phase states are predicted correctly.

The earth is a difficult laboratory within which to design and test experiments. However, simple models such as the one outlined here go some

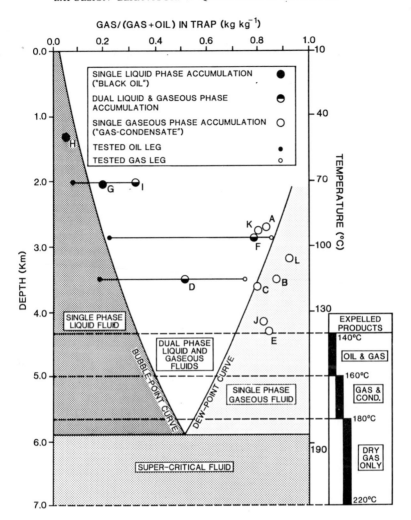

Fig. 17. Phase envelope for the basin, showing bubble point and dew-point curves which are based on a single basin-wide pressure (hydrostatic) and temperature (30°C km⁻¹) functions which allows bubble- and dew-point compositions to be related simply to depth. Curves are constructed based on the correlations of England *et al.* (1987). Also plotted are the three expulsion zones over which petroleum composition is modelled (Fig. 16). The actual observed compositions and subsurface phases of the petroleum pools are plotted at the reservoired depths.

way towards narrowing the uncertainty inherent in geochemical prospect evaluation, and ultimately impact on overall risk reduction in exploratory drilling (Tissot *et al.* 1987*b*). The results of this study, namely correct ranking of five out of the six oiliest petroleum compositions, and correct estimation of ten out of the twelve subsurface phases, shows that there is still room for thoughtful application of simple geochemical models in pursuit of this goal.

Conclusions

Currently popular expulsion models which invoke saturation of source rock pores/fracture networks as a fundamental control are based on an analogy with petroleum reservoir behaviour. Taking into account important differences in mineralogy, pore structure and water behaviour in reservoir versus source rocks, this analogy seems rather weak. Thus the widely-held view that source rock richness determines expulsion efficiency may be erroneous.

A fresh look at the problem reveals a considerable weight of evidence (chemical, petrographic) from a variety of sources (petroleum and coal geochemistry, soil science) in support of an alternative perspective: that *surface* interactions play a more important role in source rocks than does bulk petroleum flow. The implication is that kerogen composition (the ratio of generative to adsorptive material present) may be the relevant parameter in determining expulsion efficiency.

Mass balance calculations based on field data show that the 'bulk-flow-through-pores' hypothesis can not explain the behaviour of highly organic-rich source rocks whose expulsion efficiencies are low and which tend to be associated with gas rather than oil deposits in nature. The same calculations, on the other hand, do support a dominant role for kerogen composition in controlling the expulsion process.

Thus the initial hydrogen index of a source rock (reflecting the ratio of potentially generative to potentially adsorptive kerogen) can be used to forward-model expulsion efficiency, within the limits of accuracy of the mass-balance field dataset against which predictions must be compared.

Various rules of thumb arise from the prediction of expulsion efficiency in this way.

Oil expulsion is earliest and most efficient from 'Type I' source rocks (HI^0 up to 900 mg gC^{-1}).

'Type II' source rocks (HI^0 around 500–700 mg gC^{-1}) will also be early, highly efficient oil expellers.

Coal source rocks will exhibit a very wide range in expulsion behaviour, depending on their organic makeup and diagenetic history, and the resulting HI^0. Those with $HI < 200$ mg gC^{-1} are capable of expelling gas/gas-condensate only; as initial hydrogen index increases above 200 mg gC^{-1}, coals will be able to expel increasing amounts of the oil they generate, commencing at successively low levels of maturity.

Absolute petroleum potential or organic richness exerts no direct control, although source rocks with very low petroleum potential tend to have very low HI and hence low expulsion efficiency.

Gas expulsion efficiency is highly efficient (>0.9) from almost all source rocks.

A quantitative model of expulsion, based on these simple principles, can furnish realistic estimates of the various petroleum maturities, compositions and phases observed in sedimentary basins.

I would like to extend thanks to the following in connection with the production of this paper: to many BP Research colleagues who provided lively debate and discussion during formulation of these ideas during the period 1987–88, especially P. J. D. Park and R. G. Miller; T. A. Schwarzkopf and R. L. Patience for encouragement to publish; to the management of BP Research and BP Exploration for permission to publish; to the staff of the Exploration & Production Division drawing office for the Figures; to A. D. Duncan, T. A. Dodd, M. D. Lines and A. Sutter for reading an early version of the manuscript; to the two reviewers for helpful comments; and to N. C. Davis and I. C. Fuchs who proof-read the final text.

References

ANCELL, K. L., LAMBERT, S. & JOHNSON, F. S. 1980. Analysis of the coal bed degasification process at a seventeen well pattern in the Warrior Basin of Alabama. *Proceedings of 1st Annual Symposium on Unconventional Gas Recovery*, Pittsburgh, 1980, SPE/DOE 8971, 355–69.

ALLAN, J. & LARTER, S. R. 1984. Aromatic structures in coal maceral extracts and kerogens. *In*: BJOROY, M. *et al.* (eds) *Advances in Organic Geochemistry 1981*, Wiley, Chichester, 534–546.

BARKER, C. 1980. Primary migration: the importance of water-mineral-organic matter interactions in the source rock. *In*: ROBERTS, W. H. and CORDELL, R. J. (eds) *Problems of petroleum migration*. American Association of Petroleum Geologists Studies in Geology, **10**, 19–31.

BEHAR, F. & VANDENBROUCKE, M. 1989. Characterisation and quantification of saturates trapped inside kerogen: implications for pyrolysate composition. *In*: MATTAVELLI, L. & NOVELLI, L. (eds) *Advances in Organic Geochemistry 1987*. Pergamon, Oxford, 927–938.

BROSSE, E., LOREAU, J. P., HUC, A. Y., FRIXA, A., MARTELLINI L. & RIVA, A. 1988. *The organic matter of interlayered carbonates and clay sediments*. Institute Francais du Petrole, Direction de Recherches Geologie et Geochimie, Rapport ref: 35 999, March 1988.

COOLES, G. P., MACKENZIE, A. S. & QUIGLEY, T. M. 1986. Calculation of petroleum masses generated and expelled from source rocks. *Advances in Organic Geochemistry* 1985, Pergamon, 235–245.

DICKEY, P. A. 1975. Possible migration of oil from source rock in oil phase. *The American Association of Petroleum Geologists Bulletin*, **72**, 337–345.

DOLIGEZ, B., BESSIS, F., BURRUS, J., UNGERER, P. & CHENET, P. Y. 1986. Integrated numerical simulation of the sedimentation, heat transfer, hydrocarbon formation and liquid migration in a sedimentary basin: the THEMIS model. *In:* BURRUS, J. (ed.) *Thermal modelling in sedimentary basins.* Proc. 1st IFP Exploration Research Conference, Bordeaux, 3–7 June, 1985. Editions Technip, 173–195.

DURAND, B. 1983. Present trends in organic geochemistry in research on migration of hydrocarbons. *In:* BJOROY, M. *et al.* (eds) *Advances in Organic Geochemistry 1981*, 117–128.

—— & PARATTE, M. 1983. Oil potential of coals: a geochemical approach. *In:* BROOKS, J. (ed.) *Petroleum Geochemistry and Exploration of Europe* Geological Society, London, Special Publication, **12**, 285–292.

——, HUC, A. Y. & OUDIN, J. L. 1987. Oil saturation and primary migration: observations in shales and coals from the Kerbau wells, Mahakam Delta, Indonesia. *In:* DOLIGEZ, B. (ed.) *Migration of hydrocarbons in sedimentary basins.* Proc. 2nd IFP Exploration Research Conference, Carcans, June 15–19, 1987. Editions Technip, 173–196.

ENGLAND, W. A. & MACKENZIE, A. S. 1989. Some aspects of the organic geochemistry of petroleum fluids. *Geologische Rundschau*, **78/1**, 291–303.

——, ——, MANN, D. M. & QUIGLEY, T. M. 1987. The movement and entrapment of petroleum fluids in the subsurface. *Journal of the Geological Society, London*, **144**, 327–347.

ESPITALIE, J., LAPORTE, J. L., MADEC, M., MARQUIS, F., LEPLAT, P., PAULET, J. & BOUTEFEU, A. 1977. Methode raide de characterisation des roches meres de leur potentiel petrolier et de leur degre d'evolution. *Revue de l'Institut Francais du Petrole*, **32**, 23–42.

——, MADEC, M. & TISSOT, B. 1980. Role of mineral matrix in kerogen pyrolysis: influence on petroleum generation and migration. *The American Association of Petroleum Geologists Bulletin* **64**, 59–66.

FERNANDEZ, F. & QUIGLEY, R. M. 1985. Hydraulic conductivity of natural clays permeated with simple liquid hydrocarbons. *Canadian Geotechnical Journal*, **22**, 205–214.

FRIEDRICH, H.-U. & JUNTGEN, H. 1972. Some measurements of the $^{12}C/^{13}C$-ratio in methane or ethane desorbed from hard coal or released by pyrolysis. *Advances in Organic Geochemistry 1971*, Pergamon, Oxford, 639–646.

GIVEN, P. H., MARZEC, A., BARTON, W. A., LYNCH, L. J. & GERSTEIN, B. C. 1986. The concept of a mobile phase within the macromolecular network of coals: a debate. *Fuel*, **56**, 155–163.

HANBABA, P., JUNTGEN, H. & PETERS, W. 1968. Nichtisotherme instationare messung der aktivierten diffusion von gasen in festkorpern am beispiel der steinkohle. *Berichte der Bunsengesellschaft für Physikalische Chemie*, **72**, 554–562.

HINCH, H. H. 1980. The nature of shales and the dynamics of hydrocarbon expulsion in the Gulf Coast Tertiary section. *In:* ROBERTS, W. H. & CORDELL, R. J. (eds) *Problems of petroleum migration.* American Association of Petroleum Geologists Studies in Geology, **10**, 1–18.

HONDA, H. & MAGARA, K. 1982. Estimation of irreducible water saturation and effective pore size of mudstones. *Journal of Petroleum Geology*, **4**, 407–418.

HORSFIELD, B., YORDY, K. L. & CRELLING, J. C. 1989. Determining the petroleum generating potential of coal using organic geochemistry and organic petrology. *In:* MATTAVELLI, L. & NOVELLI, L. (eds) *Advances in Organic Geochemistry 1987*, Pergamon, Oxford, 121–130.

HUC, A. Y., DURAND, B., ROUCACHET, J., VANDENBROUCKE, M. & PITTION, J. L. 1986. Comparison of three series of organic matter of continental origin. *Organic Geochemistry*, **10**, 65–72.

HUNT, J. M., STEWART, F. & DICKEY, P. A. 1954. Origin of hydrocarbons of Unita Basin, Utah. *The American Association of Petroleum Geologists Bulletin*, **38**, 1671–1698.

HVOSLEV, S., LARTER, S. & LEYTHAEUSER, D. 1989. Aspects of generation and migration of hydrocarbons from coal-bearing strata of the Hitra Fm., Haltenbanken area, Offshore Norway. *In:* MATTAVELLI, L. & NOVELLI, L. (eds) *Advances in Organic Geochemistry 1987*. Pergamon, Oxford, 525–536.

JONES, R. W. 1980. Some mass balance constraints on migration mechanisms. *In:* ROBERTS, W. H. & CORDELL, R. J. (eds) *Problems of petroleum migration.* The American Association of Petroleum Geologists Studies in Geology, **10**, 47–68.

KHAN, S. U. & SCHNITZER, M. 1972. The retention of hydrophobic organic compounds by humic acid. *Geochimica et Cosmochimica Acta*, **36**, 745–754.

KHORASANI, G. K. 1987. Oil-prone coals of the Walloon Coal Measures, Surat Basin, Australia. *In:* SCOTT, A. C. (ed.) *Coal and Coal-bearing Strata: Recent Advances.* Geological Society, London, Special Publication, **32**, 303–310.

LARTER, S. 1988. Some pragmatic perspectives in source rock geochemistry. *Marine and Petroleum Geology*, **5**, 194–204.

LEYTHAEUSER, D., SCHAEFER, R. G. & RADKE, M. 1987. On the primary migration of petroleum. *Proceedings 12th World Congress, Houston, 1987, Vol. 2.* Special Paper No. 2. J. Wiley & Sons, London.

——, —— & —— 1988a. Geochemical effects of primary migration of petroleum in Kimmeridge source rocks from Brae field area, North Sea. I: Gross composition of soluble organic matter and molecular composition of C_{15+}-saturated hydrocarbons. *Geochimica et Cosmochimica Acta*, **52**, 701–713.

——, RADKE, M. & WILLSCH, 1988b. Geochemical effects of primary migration of petroleum in Kimmeridge source rocks from Brae field area, North Sea. II: Molecular composition of alkylated napthalenes, phenanthrenes, benzo- and dibenzothiophenes. *Geochimica et Cosmochimica Acta*, **52**,

2879–2891.

Low, P. F. 1976. Viscosity of interlayer water in montmorillonite. *Proceedings of Soil Science Society of America,* **40**, 500–505.

McAuliffe, C. D. 1980. Oil and gas migration: chemical and physical constraints. *In:* Roberts, W. H. & Cordell, R. J. (eds) *Problems of petroleum migration.* The American Association of Petroleum Geologists Studies in Geology, **10**, 89–107.

Mackenzie, A. S. 1984. Applications of biological markers in petroleum geochemistry. *In:* Brooks, J. & Welte, D. H. (eds) *Advances in Petroleum Geochemistry.* Academic Press, 115–215.

—— & Quigley, T. M. 1988. Principles of geochemical prospect appraisal. *The American Association of Petroleum Geologists Bulletin,* **72**, 399–415.

——, Leythaeuser, D., Muller, P., Radke, M. & Schaefer, R. G. 1987. The expulsion of petroleum from Kimmeridge Clay source rocks in the area of the Brae oilfield, U.K. Continental Shelf. *In:* Brooks, J. & Glennie, K. W. (eds) *Proceedings of 3rd Conference on the Petroleum Geology of NW Europe,* Graham and Trotman, London, 865–878.

Magara, K. 1980. Agents for primary hydrocarbon migration. *In:* Roberts, W. H. & Cordell, R. J. (eds) *Problems of petroleum migration.* AAPG Studies in Geology, **10**, 33–45.

Mann, U., Duppenbecker, S., Langen, A., Ropertz, B. & Welte, D. H. 1989. Petroleum pathways during primary migration: evidence and implications (lower Toarcian, Hils syncline, NW Germany) (abstract). *Proceedings of 14th International Meeting on Organic Geochemistry,* Paris, September 18–22, 1989.

Marzec, A., Juzwa, M., Betlej, K. & Sobkowiak, M. 1979. Bituminous coal extraction in terms of electron-donor and -acceptor interactions in the solvent/coal system. *Fuel Processing Technology,* **2**, 35–44.

Meissner, F. F. 1976. Abnormal electrical resistivity and fluid pressure in Bakken Fm., Williston Basin, and its relation to petroleum generation, migration and accumulation. *The American Association of Petroleum Geologists Bulletin,* **60**, 1403–1404.

—— 1978. Petroleum geology of Bakken Fm., Williston Basin, North Dakota and Montana. *In: Williston Basin Symposium.* Montana Geological Society, 207–227.

—— 1987. Mechanisms and patterns of gas generation, storage, expulsion-migration and accumulation associated with coal measures. Green River and San Juan basins, Rocky Mountain region, USA. *In:* Doligez, B. (ed.) *Migration of hydrocarbons in sedimentary basins.* Proc. 2nd IFP Exploration Research Conference, Carcans, June 15–19, 1987. Editions Technip, 79–112.

Murray, G. H. 1968. Quantitative fracture study-Sanish pool, McKenzie County, North Dakota. *The American Association of Petroleum Geologists Bulletin,* **52**, 57–65.

Oen, P. M., Engel-Jensen, M. & Barendregt, A. A. 1986. Skjold Field, Danish North Sea: early evaluations of oil recovery through water inhibition in a fractured reservoir. *Society of Petroleum Engineers 61st Annual Technical Conference and Exhibition of the Society of Petroleum Engineers,* New Orleans, October 5–8, 1986, SPE Paper 15569.

Pepper, A. S. 1989. Petroleum expulsion behaviour of source rocks: a novel perspective (abstract). *Proceedings of 14th International Meeting on Organic Geochemistry,* Paris, September 18–22, 1989.

Peters, W. & Juntgen, H. 1968. Einfluss der hohlraum-struktur auf gas/feststoff-reactionen an steinkohle. *Chim. Ing. Tech.,* **40**, 1039–1044.

Quigley, T. M. & Mackenzie, A. S. 1988. The temperatures of oil and gas formation in the subsurface. *Nature,* **333**, 549–552.

——, —— & Gray, J. R. 1987. Kinetic theory of petroleum generation. *In:* Doligez, B. (ed.) *Migration of hydrocarbons in sedimentary basins.* Proc. 2nd IFP Exploration Research Conference, Carcans, June 15–19, 1987. Editions Technip, 649–666.

Radke, M. 1987. Organic geochemistry of aromatic hydrocarbons. *Advances in Petroleum Geochemistry,* Academic Press, **2**, 141–207.

——, Schaefer, R. G., Leythaeuser, D. & Teichmuller, M. 1980. Composition of soluble organic matter in coals: relation to rank and liptinite fluorescence. *Geochimica et Cosmochimica Acta,* **44**, 1787–1800.

Rightmire, C. T. 1984. Coal bed methane resource. *In:* Rightmire, C. T., Eddy, G. E. & Kirr, J. N. (eds) *Coalbed methane resources of the United States.* The American Association of Petroleum Geologists Studies in Geology, **17**, 1–13.

Rullkotter, J., Leythaeuser, D., Horsfield, B., Littke, R., Mann, U., Muller, P. J., Radke, M., Schaefer, R. G., Schenk, H. J., Schwochau, K., Witte, E. G. & Welte, D. H. 1989. Organic matter maturation under the influence of a deep intrusive heat source: a natural experiment for quantitation of hydrocarbon generation and expulsion from a petroleum source rock (Toarcian shale, northern Germany). *In:* Mattavelli, L. & Novelli, L. (eds) *Advances in Organic Geochemistry 1987.* Pergamon Journals, Oxford, 847–856.

Satriana, M. 1980. *Unconventional natural gas resources, potential and technology.* Energy Technology Review No. 56, Noyes Data Corporation, New Jersey.

Schlumberger 1987. *Log interpretation principles/applications.* Schlumberger Educational Services, Houston, Texas.

Stach, E., Mackowsky, M. Th., Teichmuller, M., Taylor, G. H., Chandra, D. & Teichmuller, R. 1975. *Stach's Textbook of Coal Petrology.* Gebruder Borntraeger, Berlin.

Stainforth, J. G. 1988. Primary migration of hydrocarbons by activated diffusion through organic matter networks (abstract). *The American Association of Petroleum Geologists Conference on petroleum potential of sedimentary basins—methods, techniques and approaches.* Leesburg, Virginia,

April 1988.

—— & REINDERS, J. E. A. 1989. Primary migration of hydrocarbons by diffusion through organic matter networks (abstract). *Proceedings of 14th International Meeting on Organic Geochemistry*, Paris, September 18–22, 1989.

TALUKDAR, S., GALLANGO, O., VALLEJOS, C. & RUGGIERO, A. 1988. Observations on the primary migration of oil in the La Luna source rocks of the Maricaibo Basin, Venezuela. *Revue de l'Institut Francais du Petrole*, **43**, 59–77.

TARAFA, M. E., WHELAN, J. K. & FARRINGTON, J. W. 1988. Investigation on the effects of organic solvent extraction on whole-rock pyrolysis: multiple-lobed and symmetrical P_2 peaks. *Organic Geochemistry*, **12**, 137–149.

TEICHMULLER, M. 1974. Generation of petroleum-like substances in coal seams as seen under the microscope. *In*: TISSOT, B. & BIENNER, F. (eds) *Advances in Organic Geochemistry 1973*, Editions Technip, 379–395.

TISSOT, B. P. & WELTE, D. H. 1984. *Petroleum formation and occurrence*. 2nd Edition. Springer Verlag.

——, CALIFET-DEBYSER, Y., DEROO, G. & OUDIN, J. L. 1971. Origin and evolution of hydrocarbons in early Toarcian shales, Paris basin, France. *The American Association of Petroleum Geologists Bulletin*, **55**, 2177–2193.

——, PELET, R. & UNGERER, P. 1987a. Thermal history of sedimentary basins, maturation indices and kinetics of oil and gas generation. *The American Association of Petroleum Geologists Bulletin*, **71**, 1445–1466.

——, WELTE, D. H. & DURAND, B. 1987b. The role of geochemistry in exploration risk and decision making. *Proceedings of 12th World Petroleum Congress*, **2**, 99–112.

TOTAL 1984. *Handil field: 500 million barrels*. Total Indonesie/Inpex commemorative brochure.

UNGERER, P., BESSIS, F., CHENET, P. Y., DURAND, B., NOGARET, E., CHIARELLI, A., OUDIN, J. L. & PERRIN, J. F. 1984. Geological and geochemical models in oil exploration: principles and practical examples. *In*: DEMAISON, G. & MURRIS, R. J. (eds) *The American Association of Petroleum Geologists Memoir*, **35**, 53–57.

VAN DE WEERD, A., ARMIN, R. A., MAHADI, S. & WARE, P. L. B. 1987. Geological setting of the Kerandan gas and condensate discovery, Tertiary sedimentation and palaeogeography of the NW part of the Kutei Basin, Kalimantan, Indonesia. *Proceedings of the Indonesian Petroleum Association, Sixteenth Annual Convention*, October 1987, 317–338.

VAN KREVELEN, D. W. 1961. *Coal*. Elsevier Publishing Company.

WANG, L. & BARKER, C. 1989. Effects of organic matter content and maturity on oil expulsion from petroleum source rocks (abstract). *The American Association of Petroleum Geologists Bulletin*, **73**, 1051.

WELTE, D. H. & YALCIN, M. N. 1989. Basin modelling—a new comprehensive method in petroleum geology. *In*: MATTAVELLI, L. & NOVELLI, L. (eds) *Advances in Organic Geochemistry 1987*. Pergamon Journals, Oxford, 141–151.

WHELAN, J. K., HUNT, J. M., JASPER, J. & HUC, A. 1984. Migration of C1–C8 hydrocarbons in marine sediments. *Organic Geochemistry*, **6**, 683–694.

YALCIN, M. N. & WELTE, D. H. 1988. The thermal evolution of sedimentary basins and significance for hydrocarbon generation. *The Turkish Association of Petroleum Geologists Bulletin*, **1**, 12–26.

——, ——, MISRA, K. N., MANDAL, S. K., BALAN, K. C., MEHROTRA, K. L., LOHAR, B. L., KUMAR, S. P. & MISRA, H. S. 1987. 3-D computer aided basin modelling of Cambay Basin, India—a case history of hydrocarbon generation. *In*; KUMAR et al. (eds) *Petroleum geochemistry and exploration in the Afro-Asian region*. Balkema, Rotterdam, 417–450.

YOUNG, A. & McIVER, R. D. 1977. Distribution of hydrocarbons between oils and associated fine grained sedimentary rocks—physical chemistry applied to petroleum geochemistry II. *The American Association of Petroleum Geologists Bulletin*, **61**, 1407–1436.

Expulsion of petroleum from type III kerogen source rocks in gaseous solution: modelling of solubility fractionation

D. LEYTHAEUSER & H. S. POELCHAU

Institute of Petroleum and Organic Geochemistry, Research Center Jülich (KFA),
Germany

Abstract: The composition of n-alkane mixtures expelled from type III kerogen-bearing source rocks at moderate maturity levels shows strong signs of fractionation according to molecular chain length. This effect is attributed to the process of expulsion in gaseous solution. The degree of fractionation varies with the pressure and temperature conditions of primary migration. Fractionation effects for C_{15+}-n-alkanes have been simulated with the help of equations developed from experimental data of Price *et al.* (1983) for gaseous solution of oil. For known compositions of expelled n-alkanes our simulation model allows the estimation of pressure and temperature conditions during primary migration. Expulsion of petroleum in gaseous solution is a particularly effective migration mechanism for overpressured type III kerogen-bearing source rocks.

Two geological case histories are presented which demonstrate the compositional effects of petroleum expulsion in gaseous solution from such source beds. The observed n-alkane distributions expelled can be closely matched by simulated distributions, which were calculated for *P–T* conditions estimated from reconstructions of the burial histories of both case histories.

Traditionally, the composition of the hydrocarbon products generated in, and expelled from, shale source rocks was believed to be controlled only by its kerogen quality and maturity (e.g. Tissot & Welte 1984; Hunt 1979). Only recently has it become apparent that additional important controls are exercised by the type of expulsion mechanism and by PVT-related changes during secondary migration (e.g. Larter 1988; England & Mackenzie 1989). Based on detailed analysis of geochemical effects of expulsion from molecular composition of the residual bitumen remaining in the source rock pore system after expulsion, Mackenzie *et al.* (1987) and Leythaeuser *et al.* (1987) have shown the relationship between type of expulsion mechanism and kerogen quality and richness.

Rich, mature source rocks bearing type II kerogens expel petroleum as a separate fluid phase driven by pressure gradients resulting from a combination of shale compaction, volume expansion with kerogen conversion (e.g. Tissot & Welte 1984; Hunt 1979) and in addition, near the edges of thick source rock units, by capillary forces (Hubbert 1953; Mackenzie *et al.* 1987; Leythaeuser *et al.* 1988). Under equivalent maturity conditions, moderate-quality source rocks bearing hydrogen-leaner kerogens of a type III nature expel their oil in a different way. Such source rocks generate abundant gas and relatively little oil, which is expelled in gaseous solution. Since solubilities of petroleum compounds in gas depend on molecular size and type (e.g. Zhuze *et al.* 1973), fractionation effects can be associated with this type of expulsion. This condition applies as long as the gas acting as a solvent is in limited supply. In a recent review paper Price (1989) has stressed the role of gaseous solution as a migration mechanism for such source rocks.

In the present paper an integrated approach was applied to assess the role and the effects of gaseous solution as an expulsion mechanism. Based on published experimental data (Price *et al.* 1983) a simple numerical model was established to simulate the compositional fractionation effects due to expulsion of petroleum hydrocarbons by gaseous solution for any pressure–temperature (*P-T*) combination reached during source rock burial. For two source rock examples, where expulsion by gaseous solution had previously been demonstrated as the main migration process, a close match is shown between the observed and the calculated distribution envelopes for the C_{15+}-n-alkanes. Also, the calculated *P-T* conditions for both case histories appear to match reasonably well with those which, based on geological reconstructions, must have prevailed during maximum burial while these source rocks generated and expelled petroleum hydrocarbons. Based on these findings, it is shown that expulsion of petroleum hydrocarbons in gaseous solution can be an important process for thick, over-pres-

From England, W. A. & Fleet, A. J. (eds), *Petroleum Migration*
Geological Society, Special Publication No. 59, pp. 33–46.

sured sequences with type III kerogen-bearing source strata.

Geochemical migration effects revealing expulsion mechanisms

The molecular composition of the residual petroleum remaining in the source rock pore system after expulsion, has previously been demonstrated to be a key for recognizing the effects of migration and, in particular, for revealing the type of expulsion mechanism (Leythaeuser & Schaefer 1984; Leythaeuser *et al.* 1984*a, b* 1987, 1988; Mackenzie *et al.* 1987). The basic approach involved hydrocarbon mass balances between samples from the centre of thick source rock units and their margins adjacent to sandstone reservoir rocks. The most pronounced migration effects, however, were observed for comparisons of samples from thin shale layers interbedded in reservoir sands with samples of equal kerogen quality from the centre of adjacent thick source rock units. As an example, Fig. 1 shows this approach for a sample pair from a mature (0.82% R_m), type III kerogen-bearing source rock sequence from the Palaeocene Firkanten Formation of Svalbard, Norway. Although both source rock samples have had the same conditions for hydrocarbon generation (i.e. similar kerogen quality and same temperature history) the composition of their saturated C_{15+}-hydrocarbons differs drastically (Leythaeuser *et al.* 1984*a*). Sample 127.0 m taken from the centre of a thick source rock unit experienced little or no expulsion of oil. Therefore, the compositional, signature of the originally generated hydrocarbon mixture (broad, bimodal distribution of n-alkanes; predominance of n-alkanes over isoprenoids) was largely preserved. By contrast, sample 62.5 m taken from a 5 cm thin shale layer reveals a unimodal, heavy-end biased n-alkane distribution with a pronounced predominance of pristane. Obviously, this source rock has expelled hydrocarbons to a much higher degree than sample 127.0 m, and it has preferentially expelled lower molecular weight n-alkanes. This qualitative difference, indicated schematically in Fig. 1 by superposition of both n-alkanes distribution envelopes, has been quantified based on determination of absolute concentrations for each compound shown. The n-alkanes expelled from the thin shale bed 62.5 m, as determined by this mass balance (curve A in Fig. 2), reveal a front-end biased composition, which resembles that of a light oil or gas condensate. This feature is remarkable in view of the moderate maturity and the type III kerogen

comprising this source rock, characteristics which are generally believed to generate oils with heavy-end biased n-alkane distributions. Taking the n-alkane concentrations of sample 127.0 m as reference, relative expulsion efficiencies can be determined for each molecule (difference in concentration as percentage of the reference concentration). The molecular dependency of this relative expulsion efficiency shown in Fig. 3 (trend A) is a measure of the degree of fractionation which was associated with primary migration in this source rock. Moreover, this fractionation according to molecular chain length clearly indicates transport of these n-alkanes in gaseous solution, which is known to decrease sharply with increasing carbon number (Mackenzie *et al.* 1987).

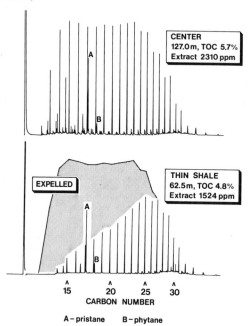

Fig. 1. Capillary gas chromatograms indicating composition of saturated C_{15+}-hydrocarbons of two source rock samples from the Palaeocene Firkanten Fm. (RD) from Svalbard, Norway (modified after Leythaeuser *et al.* 1984*a*). Sample 127 m is from the centre of a thick source rock unit, while sample 62.5 m is from a 5 cm thin shale layer of the same kerogen quality interbedded between reservoir-sands. Basic geochemical data for both samples are indicated. Prominent peaks are n-alkanes of indicated carbon number range, while A and B denote pristane and phytane, respectively. Since hydrocarbon generation conditions were the same for both source rock samples, the compositional difference is attributed to differential expulsion: the composition of n-alkanes expelled from sample 62.5 m is indicated schematically by the difference between the two n-alkane distributions (stippled area).

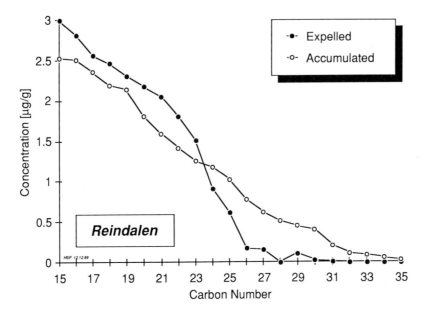

Fig. 2. Concentrations (µg g⁻¹)versus carbon number for two n-alkane mixtures from the Palaeocene Firkanten Formation (RD), Svalbard (modified after Leythaeuser *et al.* 1984a): (**A**) 'Expelled' n-alkanes from thin source rock layer at 62.5 m, as determined by mass balance between the two samples shown in Fig. 1. (**B**) 'Accumulated' n-alkanes obtained by solvent extraction of a nearby, impregnated sandstone.

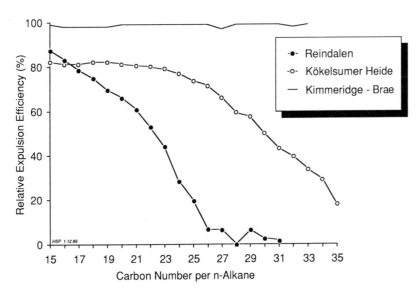

Fig. 3. Relative expulsion efficiencies (%) of n-alkanes expelled from thinly interbedded shale source rock layers.
(**A**) Reindalen, Palaeocene Firkanten Fm., sample 62.5 m from Reindalen (RD) corehole Svalbard (kerogen type III, 0.82% R_m); modified from Leythaeuser *et al.* (1984a). (**B**) Kökelsumer Heide, Upper Carboniferous Westphalian B, sample 1126.7 m from Kökelsumer Heide (KH) corehole (kerogen type III, 1.1% R_m); modified from Leythaeuser & Schaefer (1984). (**C**) Kimmeridge, Brae Area, Upper Jurassic Kimmeridge Clay Fm., sample 4072 m from well 1 in the Brae area, British North Sea (kerogen type II, 0.67% R_m); modified from Leythaeuser *et al.* (1988).

Table 1. *Normalized carbon-range distributions (in %) of Spindle field crude oil samples dissolved in methane at experimental P–T conditions, and the whole oil solubility.*

T(°C)	P(bar)	C_4–C_{10}	C_{11}–C_{15}	C_{16}–C_{20}	C_{21}–C_{25}	C_{26}–C_{30}	C_{31+}	oil sol. (g l^{-1})
49.4	327	33.60	53.55	11.27	1.59	0.00	0.00	0.060
49.2	356	40.67	44.00	11.72	2.93	0.69	0.00	0.130
49.2	413	35.48	47.28	11.82	4.05	1.37	0.00	0.153
49.2	529	31.32	38.06	18.25	8.47	2.19	1.70	0.207
49.5	621	27.08	38.16	20.80	8.68	3.39	1.59	0.282
49.5	626	25.82	37.62	20.36	9.10	4.09	3.01	0.284
49.8	688	20.47	37.18	22.77	10.75	5.24	3.38	0.360
49.4	807	17.20	30.14	22.04	13.91	8.25	8.47	0.485
98.3	249	35.81	51.52	9.92	2.23	0.57	0.00	0.103
97.8	343	36.08	46.17	13.51	3.16	1.08	0.00	0.184
99.2	468	36.30	39.22	16.57	5.30	1.81	0.80	0.267
99.2	588	29.13	31.40	18.75	9.92	5.50	5.30	0.393
99.8	701	17.88	25.97	19.62	12.91	8.99	14.63	0.530
99.4	702	16.88	27.98	19.39	12.96	9.10	13.69	0.530
99.8	811	18.71	27.77	19.27	13.08	8.83	12.34	0.757
99.4	1008	19.51	25.66	18.46	13.11	8.61	14.64	1.362
149.2	76	49.45	40.80	6.73	2.25	0.77	0.00	0.123
150.2	255	38.57	41.91	12.62	4.70	2.20	0.00	0.190
150.2	364	34.00	43.53	15.33	4.81	2.01	0.32	0.261
150.5	495	31.76	33.44	20.12	8.57	3.81	2.20	0.391
150.2	635	21.34	26.81	18.90	12.25	8.15	12.55	0.653
150.0	798	18.27	26.75	19.57	12.57	9.28	13.56	1.030
150.0	909	20.59	26.22	19.07	12.76	8.12	13.24	1.450
205.8	88	59.48	33.71	5.60	1.01	0.20	0.00	0.200
204.8	163	38.76	45.23	12.04	3.51	0.46	0.00	0.270
203.3	197	38.52	45.70	12.98	2.46	0.34	0.00	0.330
205.8	340	38.14	37.45	15.77	5.51	2.08	1.05	0.434
205.2	458	23.83	30.20	24.57	12.21	6.81	2.38	0.600
205.8	609	19.31	26.95	19.78	12.97	8.82	12.18	0.900
251.1	63	34.61	54.93	8.55	1.91	0.00	0.00	1.120
248.9	90	38.44	49.34	10.44	1.77	0.00	0.00	0.699
250.2	161	38.63	44.36	13.42	3.03	0.55	0.00	0.489
251.1	259	30.79	41.48	17.37	6.26	2.57	1.52	0.507
251.3	452	14.87	31.77	22.12	13.09	9.26	9.00	0.828
251.1	523	10.94	28.91	21.24	14.39	9.93	14.59	1.150
25.0	1	29.30	23.32	17.03	11.13	7.29	11.94	

Selected from data in Price *et al.* (1983) as database for the simulation model. The last line gives the composition of the crude oil used in the experiments, also shown in Fig. 4.

The same approach was applied to a more mature (1.1% R_m), but otherwise similar sample pair from an Upper Carboniferous sequence in the Ruhr area, Germany (well Kökelsumer Heide, KH; Leythaeuser & Schaefer 1984). It revealed uniformly high relative expulsion efficiencies up to about n-C_{25} and fractionation effects limited to the molecular range beyond. This decrease in the degree of molecular fractionation with increasing maturity, as illustrated by comparison of trends A and B in Fig. 3, was interpreted to reflect a progressive evolution of the relationships between amount of gas available for dissolution and quantity of C_{15+}-n-alkanes generated. As more and more gas is generated and expelled, all but the very long-chain n-alkanes are mobilized by gaseous solution. In summary, the n-alkane mixture expelled from low to moderate-maturity source rocks is in compositional disequilibrium with the residual fraction left in the source rock. The composition of this initially expelled mixture is not controlled by the nature of the kerogen in the source rock, but by the fractionation effects during primary migration, due to expulsion of petroleum in gaseous solution. If these initially expelled hydrocarbons accumulate in a trap, a light oil or gas condensate deposit is formed even though the hydrocarbons are derived from a type III source rock. With progressing maturity compositional equilibrium is gradually reached between the expelled and the residual hydrocarbon fractions, which ultimately represents a steady-state situation. At this stage the composition of the expelled and accumulated hydrocarbons indeed reflects the nature of the native kerogen.

Mature source rocks bearing a hydrogen-rich type II kerogen generate predominantly oil and only small quantities of gas over most of the liquid window interval. Gas is expelled by dissolution in oil, i.e. no separate gas phase evolves during expulsion and all hydrocarbons are expelled as a single phase fluid. Therefore, as shown by trend C in Fig. 3 for a similar comparison of type II kerogen-bearing samples, no molecular fractionation effects are observed for the C_{15+}-n-alkanes. Although a separate phase flow has been proposed by many authors to explain petroleum expulsion from rich, type II kerogen-bearing source rocks (e.g. Tissot & Welte 1984; Durand 1988), the data shown in trend C in Fig. 3 are believed to represent, for the first time, direct chemical evidence in support of this conclusion.

Development of a model to simulate fractionation effects due to petroleum expulsion in gaseous solution

Numerical simulation models, based on basic laws of physics and chemistry and calibrated with real data, permit quantitative assessment of the role of geological and geochemical processes. Moreover, they permit the reconstruction of processes and evolution of source rocks with geological time. A model to simulate compositional fractionation effects resulting from expulsion of oil in gaseous solution should ideally be based on the equation of state for the PVT relationships of all oil and gas components and mixtures. This approach has been attempted by Nogaret (1983) but has been unsuccessful mainly because the necessary characteristic parameters of most components are unknown. Instead, this study presents a more pragmatic approach by generating a solubility equation fitted to existing experimental solubility measurements. Application of this model to the case histories introduced in the

previous chapter allows prediction of the geological pressure and temperature conditions at which expulsion of n-alkanes in gaseous solution may have occurred.

Experimental input data

The data on which the model is based come from the experiments of Wenger and Price (Price et al. 1983) who determined the solubility in methane of crude oil from the Cretaceous Sussex Sandstone reservoir of the Spindle field (Denver Basin, Colorado) at varying P–T conditions up to 250°C and 1000 bar. Figure 4 shows the range of experimental P–T conditions used and the trend of whole oil solubility in methane with pressure and temperature. The contours are drawn by hand and reflect a reversal in solubility shown in their data for high temperature and low pressures. The cause of the reversal is uncertain. Price et al. (1983) comment that these reversals 'were actually experimental artifacts' caused by 'selective solution of the lower molecular weight hydrocarbons at lower pressures and depletion of those hydrocarbons in the hydrocarbon reservoir of the pressure vessel'. Another cause could be the relatively high water content listed in table 1 of Price et al. (1983) for these samples. However, a similar reversal is obvious in the solubility curve of Moiseev et al. (1982, fig. 2).

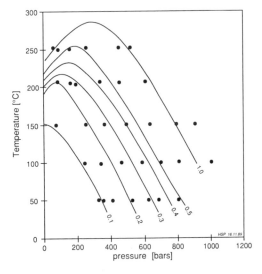

Fig. 4. Whole oil solubility (g l⁻¹) of Spindle field crude in methane as a function of experimental temperature and pressure, contoured in 0.1 mg/l concentration. Selected data from Price et al. (1983).

In addition to determination of dissolved concentrations, Price *et al.* (1983) analysed the composition of the dissolved hydrocarbon fractions by gas chromatography and reported some 45 analyses in terms of carbon number distributions in normalized percentages. 35 of these analyses were selected for this study (Table 1 and Fig. 4), the rest were found unsuitable either because of poor fit in the temperature or pressure trends or errors in the data.

These experimental data were prepared and transformed for regression as follows: the compositional percentage values were divided by the range of carbon numbers in the class (usually 5) to generate a midpoint value, and assigned to the middle carbon number of the class (see Fig. 5). For instance, given a C_{11}–C_{15} fraction of 25%, a value of 5% was assigned to C_{13}. This percent value was then multiplied with the whole oil solubility (last column in Table 1) at the given pressure and temperature to obtain the amount, or concentration (in mg l^{-1}), of all the compounds with this same carbon number. In some cases the whole oil solubility values had to be interpolated because the *P–T* values listed by Price *et al.* (1983) did not always coincide with those of the analyses.

dent variable 'dissolved concentration' was log-transformed to improve the fit and avoid negative concentrations.

Several regressions were run with various variable and data combinations. In some cases it was tried to force certain variables to remain in the equation. Most of the resulting fits had correlation coefficients around 0.97 with the number of variable terms included ranging from 8 to 11.

Selected equations were then tested in two ways. First, for a representative range of pressures and temperatures dissolved carbon number distributions were calculated, reduced to percentages, and lumped into the same groups as the original data. The shape of these distribution curves was then compared with the original experimental data at the same *P–T* values. Secondly, the calculated dissolved concentrations were added up to give whole oil solubility and compared with the corresponding experimental solubilities at selected *P–T* values.

Based on these comparisons the regression equation giving the best fit (here named P/W C1) was chosen in this study for simulations of actual case histories. The regression coefficients and associated variables for this equation are given in Table 2. The fit is best for intermediate temperatures (50–200°C) and medium to high pressures (250–600 bar).

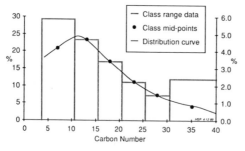

Fig. 5. Carbon number distribution of hydrocarbons of the original Spindle field oil used by Price *et al.* (1983) for solubility experiments. Shown are groupings of compounds into classes of 5 or more (bars, left scale), the class mid-point values (dark dots), and an interpreted continuous distribution (right scale).

Table 2. *Coefficients for solubility simulation equation P/W C1*

Variable	Coefficient
Intercept	1.4100×10^{-1}
C	1.1800×10^{-1}
C^2	-1.2000×10^{-3}
C^3	1.5734×10^{-4}
T^3	4.8373×10^{-8}
CT	2.1058×10^{-4}
CP	2.1000×10^{-4}
CP^2	-7.5888×10^{-8}
PT^2	-2.1544×10^{-15}
standard error	0.171
$R =$	0.971

$\log_{10} X = \Sigma(\text{coeff.} \times \text{var.}) + \text{intercept}$ where $C =$ carbon number; $P =$ pressure in bars; $T =$ temperature in °C; $X =$ dissolved concentration.

Regression analysis

Standard step-wise multiple regression (Krumbein & Graybill 1965) was applied to the data set to generate predictive equations for solubilities of each carbon number group at any given temperature and pressure. The analysis considered the independent variables *P*, *T*, and *C* (the mid-point carbon number of each group), as well as squares and cubes of each variable and interaction terms between variables. The dependent

Solubilities v. dissolved composition

The experimental data of Price *et al.* (1983) used for this model are the compositions of the dissolved fraction of crude oil at experimental pressure and temperature conditions. Hence, the regression equation based on these data reproduces only the composition of dissolved Spindle field crude oil at a given pressure and temperature combination.

This means that the model (equation P/W-C1) should only be applied to Spindle field oil or oils that are very similar in composition. However, since the experimental distributions all show an imprint of the original oil distribution, the assumption can be made that the amount of a component in gaseous solution depends not only on its specific solubility but also on the relative original concentration available to be dissolved. If that is the case a relative solubility (independent of the oil used in the experiment) may be derived by simply dividing dissolved distribution with the relative crude oil distribution of the original Spindle field oil. This is illustrated in schematic fashion in Figure 6.

This set of new parameters is essentially ana-

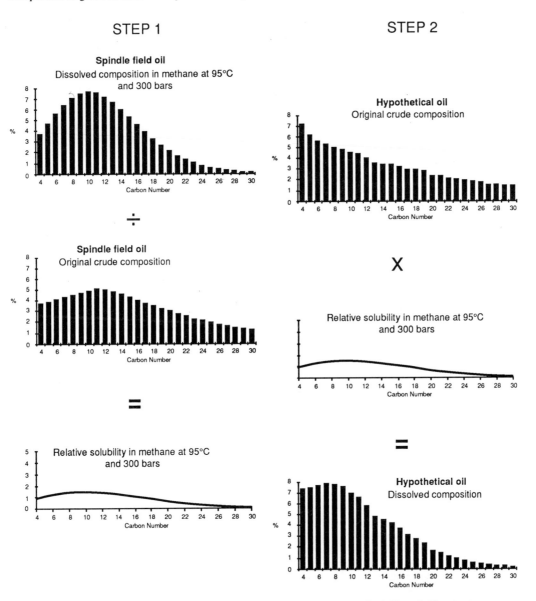

Fig 6. Schematic illustration of procedure used to normalize solubility of Spindle field crude in order to obtain a relative solubility distribution which can be applied to any other oil. Step 1 (left column) shows how the normalized relative solubility is derived. Step 2 (right column) demonstrates the computation of the dissolved fraction of another oil for the same P–T conditions.

logous to a series of vapour-liquid equilibrium constants (Thompson 1987) for each component in the system. The vapour–liquid equilibrium constant is a ratio describing the partitioning of a compound between the vapour and the liquid phase, and depends on the fugacities, the activity and fugacity coefficients, and therefore on P, T, and composition. As Thompson (1988) has shown for his evaporative fractionation mechanism, the composition of the condensate oils, which result from oil transport in the vapour phase, is indeed strongly dependent on the composition of the parent oil.

This normalized relative solubility can be applied to any crude oil to derive the distribution of dissolved composition at given pressure and temperature by simply multiplying the percentage or amount of the corresponding carbon number class of the oil with the relative solubility values. It should be emphasized, however, that the solubilities so derived cannot be used to calculate absolute concentrations of dissolved compounds (e.g. mg l^{-1}) but rather to give the amount of one carbon number class relative to the neighbouring carbon number classes. The result is simply a relative distribution curve.

Assumptions

Although the numerical method is straightforward there are a number of assumptions inherent in the experimental data and in the way they are employed here which need to be kept in mind when using and applying this simulation model.

(1) The distribution of data, or the mathematical hypersurface described by the data, is continuous. This assumption appears warranted as solubility plots of the data (e.g. Fig. 4) show no discontinuities. In addition, ill fitting data have been removed before the regression.

(2) The carbon number ranges can be represented by the central carbon number of each range and assigned a linear proportional percentage of the fraction. This is, or course, a first approximation which works well for smooth, uniformly changing segments of the distribution, but could be inaccurate where large carbon number predominances occur. However, the available data do not allow a better fit.

(3) The experimental data can be extrapolated to other crude oils. The solubility experiments were performed on one specific crude oil (Spindle field crude). It is assumed that this bias can be removed by normalizing the solubilities against the crude composition (see above). How-

ever, the predictive equation based on these data may be valid only for oils of similar composition.

(4) The whole oil solubilities are representative for the individual hydrocarbon compounds, i.e. the effect of the presence of other compounds on the solubility of a given compound can either be ignored, or is already inherent in the experimental data. Since the end results of the simulation consist of relative percentages, not of absolute concentrations, the assumption reduces to the statement that the ratios of solubilities between individual compounds of different carbon-numbers should be similar to solubility ratios of corresponding groups of compounds. This assumption is crucial when the method is applied not to whole oil composition but to n-alkane distributions as in this study.

(5) Laboratory data can be applied and extrapolated to natural, geological systems. The solvent in the experiments was pure methane (with varying amounts of water). While presence of water according to Price et al. (1983) enhances the capacity of methane to dissolve oil, it is also known (Nogaret 1983) that natural gas which includes admixtures of ethane, propane, etc. is capable of dissolving twice as much oil as pure methane. This means that the solubilities of this model probably lie below those expected in nature. It is not clear, however, if the increased solubility due to natural gas will, similar to solubility increased by temperature or pressure, shift the dissolved composition patterns towards higher carbon-numbers. If that were true the overpressures required in the case histories discussed below could be reduced to some degree.

Application of the solubility simulation model: comparison of measured v. calculated n-alkane distributions

The simulation model of hydrocarbon fractionation by solution in gas during primary migration can be applied in two ways: (i) reliable constraints can be placed on the estimates of geological P–T conditions under which primary migration took place for the particular case histories considered, or (ii) if P–T conditions are known from geological reconstructions, modeling results can be used as an argument in favour of petroleum migration in gaseous solution. Also, the n-alkane distribution of an expelled oil can be predicted for any geologically reasonable combination of P–T conditions.

For either application of the model the composition of the original oil prior to migration as well as that of the oil expelled in gaseous solution has to be known. This is the case for the two examples to which the model was applied, i.e., source rocks of the Palaeocene Firkanten Formation from Svalbard and Upper Carboniferous strata from the Ruhr area, Germany. The molecular evidence indicating expulsion of oil in gaseous solution as well as basic geochemical data were given in the first part of this paper. In each case the composition of the 'original oil' prior to expulsion is assumed to be the same as that of the bitumen extract of the sample from the centre of the thick source rock unit (termed 'reference sample' above) which is presumed to have expelled little or no oil. The composition of the 'expelled oil' was obtained, as explained above, by the mass balance between reference sample and thin shale sample. In the case from the northwest German Upper Carboniferous (KH), a range of reasonable pressure and temperature values can be estimated based on published information about the geological history and vitrinite reflectance measurements. The simulation model can therefore be tested against a real case history. In the second case (RD) from the Palaeocene of Svalbard the model is used to estimate the subsurface conditions of oil expulsion.

The core samples from the well Kökelsumer Heide (KH) are from dark gray shales of Upper Carboniferous Westphalian B age (Leythaeuser & Schaefer 1984). It is known that the coalification process of Carboniferous strata in the Ruhr area had advanced to nearly present-day rank levels by the beginning of Permian times (Teichmüller & Teichmüller 1971). This was achieved by rapid subsidence and burial during the Westphalian and Stephanian followed by strong uplift during uppermost Stephanian times. Using published data on depth vs. vitrinite reflectance trends for Upper Carboniferous strata (Scheidt & Littke 1989), a maximum burial depth of 3–3.5 km can be estimated for this KH shale unit. This rather rapid burial of the predominantly shaly sequence in less than 12 Ma (Hess & Lippold 1986) must have caused pore fluid pressures to rise well above hydrostatic. Scheidt & Littke (1989) concluded that vitrinite reflectance gradients with depth for Upper Carboniferous strata in the general area of the KH well are comparable to those in the Upper Rhine Graben, where rather high geothermal gradients in the range 40–80 K km^{-1} are observed at present. Therefore, 48.5 and 60 K km^{-1} were selected as lower and upper limits for geothermal gradients which must have prevailed during the time of hydrocarbon generation and migration. Based on published vitrinite reflectance vs. temperature relationships (Quigley et al. 1988) the measured

vitrinite reflectance of 1.1% R_{m} is converted into a temperature estimate of 160–180°C. These boundary conditions for pressure and temperature at the time of maximum burial of the KH shale unit are displayed in Fig. 7 as ranges for the 3000 m and 3500 m depth levels.

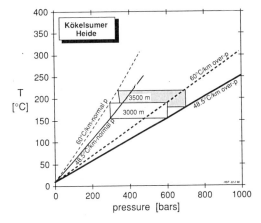

Fig. 7. Pressure and temperature conditions of shale unit studied at Kökelsumer Heide (KH) which are considered to be geologically possible and reasonable. Shown are *P–T* curves for normal and over-pressures and geothermal gradients of 48.5 and 60°C km^{-1} which limit the range of pressure and temperature considered possible at the two depths, 3000 m and 3500 m.

Using this range of geologically realistic *P–T* conditions the simulation model can generate any number of n-alkane distributions resulting from transport in gaseous solution. From these, the calculated distribution which reveals the best match with the observed one can be selected. In Fig. 8 three n-alkane distributions were chosen which cover the extremes of the range from low temperature and normal pressure at 3000 m to the highest temperature and over-pressure at 3500 m depth. Clearly there is an excellent match between the observed n-alkane distribution and the one calculated for dissolution of n-alkanes in gas at temperatures of 160°C and 585 bar pressure. This perfect match in conjunction with the fact that these *P–T* conditions are realistic for the sequence is interpreted as an idication that the mechanism of primary migration of oil had indeed occurred in gaseous solution.

However, a match between the calculated and the observed n-alkane distribution is not unique for this particular *P–T* combination. Raising the temperature of the simulation an equally good match can be found for correspondingly lower pressures. The locus of all these *P–T* pairs is the curve of equal dissolution character (i.e., same

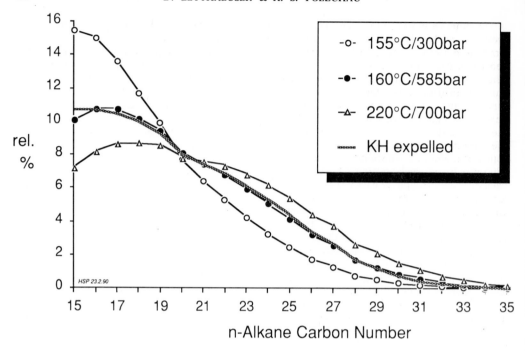

Fig. 8. n-Alkane distribution for the expelled KH oil observed by Leythaeuser & Schaefer (1984) compared to three simulated curves of dissolved oil composition for different pressures and temperatures. Note the excellent match between the expelled oil and the simulated oil at 160°C and 585 bars.

n-alkane distribution patterns) called 'isolyse' which is shown in Fig. 9 for the KH series. Any temperature–pressure value combination along the isolyse is a possible subsurface condition to match the observed n-alkane distribution envelope with a simulated distribution. Thus, the isolyse narrows the choice of P–T combinations within the field of geologically reasonable conditions. Values to construct the isolyse can be generated by running a least-squares fit approximation for a number of fixed temperature values with varying pressures (or vice versa).

The P–T-isolyse diagram (Fig. 9) can also be used to enter a geologically preferred temperature, say 160° C, determine the intersection with the isolyse, and read off the corresponding pressure, in this case 585 bars. In this way it is evident that for the temperatures equivalent to the observed vitrinite reflectance value (1.1%) overpressures are required to explain the calculated solubilities. This is an indirect indication that overpressures must have existed in parts of the Upper Carboniferous sequence. Similar overpressure conditions have been postulated by Radke *et al.* (1982), based on molecular compositions of polycyclic aromatic hydrocarbons from coals. Figure 9 shows furthermore that a depth of 3000 m was sufficient for mobilization

of n-alkanes in gaseous solution and that for the temperatures indicated geothermal gradients ranging from 48.5 to 60 K km^{-1} are appropriate.

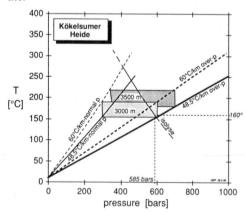

Fig. 9. Pressure and temperature conditions of shale unit studied at Kökelsumer Heide (KH), as in Fig. 7 with isolyse added. The isolyse is the curve which follows the P–T values at which the calculated dissolved composition is the same and matches the expelled KH oil. Indicated is the temperature (160°C) and corresponding pressure considered most likely to have caused the observed fractionation effects due to expulsion in gaseous solution.

The other case history to which the solubility model is applied is the Palaeocene Firkanten Formation from Svalbard (RD sequence), the geochemical data of which were discussed above. The lithology consists of dark shales and siltstones with occasional interbedded sandstones. Unfortunately, the conditions for petroleum generation cannot be reconstructed as well as for the KH sequence. This is due to major uplift of the Firkanten Formation at the RD site. A rough estimate of a maximum burial depth of 2800 m is based on the geological setting (Leythaeuser *et al.* 1984*a*). The measured vitrinite reflectance level of 0.82% R_m would, according to Quigley *et al.* (1988), suggest a maximum temperature in the range of 110 to 130°C for a low to intermediate heating rate. For the following calculations a value of 115°C was assumed. The observed fractionation effects of the n-alkanes according to chain length in the expelled petroleum indicated gaseous solution as the predominant migration mechanism. The geochemical input data for modelling include the n-alkane composition of the petroleum expelled as well as the very similar composition of an oil (curve B in Fig. 2) extracted from a nearby impregnated sandstone (Leythaeuser *et al.* 1984*a*). This oil is assumed to have accumulated due to expulsion from the source rock in gaseous solution. Thus,

it is an ideal case to attempt a match with a simulated n-alkane distribution.

As shown in Fig. 10, the measured n-alkane distribution can be matched fairly closely with the one simulated for the *P–T* combination of 115°C and 421 bar. Again, as in the previous case, an isolyse line, specifically for this n-alkane distribution, can be constructed for a series of pressure and temperature combinations. Figure 11 shows this isolyse as a function of temperature and pressure for various geothermal regimes. Using the temperature estimate 115°C, the pressure required for mobilization of this n-alkane distribution in gaseous solution (421 bars) can be read off the *x*-axis. The intersection with the isolyse falls on a geothermal gradient of 25 K km^{-1}. Given these data either a depth greater than 4000 m, or an overpressure gradient of about 150 bar km^{-1} has to postulated. Thus, in case of an oil known or suspected to have migrated in gaseous solution the simulation model can help to estimate temperature and pressure conditions of oil generation and expulsion.

Discussion of the geological significance

The geological significance of the role of gaseous solution for expulsion of oil from type III kero-

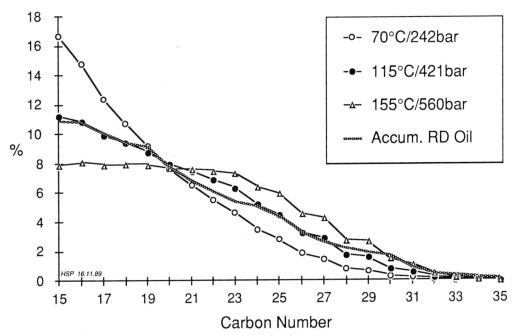

Fig. 10. n-Alkane distribution for the accumulated RD oil (see Fig. 2) in comparison with three simulated curves of dissolved oil composition for different pressures and temperatures. Note the good match of the oil mobilized in gaseous solution at 115°C and 421 bars with the accumulated oil.

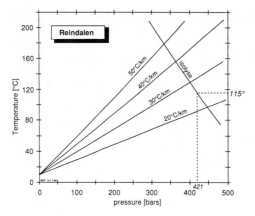

Fig. 11. Range of pressure and temperature conditions which may have prevailed for the Reindalen sequence during oil generation and migration. The isolyse shows the *P–T* conditions for which the best match of dissolved oil simulation curves with the observed oil is achieved. For the chosen temperature value of 115°C a pressure of 421 bars would be required (dotted line).

and thermal cracking of oil during advanced catagenetic stages (R_m around and above 1.3%), but result alternatively also from fractionation-controlled expulsion due to gaseous solution occurring at moderate maturity levels.

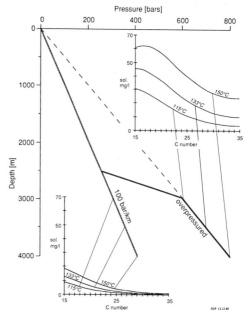

Fig. 12. Application of the solubility model (P/W C1) to a hypothetical basin with type III kerogen-bearing source rocks. Comparison of amount and composition of petroleum expelled in gaseous solution at three depth levels for hydrostatic and overpressure conditions (200 bar km⁻¹) and a geothermal gradient of 35°C km⁻¹.

gen-bearing source rocks can be best illustrated by applying the solubility model to a hypothetical basin (Fig. 12). Hydrostatic and overpressure regimes are compared at a mean geothermal gradient of 35°C km⁻¹. It is assumed that there are mature type III kerogen-bearing source rocks below 3000 m depth, which have generated oil similar in composition to the Spindle field oil. Application of the model allows the prediction of the amount and composition of petroleum expelled in gaseous solution at the indicated depth levels for hydrostatic and overpressure conditions. Figure 12 clearly shows that over-pressures greatly increase the total amount of oil mobilized in gaseous solution compared to normal pressure regimes. Also shown is the evolution of compositional features in the expelled oil with increasing depth, temperature and pressure. The oil expelled in gaseous solution from the overpressured source rock (e.g. at 4000 m depth) reveals a front-end biased distribution, which is typical for light oils or gas/condensates. This feature is, however, in disagreement with the type III kerogen nature of the source rocks which usually lead to generation of waxy oils with heavy-end biased distributions (Hedberg 1968). Thus, these modelling results support earlier conclusions (Leythaeuser *et al.* 1984*a*) of an alternative mode of origin of light oils and gas-condensates. They originate not only from very deeply buried, high maturity source rocks

The principal fractionation mechanism involved is the same as for a process termed 'evaporative fractionation' by Thompson (1987, 1988) which occurs, however, in reservoir accumulations. Using observational and experimental evidence, Thompson demonstrated convincingly, that differential uptake of light hydrocarbons from oil into gaseous solution occurs under certain geological conditions. Combined with later migration of this gas, this leads to formation of new gas-condensate deposits on the one hand, and residual oils of altered light hydrocarbon composition on the other.

In summary, compositional fractionation associated with hydrocarbon transport in gaseous solution can occur in two different subsurface situations: during hydrocarbon expulsion from certain source rocks (as shown in this paper) and during tectonically-induced pressure changes in gas-saturated oil reservoirs (Thompson 1987, 1988). Both processes are able to produce gas-condensate deposits.

Conclusions

Based on an integrated approach using observed molecular compositions of bitumen extracts from source rocks, experimental data on oil solubility in methane from the literature, and numerical modelling the following conclusions have been reached.

(a) Observed distribution patterns of C_{15+}-n-alkanes expelled from type III kerogen-bearing source rocks can be matched closely by simulated distributions calculated for geologically reasonable pressure and temperature conditions. Modelling results indicate that in many instances overpressures must be postulated to accomplish significant oil expulsion by gaseous solution.

(b) Gaseous solution of oil represents an efficient mobilization process during primary migration for type III kerogen-bearing source rocks.

(c) Mobilization of n-alkanes in gaseous solution is enhanced by increased pressure and temperature and is associated with fractionation effects according to molecular chain length.

(d) The degree of fractionation of the expelled n-alkanes varies from high at early stages of migration to very low at advanced stages of migration and maturity. Ultimately, with an oversupply of gas, fractionation effects will no longer be apparent as more and more of the limited amount of oil generated by such source rocks is dissolved and expelled.

(e) For known compositions of n-alkanes expelled from type III source rocks it is possible to estimate the pressure and temperature conditions of primary migration.

(f) The model supports our earlier hypothesis for an alternative mechanism to create gas condensates from type III kerogen source rocks in overpressured settings similar to the Gulf Coast.

The concepts developed here were discussed at various stages with A.S. Mackenzie (BP International Ltd., Sunbury Research Centre) and with several of our colleagues at the Institute of Petroleum and Organic Geochemistry. We are grateful for feedback and input especially from R.G. Schaefer, B. Krooß, R. Littke and S. Düppenbecker. We appreciate the stylistic editing by S. Poelchau and the constructive comments by the reviewers, L.C. Price and especially S. Larter. We also thank D.H. Welte for his support.

References

DURAND, B. 1988. Understanding of hydrocarbon migration in sedimentary basins (present state of knowledge). In: *Advances in Organic Geochemistry 1987. Organic Geochemistry* **13**, 445–459.

ENGLAND, W.A. & MACKENZIE, A.S. 1989. Some aspects of the organic geochemistry of petroleum fluids. *In*: POELCHAU, H. S. & MANN, U. (eds) *Geological Modeling—Aspects of Integrated Basin Analysis. Geologische Rundschau*, **78**, 291–303.

HEDBERG, H. D. 1968. Significance of high-wax oils with respect to genesis of petroleum. *American Association of Petroleum Geologists Bulletin* **52**, 736–750.

HESS, J. C. & LIPPOLT, H. J. 1986. $^{40}Ar/^{39}Ar$ ages of tonstein and tuff sanidines: New calibration points for the improvement of the Upper Carboniferous time scale. *Isotope Geoscience*, **59**, 143–154.

HUBBERT, M. K. 1953. Entrapment of petroleum under hydrodynamic conditions. *American Association of Petroleum Geologists Bulletin* **37**, 2026.

HUNT, J. M. 1979. *Petroleum geochemistry and geology*. W. H. Freeman and Co., San Francisco.

KRUMBEIN, W. C. & GRAYBILL F. A. 1965. *An introduction to statistical models in geology*. McGraw-Hill Book Co., New York.

LARTER S. 1988. Some pragmatic perspectives in source rock geochemistry. *Marine and Petroleum Geology*, **5**, 194–204.

LEYTHAUSER, D. & SCHAEFER, R. G. 1984. Effects of hydrocarbon expulsion from shale source rocks of high maturity in Upper Carboniferous strata of the Ruhr area, Federal Republic of Germany. *Organic Geochemistry* **6**, 671–681.

——, —— & RADKE, M. 1988. Geochemical effects of primary migration of petroleum in Kimmeridge source rocks from Brae field area, North Sea. I: Gross composition of C^{15+}-soluble organic matter and molecular composition of C^{15+}-saturated hydrocarbons. *Geochimica et Cosmochimica Acta*, **52**, 701–713.

——, —— & —— 1987. On the primary migration of petroleum. *Special Paper No. 2, Proceedings of the 12th World Petroleum Congress*, John Wiley & Sons Ltd., Chichester, 227–236.

——, MACKENZIE, A. S., SCHAEFER, R. G. & BJORØY, M. 1984a. A novel approach for recognition and quantification of hydrocarbon migration effects in shale-sandstone sequences. *American Association of Petroleum Geologists Bulletin*, **68**, 196–219.

——, RADKE, M. & SCHAEFER, R. G. 1984b. Efficiency of petroleum expulsion from shale source rocks. *Nature*, **311**, 745–748.

MACKENZIE, A. S. PRICE, I., LEYTHAEUSER, D.,

MULLER, P., RADKE, M. & SCHAEFER, R. G. 1987. The expulsion of petroleum from Kimmeridge Clay source rocks in the area of the Brae Oilfield, U.K. Continental Shelf. *In:* BROOKS, J. & GLENNIE, K. W. (eds) *Petroleum Geology of North West Europe* Graham & Trotman, London, 865–877.

MOISEEV, V. D., LEONOV, V. I. VASHURKIN, A. I. & LOZHKIN, G. V. 1982. Effect of pressure on the solubility of petroleum in a dry hydrocarbon gas. *Neftyanoe Khozyaistro*, **8**, 33–35.

NOGARET, E. 1983. *Solubilité des hydrocarbures dans le gaz naturel comprimé. Application à la migration du pétrole dans les bassins sédimentaires.* Dissertation, École Nationale Superior des Mines Paris. (Report IFP-31-115).

PRICE, L. C. 1989. Primary petroleum migration from shales with oxygen-rich organic matter. *Journal of Petroleum Geology*, **12**, 289–324.

——, WENGER, L. M., GING, T. & BLOUNT, C. W. 1983. Solubility of crude oil in methane as a function of pressure and temperature. *Organic Geochemistry*, **4**, 201–221.

QUIGLEY, T. M., MACKENZIE A. S. & GRAY, J. R. 1988. Kinetic theory of petroleum generation. In: DOLIGEZ B. *et al.* (eds) *Proceedings of Conference on Migration of hydrocarbons in sedimentary basins— Bordeaux, June 1987.* Editions Technip, Paris, 649–665.

RADKE, M., WILLSCH, H., LEYTHAEUSER, D. & TEICHMULLER, M. 1982. Aromatic components of coal: relation of distribution pattern to rank. *Geochimica et Cosmochimica Acta*, **46**, 1831–48.

SCHEIDT, G. & LITTKE, R. 1989. Comparative organic petrology of interlayered sandstones, siltstones, mudstones and coals in the Upper Carboniferous Ruhr Basin, Northwest Germany, and their thermal history and methane generation. *In:* POELCHAU, H. S. & MANN, U. (eds) *Geologic Modeling — Aspects of Integrated Basin Analysis and Numerical Simulation. Geologische Rundschau*, **78**, 337–318.

TEICHMULLER, M. & TEICHMULLER, R. 1971. Inkolung. *In: Die Karbon-Ablagerungen in der Bundesrepublik Deutschland: Das Rhein-Ruhr-Revier. Forschritte in der Geologie von Rheinland und Westfalen*, **19**, 47–56, Krefeld.

TISSOT, B & WELTE, D. H. 1984. *Petroleum Formation and Occurrence, 2nd. ed.*, Springer Verlag, Berlin.

THOMPSON, K. F. M. 1987. Fractionated aromatic petroleums and the generation of gas-condensates. *Organic Geochemistry*, 573–590.

—— 1988. Gas-condensate migration and oil fractionation in deltaic systems. *Marine and Petroleum Geology*, **5**, 237–246.

WENGER, L. M., Jr. 1982. *Solubility of crude oil and heavy petroleum distillation fractions in methane (with water present) at elevated temperatures and pressures as applied to the primary migration of petroleum.* MS Thesis, Idaho State Univ., Pocatello.

ZHUZE, T. P., YOUCHKEVITCH, G. N. & TCHAKHMAKHTCHEV, B. A. 1973. L'influence de la température et de la pression sur les compositions des phases dans les systèmes huile-gaz naturels. *Advances in Organic Geochemistry, 6th International Congress, Ruel-Malmaison, Proceedings*, 463–469.

Numerical modelling of petroleum expulsion in two areas of the Lower Saxony Basin, Northern Germany

S. J. DÜPPENBECKER[1,3], L. DOHMEN[2], D. H. WELTE[1,2]

[1] *Institute of Petroleum & Organic Geochemistry (ICH-5), KFA Jülich, PO Box 1913, D-5170 Jülich, Germany*

[2] *Integrated Exploration Systems (IES), Bastionsstraße 11–19, D-5170 Jülich, Germany*

[3] *Present address: BP International Ltd, Sunbury Research Centre, Chertsey Road, Sunbury-on-Thames TW16 7LN, UK*

Abstract: The process of petroleum expulsion is mainly determined by the initial quality of the source rock and its sedimentary facies and by the temperature and pressure histories during burial. A numerical model of the complex processes of expulsion was developed based on observations of expulsion phenomena, detailed petrophysical and organic geochemical investigations and data provided by basin modelling. In this model petroleum expulsion is treated as a pressure driven transport in a separate phase using the existing pore system and newly formed microfractures as migration pathways. Microfracturing predominantly parallel to the bedding plane, as it is observed in oil generating source rocks, is initiated when the pore pressure build-up exceeds the rock strength. Major sources of overpressure in source rocks are considered to be volume expansion of organic matter during petroleum generation and compaction disequilibrium. Experimental data were obtained from detailed investigations of the Posidonia Shale source rock of the Lower Saxony Basin. An application of expulsion modelling is presented for two regions of the Lower Saxony Basin, where timing and quantity of petroleum generation and expulsion from Posidonia shale was studied with respect to different heating rates. The efficiency of transport mechanisms and migration pathways through both the pore network and newly-formed fractures could be determined by this quantitative treatment. The modelling results were in good agreement with geological observations and a mass balance based on organic geochemical analysis.

Over the last decade it has become widely accepted that expulsion of petroleum from source rocks takes place in a separate phase. Evidence for this has been presented in the work of Snarsky (1962), Tissot & Pelet (1971), Dickey (1975), Meissner (1978), Momper (1978), Magara (1978) and Durand (1983, 1988). As a consequence, primary and secondary migration can basically be treated as a multiphase fluid flow driven by petroleum fluid potential gradients (England *et al.* 1987) but on different scales so that acting forces have different effects. In clastic oil-prone source rocks low permeability, capillary effects and the presence of a significant amount of kerogen and its conversion to petroleum characterise the special conditions for primary migration. Two basic concepts have been proposed for primary migration: the organic network concept (Yariv 1976; McAuliffe 1980; Durand 1988) and the microfracturing concept (Snarsky 1962; Tissot & Pelet 1971; Meissner 1978; du Rouchet 1981; Talukdar *et al.* 1986, 1987). Besides this, two-phase fluid flow has been used to describe not only petroleum expulsion but also secondary migration and petroleum accumulation (England *et al.* 1987; Ungerer *et al.* 1987).

For a comprehensive description of the expulsion process a variety of controlling factors determined by the depositional environment of the source rock and the development of the basin have to be taken into account (Fig. 1). The quality and quantity of kerogen embedded in the source rock are controlled mainly by the sedimentary environment and are essential parameters in studying migration. The mass balance calculations of Cooles *et al.* (1986) showed clearly, that organic-rich, oil-generative source rocks with an initial petroleum potential exceeding 0.01 kg/kg rock can expel 60% to 90% of the generated petroleum, whereas below 0.005 kg/kg the expulsion process is relatively inefficient. Organic matter type ultimately governs for given heating conditions the rate of petroleum formation and the compositional variations of generated products. Details regarding the geometric distribution of the organic matter inside the rock matrix are important with respect to the exis-

From England, W. A. & Fleet, A. J. (eds), *Petroleum Migration*
Geological Society, Special Publication No. 59, pp. 47–64.

47

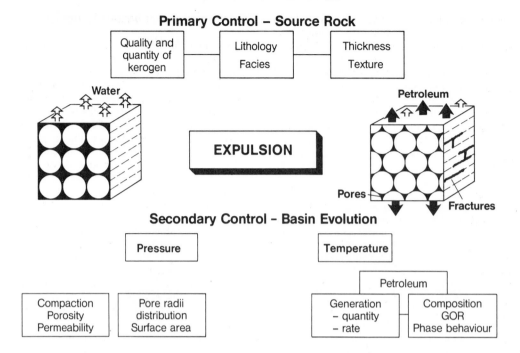

Fig. 1. Controlling factors of petroleum expulsion.

tence of a continuous oil-wet kerogen network and with respect to compaction effects while kerogen conversion to petroleum and expulsion are taking place. Lithology and lithofacies determine the physical properties of the sediment regarding its evolution of porosity and permeability as well as the wettability of its mineral surfaces and its fracture resistance.

Based on these inherent controls, the pressure and temperature conditions experienced by a source rock during basin evolution are crucial controlling factors for the process of petroleum expulsion. Increasing overburden pressure during burial results in increasing compaction and decreasing porosity and permeability as well as changes in pore radius distribution and a reduction in the specific surface area of the rock. All these processes affect fluid flow conditions inside the source rock. The temperature history determines the kinetically controlled process of petroleum generation from the source rock kerogen. The compositional variation of the generated petroleum is the most important factor influencing both, the degree of volume expansion due to formation of petroleum and the physical behaviour of the petroleum phase during expulsion as well as during secondary migration.

Abnormally high pore pressures in source rocks control the fluid flow and can cause fracturing of the rock by exceeding its mechanical strength. From numerous possibilities, the authors consider compaction disequilibrium and volume expansion of organic matter during petroleum generation to be the dominating sources of pressure, which act as a driving force for fluid expulsion from the source rock. Additionally, diagenetic transformation of clay minerals may contribute to pressure build-up (Powers 1967; Burst 1969; Dutta 1986) while the phenomena of thermal expansion of water (Barker 1972) might not be important under natural geological conditions (Daines 1982). Fractures, opened and propagated by overpressure, can serve as new pathways for the pressure-driven transport of petroleum.

Numerical modelling of petroleum expulsion

Any numerical simulation of the petroleum expulsion process should take into account all the above mentioned considerations. Parameters controlled by the depositional environment can be determined by organic geochemical, petrographical, and petrophysical investigations of source rock samples. The variations of pressure and temperature in a source rock during basin evolution can only be obtained by numerical simulation of its geological history. Therefore expulsion modelling must be integrated in comprehensive basin modelling systems. Coupled with kinetic modelling this leads to the determination of the timing of the process and the quantification of expelled petroleum. The PVT behaviour of migrating petroleum can be taken into account by defining compositional changes of generated petroleum during kerogen maturation.

Having this in mind, an expulsion program was developed as an integrated part of the Predrilling Intelligence (PDI) Basin Modelling System. The work was part of a research project on the generation and expulsion of petroleum from Posidonia Shale in the Lower Saxony Basin (Düppenbecker 1991). Although all source rock-specific parameters were taken from the marine Posidonia shale, we think that the method is applicable without major modifications to other clastic source rocks. The aim of this paper is to outline the recently developed petroleum expulsion program and demonstrate its application on the example of the Posidonia Shale in two areas of the Lower Saxony Basin.

The main features of the numerical simulation are as follows:

(1) The continuous mass and volume balancing of the source rock components under changing PVT conditions. The overall volume expansion of the organic matter due to petroleum generation (Tissot & Pelet 1971; Momper 1978) is considered to be a major source for overpressure build-up inside the pore system. Theoretical considerations of Ungerer et al. (1983) showed that gas generation results in significant volume expansion of organic matter.

(2) The build-up of pore pressure due to volume expansion and compaction disequilibrium. This leads to the formation of microfractures, if the mechanical strength of the rock is exceeded. Open microfractures enhance the conditions for petroleum migration both within and out of the source rock.

(3) The pressure driven transport of a separate petroleum phase along a preferentially water-wet pore system and newly formed microfractures, if present. The concept of petroleum transport does not take into account a continuous organic network. Up to now, our microscopical observations on the Posidonia Shale do not support a continuous 3D organic system. Therefore, in the Posidonia shale the particulate organic matter can serve only as a distributional system for petroleum on a small scale in the source rock.

Volume expansion of organic matter

The petrophysical behaviour of most sedimentary rocks is not affected by changes of the chemistry and physics of their organic matter contents because it is generally low. However, these changes are crucially important for organic rich rocks such as oil source rocks (Goff 1983; Larter 1988). For immature Posidonia Shale with 10 wt % of organic carbon, a mineral grain density of $2690 \, kg/m^3$ and an initial kerogen density of $1200 \, kg/m^3$ the matrix density is $2390 \, kg/m^3$ and the organic matter occupies 20% of the solid volume. Variations in structure and composition of organic matter with increasing maturity causes significant volumetric variations. The immobile parts of a source rock are mineral matrix, irreducible water bound to the mineral surface and kerogen. The mobile parts are free pore water and petroleum, which together represent porosity. During maturation of organic matter the conversion of solid kerogen to petroleum leads to an increase in porosity. This is modified by the assumption, that the kerogen structure has a certain adsorption capacity for generated petroleum compounds, which decreases continuously with kerogen breakdown. The overall rock compressibility is affected by this porosity increase and gives rise to further compaction if expulsion takes place. Collapsed algal structures (metaalginite) as observed at advanced maturity stages (Littke & Rullkötter 1987) are thought to indicate this compaction.

A kinetic model of kerogen degradation was used to determine the timing and quantity of petroleum generation. The process of so-called primary cracking reactions is described by a set of parallel reactions each with a specific initial petroleum potential. Direct calibration of the kinetic parameters for the Posidonia Shale resulted in a distribution of activation energies and one single frequency factor (Düppenbecker & Horsfield 1990). The consecutive cracking of oil (C_{6+}) to gas (methane) is treated as a single reaction characterized by a single activation energy and one frequency factor. The transformation ratio (TR) describes the portion of kerogen that has been converted to petroleum by primary cracking reactions and defines the stage of kerogen maturity by values between 0 (no conversion) and 1.0 (complete conversion).

Applying this bulk oil generation approach to given temperature histories the output is in the form of generated masses of petroleum with unspecified composition. Masses of educt (kerogen) and product (petroleum) must then be converted to volumes using densities. The major compositional effect on subsurface liquid petroleum density and therefore on volume is the quantity of petroleum of the C_1–C_5 range (gas) dissolved in the heavier C_{6+} fraction (oil) (England *et al.* 1987). From closed system pyrolysis experiments on Posidonia Shale kerogens a function describing the variation of the gas/oil ratio (GOR) with increasing TR values could be established, which is shown in Fig. 2 (Düppenbecker & Horsfield 1990). On the basis of these experimental data, physical properties of the generated petroleum fluids can be determined using correlations from the reservoir engineering field, which were published by Glaso (1980) and further discussed by England *et al.* (1987). The variation of saturation GOR with subsurface pressure and temperature conditions, predicted from the basin modelling, is determined according to Glaso (1980). Compared with the actual GOR of the generated product composition, given for the calculated TR, the phase behaviour of petroleum in the source rock can be determined. For the volumetric balancing the variation of liquid petroleum density can be defined as a function of the predicted actual GOR using Fig. 3. Density changes of petroleum in a single liquid phase at pressures above saturation pressure are neglected because of the low compressibility factors of 10^{-6}/psi to 15^{-6}/psi for produced oils and because of the minor changes in formation volume factors of oils above bubble point pressures (McCain 1973, pp. 157; Dake 1978, pp. 20,51,82).

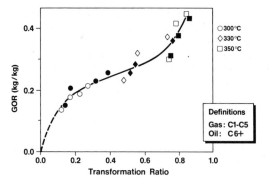

Fig. 2. Variation of the GOR as a function of experimentally simulated maturity for Posidonia Shale.

The progressive loss of functional groups and substituent entities from kerogen during catagenesis in the form of petroleum compounds changes the chemical and physical properties of the kerogen. A progressive increase in density with maturation is known for liptinitic coal macerals (Allan 1975). For the marine type II Posidonia Shale kerogen, approximate density values have been found to vary from 1200 kg/m^3 for very immature stages (TR of 0–10%) up to 1600 kg/m^3 for the overmature kerogen (TR over 90%) (Littke, pers. comm.). On the basis of these data, kerogen density increase could be determined as a function of increasing transformation ratios, which is manifested as a relative volume reduction. The amount of coke that is formed besides dry gas by secondary

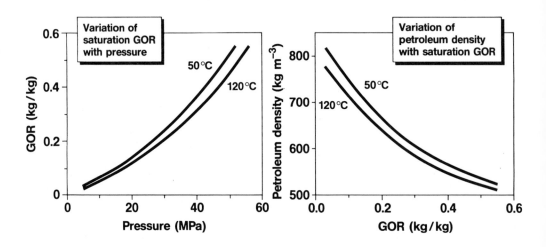

Fig. 3. Application of GOR for determination of (*a*) phase behaviour of petroleum and (*b*) subsurface petroleum densities. Correlations given by Glaso (1980) with oil density (STP) = 35° API and gas density (STP) = 0.75 kg/m^3.

cracking reactions of oil compounds at higher maturity stages is added to kerogen volume in the model.

Compositional and structural changes in organic matter during maturation cause significant volume changes in the source rock system. Therefore, a detailed balancing of masses and volumes of all compounds is carried out in our petroleum expulsion model. The increasing density contrast between kerogen and its product during progressive maturation results in an increase of total volume of organic matter, that is sufficient to exceed the available rock volume. This volume expansion of organic matter can be quantified as a function of TR, if we assume a mass conservation of the system. The total volume of organic matter V_{om} can be expressed as

$$V_{om} = M_{in} \times ((1 - TR)/D_{ker} + TR/D_{pet}).$$

The initial mass M_{in} is the initial petroleum potential of the kerogen, which is for Posidonia Shale 70% of the initial organic carbon mass. D_{ker} is the actual calculated density of kerogen and D_{pet} is that of the petroleum generated throughout catagenesis. The volume increase relative to the initial volume V_{rel}, can also be expressed without the masses as

$$V_{rel} = (D_{kin}/D_{ker}) \times (1 - TR) + (D_{kin}/D_{pet}) \, TR$$

D_{kin} is the assumed initial density of the kerogen. Consequently the reduced volume of the kerogen V_{ker} and the increased volume of petroleum phase V_{pet} is

$$V_{ker} = M_{ini} \times (1 - TR)/D_{ker}$$

$$V_{pet} = M_{ini} \times TR/D_{pet}.$$

Please note that D_{pet} is a cumulative value for the petroleum density inside the source rock being derived from the generated petroleum portions whose densities are continuously decreasing with increasing TR.

At high maturity stages the GOR of the petroleum composition generated by kerogen degradation is affected by the generation of dry gas due to secondary cracking of oil (C_{6+}). The total GOR_t of the petroleum composition is to be calculated based on the GOR from primary cracking reactions of kerogen (GOR_{pr}) as a function of TR (Fig. 3) and on the GOR from secondary cracking reactions (GOR_{sc}) based on the masses of generated gas G_s and remaining oil O_s from oil cracking.

$$GOR_t = GOR_{pr} + GOR_{sc} \times (1 + 2GOR_{pr}).$$

Due to stoichiometric considerations only about 50% of the initial petroleum potential can be taken into account for the total gas generation potential by cracking of oil to gas (Welte et al. 1988; Ungerer et al. 1988). Therefore the mass of oil degraded by secondary cracking must be multiplied by a factor of 0.5 to result in the masses of the two products dry gas (G_s) and coke (C_s).

These masses can be expressed in terms of parameters introduced earlier, namely

$$C_s = G_s = GOR_{sc} \times (M_{in} \times TR)/$$
$$((1 + GOR_{pr}) \times (1 + 2GOR_{sc})).$$

Coke is solid organic matter; its volume is calculated by using the maximum kerogen density and added to the kerogen volume.

The considerable overall volume expansion of organic matter represents an excellent potential for a pressure build-up in the source rock pore system, which is then the driving force for expulsion. Generally source rocks cannot be treated as isolated systems and consequently volume increase is compensated by a volume flow of water and/or petroleum out of the system. However, this depends on transport conditions for the different phases and quantification and the pressure build-up. Under certain conditions this pore pressure build-up can exceed the rock strength and thus form fractures. Balancing the volumetric changes in petroleum expulsion modelling leads to quantitative assessment of the reduction in overall volume and source rock porosity with time.

Fracturing concept

The expulsion of petroleum along microfractures out of the source rock was originally proposed by Snarsky (1962) and later by Tissot & Pelet (1971). The microfracturing concept considers that pore pressures may exceed the mechanical strength and newly formed fractures serve as transportation avenues for a fluid flow out of the system resulting in a pressure release. Microfractures containing either bitumen or diagenetic mineral-fill such as calcite have been observed in a number of prominent organic-rich oil source rocks. Meissner (1978) described oil-filled fractures in combination with overpressure in mature areas of the Bakken Shale Formation (Williston Basin) where oil is actually produced from the source rock. He proposed that fracturing had occured at some high pore pressure due to petroleum generation, and that an outward flow of oil could take place through the spontaneously generated fracture system. Lindgreen (1987) described diagenetic mineral-bearing microfractures in the Kimmeridge Clay (North Sea, Central Graben) as being up to 0.1 mm in width and orientated both parallel to the bedding plane as well as inclined and normal to it. He suggested that primary oil migration had taken place along those microfractures and that microfractures parallel to the bedding plane had been created earlier. Fractures filled with bitumen and partly with calcite are reported from the organic-rich shales of the Woodford Formation in Oklahoma and Arkansas by Comer & Hinch (1987). The provinces where mature source rocks commonly contain oil-filled fractures coincide with regions where large reserves of Woodford type oil occur. For the predominantly carbonate source rocks of the La Luna Formation (Maracaibo Basin) Talukdar *et al.* (1986, 1987) proposed that the dominant mode of oil expulsion took place in a separate phase through microfractures, driven by excess fluid pressure. Some production is recorded from the fractured and overpressured mature source rocks, and the bitumen-filled fractures in the clay-rich sections are predominantly orientated parallel to the bedding plane. Mature samples ($>0.7\%$ R_o) of the Posidonia Shale from the SE region (Hils Syncline) of the Lower Saxony Basin showed micro and macrofractures ($>200\,\mu m$) generally orientated parallel to the bedding plane (Jochum 1988; Leythaeuser *et al.* 1988; Littke *et al.* 1988). The fracture fill is usually calcite, which contains fluorescing oil inclusions between the crystal margins and, less commonly, non-fluorescing bitumen. This indicates, that these fractures served as migration avenues for the expulsion process. No comparable fractures were observed in the over- and underlying shales, which contain only little organic matter of poorer quality.

In view of this widespread coincidence of fractured mature source rocks, fracturing must be included in expulsion modelling. Concepts for the development of fractures for expulsion in oil generating source rock were presented by Meissner (1978) and du Rouchet (1981). They followed the basic failure theory for isotropic brittle rocks, in which the fracture criterion can be obtained by Mohr's circles describing the stress field by normal and shear stresses. For the formation of open extension fractures tensile stress conditions for at least one principle stress must be established. According to Hubbert & Rubey (1959) the stress field for fluid-filled po-

rous rocks with pore pressures P_{fl} above hydrostatic pressure is related to the effective stress S_{eff}, which allows tensile stress conditions to be present under a maximum compressive stress S_1 and a least compressive stress S_3.

$$S_{eff} = S_i - P_{fl} \ (i = 1\text{--}3).$$

The shift of the stress field due to increasing pore-fluid pressures can cause tensional fracturing for petroleum expulsion if the least compressive stress S_3 and the tensile strength of the rock is exceeded.

As long as the least compressive stress is oriented horizontally, which is true under the overall tensional tectonic conditions in a subsiding basin, the above fracture model describes the fracture criterion only for vertical fractures (Fig. 4a). But microfractures in source rocks are more frequently oriented parallel to the subhorizontal bedding plane, which in most cases is parallel to S_3. This ubiquitous phenomenon in source rocks is a most uncommon feature in classical rock fracture mechanics and cannot be explained by the concept of Meissner (1978) and du Rouchet (1981).

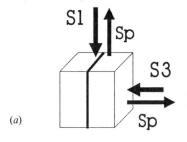

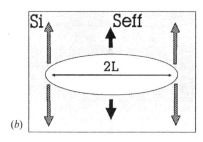

Fig. 4. Concepts for the development of fractures. (*a*) Tension against the least compressive stress. (*b*) Intensification of stress according to the Griffith fracture theory.

Our new tensile failure criteria are gained by the application of the Griffith theory of fracture and linear elastic fracture mechanics. The Griffith theory is based on the assumption that cracks are present in every material and can be approximated as flat elliptical holes (Fig. 4b). Stress fields are greatly magnified at the tip of each crack and this intensification depends on the geometry and the orientation of the ellipsoid in the overall stress field. For a 2-dimensional stress field that can be expressed as

$$S_i = 2S_{eff} \times (L/r_{min})^{1/2}$$

where:
S_i = intensified stress [MPa]
L = half length of the ellipsoid [nm]
r_{min} = radius of the curvature at the tip of the ellipsoid [nm].

Macroscopic failure can develop when the maximum tensile stress near the tips of the ellipsoids exceeds the rocks strength and the resulting fractures become connected laterally. In our expulsion model the pore geometrics of clastic fine-grained source rocks are assumed to be ellipsoidal and oriented with their long axis predominantly parallel the bedding plane.

From linear elastic fracture mechanics, which is directly related to the Griffith theory, fracture growth occurs when the stress intensity at the crack tip reaches a critical value. The critical stress intensity factor K_c, which is known as the fracture toughness, is a directly measurable material constant with units of MPa m$^{1/2}$ (Schmidt & Rossmanith 1983). If the fracture toughness and the geometry of existing cracks or ellipsoids are known, then the critical stress S_c required to initiate fracture extension can be determined from:

$$S_c = K_c/(\pi \times L)^{1/2},$$

Schmidt (1976) and Costin (1981) measured the fracture toughness of oil shales (Green River Shale) with various techniques and determined the values as a function of orientation with respect to bedding plane and kerogen content. The values varied between $K_c = 0.3$–1.2 MPa m$^{1/2}$. The low values were obtained parallel to the bedding plane and fracture toughness was found to decrease with increasing kerogen content.

Pore geometry and pore classes

The application of this fracture concept enables us to describe the formation of microfractures as a function of pore pressure build-up, provided that pore geometries are known. Mercury porosimetry and nitrogen adsoprtion techniques can

be applied to estimate the pore volume in terms of pore radii distributions and specific surface area. Pore radii distributions of a variety of Posidonia Shale source rock samples were determined with those techniques (Mann et al. 1990). Thereby a continuous measurement of mercury injection capillary pressure curves up to 50 000 psi (345 MPa) was applied. The injection pressures can be converted to pore radius equivalents, which represent the radius of the pore throats (pore entry radius). The curves in Fig. 5 represent the shift of pore entry radius distributions to lower values as a result of increasing maximum burial for different sampling locations of Posidonia shale in the Lower Saxony Basin. The two bar diagrams of Fig. 6 a & b show the frequency distribution of four defined pore classes derived from the continuous cumulative curves A and B of pore volume v. pore entry radius in Fig. 5. Pore radii from mercury porosimetry measurements are taken to represent the short axis of the assumed pore ellipsoids. The length L of the long axis as used for the fracturing concept is determined by taking the decompacted value of the short axis and a defined relation between the two axes at the time of sediment deposition.

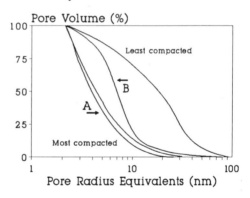

Fig. 5. Pore radii distribution for four Posidonia Shale samples with different compaction stages.

These class distributions of pore volumes for specific ranges of pore radii are used to characterise the specific variations in pore geometries of various source rocks. Nevertheless, it should be kept in mind that the ranges describe only pore throat radii and that the radii of pores *sensu strictu* having this entry radius may be larger. Our concept for the pore system states that all the defined pore classes are connected in ascending order of pore entry radius and that mass and volume flow of petroleum takes place between the classes according to pressure gradients and resisting capillary forces. Thereby we assume the class of pore throats with 5 nm and 10 nm radius to belong to pores (*s. s.*) with a radius of one order of magnitude higher. All the petroleum flow out of the system is defined to go along the largest pore class. Mass and volume balancing of the petroleum increase in a pore class by generation and migration into it from smaller pore classes and of the petroleum decrease by expulsion to the next larger class or out of the system has to be done for each class continuously. Hence, all volume changes due to in- and outflow of petroleum between all pore classes are taken into account.

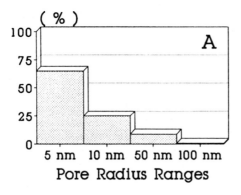

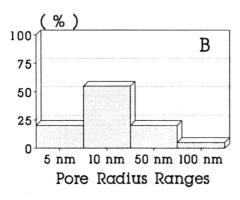

Fig. 6. Classification of the pore radii distribution A and B in four different classes of pore ranges.

Transport of petroleum in source rocks

The movement of a petroleum phase in the porous system of a fine-grained source rock is strongly controlled by capillary forces, resisting the migration through a water wet pore system (Schowalter 1979; England *et al.* 1987). The specific surface area of clastic source rocks like the Posidonia Shale can be considered to be

composed of oil-wet portions, namely kerogen surface, while the rest, the mineral portion, is initially water wet. Such a system with so-called fractional wettability, which is assumed to exist on the scale of pores, does not provide a continuous oil wet path out of the source rock. Experimental results on reservoir rock systems with fractional wettability (Amott 1959; Fatt & Klikhoff 1959) suggest that even in organic-rich source rocks, where the kerogen represents up to 20 volume % (such as the Posidonia Shale), the system should be treated as uniformly water wet.

The resisting force, the capillary pressure P_c, is given by the La Place equation

$$P_c = 2\gamma \times \cos \theta / r,$$

where γ is the interfacial tension between petroleum and water, θ the contact angle of oil/water against the pore walls as an expression of the wettability of the system and r the pore entry radius. Besides the oil/water interfacial tension, whose values are in the range of 13–36×10^{-3} N m^{-1} and vary with temperature (Schowalter 1979), the size of the pore entry radius is the critical factor controlling transport conditions for oil. Assuming that the La Place Equation is valid for the smallest defined pore throat radius of 5 nm, capillary pressures in the order of several MPa must be exceeded by petroleum pressure. It must be kept in mind that the radius is reduced additionally by the width of the irreducible water bound to the mineral surface. In our model two molecular layers of water, each with 0.3 nm width, are taken into account. This is based on considerations regarding the DLVO theory for adsorbed hydration water layer thickness (Hirasaki 1988).

Reservoir engineering experiments have demonstrated that for uniformly water wetted rocks, with contact angles in the range of 0–50°, the influence of contact angles can be neglected (Anderson 1987). In our model we considered a source rock to be an overall water wet system. It should however be noted that the initial wettability of the minerals can be altered during petroleum migration because organic compounds become adsorbed to the mineral surface, thereby changing the mineral–oil–water system (Clementz 1982). Such a wettability alteration of mineral surfaces from water wet to oil wet might explain the increase in resistivity of source rocks in the zone of petroleum formation and migration as reported by Meissner (1978) for the Bakken Shale. However, for the *onset* of expulsion and the *timing* of pressure build-up the source rock should be treated as a water wet system.

If the capillary pressures in the defined pore classes are exceeded by the pressure build-up in the petroleum phase or fractures are formed, it is assumed in our model that petroleum is the only mobile phase. A flow of petroleum into the next lager pore class or out of source rock takes place according to pressure gradients. The flow velocity v of a single phase flow in a porous rock through a cross section of an area A is defined as flow rate Q over A:

$$v = Q/A \text{ and } Q = \mathrm{d}V/\mathrm{d}t$$

where V is the volume transported per time step. For small flows of a liquid petroleum phase the velocity v_p can be described by Darcy's law

$$v_P = -K/\mu \times \mathrm{grad}(P + P_c - P_h)$$

where K is the intrinsic rock permeability, μ the dynamic viscosity of oil, P the pressure in the oil phase and P_h the hydrostatic pressure at depth z relative to sea level. The rate of flow Q out of each pore classes is then given by

$$Q = -K/\mu \times A \times \mathrm{grad}(P + P_c - P_h).$$

By this simple approach the volume of a single petroleum phase flowing out of the pore classes per time step of numerical simulation is calculated. The viscosity of liquid petroleum is determined as a function of temperature and the GOR (McCain 1973, pp. 196). The permeability is assumed to be constant and according to Rieke & Chillingarian (1974) values of 10^{-21} m^2 and less were used for compacted shale. In our model we have used only a single phase fluid flow of liquid petroleum. This is because it is suggested from different case histories of the Posidonia Shale in the Lower Saxony Basin that pressure conditions at that time of maximum expulsion exceed the oil saturation pressure.

For the largest pore class, which represents the smallest portion of the pore volume and receives the petroleum expelled from the smaller classes, the contribution of capillary pressures is no longer significant. The pressure gradient for the flow out of this class is calculated with respect to the pore pressures of the over- and underlaying layer. If the critical pressure for fracturing is reached in a pore class the mobile volume of petroleum in this pore class is transported to the largest pore class governed by a flow with increased permeability of two orders of magnitude.

Application: case study of two regions from within the Lower Saxony Basin

Geological setting

The Lower Saxony Basin (LSB) is a Mesozoic trough, 300 km long and 65 km wide, and represents the most petroliferous region in Germany. In this basin the Posidonia Shale is the major prolific and regionally extensive source rock. As a central part of the NW-European Basin the LSB underwent a complex geological history from the late Jurassic till present reflecting the changes in the megatectonic regime of NW-Europe (Betz et al. 1987). Phases of subsidence in Upper Jurassic and Cretaceous time were followed by basin inversion, which caused strong uplift and the formation of oil-trapping structures in the northern and western part of the LSB. Before this inversion, several igneous massifs intruded the southern and central part of the basin. These intrusions had a strong influence on the thermal regime at that time, and produced several maturity anomalies. They are indicated by the isoreflectance contours on Fig. 7.

Numerical basin modelling has been applied to find out how and to what extent this thermal event associated with the deep seated magmatic intrusions has influenced petroleum generation and expulsion from the Posidonia Shale in different parts of the basin. The study areas were located in the western and southeastern parts (Hils Syncline) of the basin. Core samples from the Posidonia Shale and overlaying strata were taken from several well locations (Fig. 7). This paper focusses mainly on the aspects of expulsion while the entire study is presented elsewhere (Düppenbecker 1991).

Characterization of the Posidonia Shale

The Posidonia black shales were deposited in an epicontinental sea under restricted environmental conditions with water stratification and anoxic bottom waters. A most characteristic sedimentological feature is the fine parallel lamination; nevertheless organic matter is uniformly and finely disseminated throughout (Littke et al. 1988). Comprehensive organic geochemical and micropetrographical investigations were carried out on a large number of samples. Those samples from the SE region (Hils Syncline) were taken from an extensive collection used for detailed investigations on hydrocarbon generation (Rullkötter et al. 1988 and references therein).

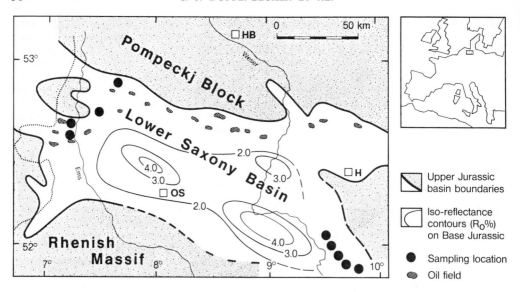

Fig. 7. Delineation of the Lower Saxony Basin and the sampling locations.

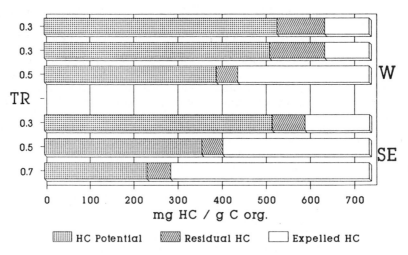

Fig. 8. Results of mass balance calculations of generated and expelled petroleum for three mature sampling locations with various transformation ratio (TR) in the western and the southeastern region according to the algebraic scheme from Cooles *et al.* (1986).

The total organic content varies between 6 and 12 wt%, whereas immature samples have values always between 9 and 12 wt%. About 80% of the microscopically visible organic matter consists of marine alginite; less than 10% is of terrestrial origin. Typical total pyrolysis yields of about 700 mg HC/g $C_{org.}$, atomic H/C ratios of around 1.35 and O/C ratios of about 0.16 for immature samples. These values identify the Posidonia Shale as an excellent potential oil source rock. Specific parameters for kinetic modelling were determined. In this context compositional changes of petroleum products were determined as a function of increasing maturity (Düppenbecker & Horsfield 1990). The maturity of the Posidonia Shale in the sampling locations of the two regions ranged from 0.45% R_o up to 0.88% R_o, corresponding to a maximum kerogen transformation ratio of 0.7.

Micro- and macrofractures, predominantly

parallel or subparallel to the bedding, were found to be most abundant in the mature Posidonia Shale units of the SE region (Jochum 1988; Leythaeuser *et al.* 1988). These fractures served as migration pathways for petroleum. Because such fracture phenomena were not observed in the west region it can be suggested that other mechanisms of expulsion must have been active there.

Organic geochemical mass balance

On the basis of organic geochemical investigations mass balance calculations of generated and expelled petroleum was carried out according to the algebraic scheme published by Cooles *et al.* (1986). The starting ratio of initial oil/ reactive kerogen/inert kerogen for an immature reference was defined by taking representative immature samples from six locations. Results for three mature sampling locations in the west and southeast regions are presented in Fig. 8. The total bar length represents the initial petroleum potential, which is subdivided into portions of already expelled petroleum, residual petroleum and the remaining petroleum generation potential. For transformation ratios (TR) of about 0.3, expulsion efficiencies of 40% were observed in the west region, whereas values of 70% expulsion efficiency were observed in the SE region. For transformation ratios of about 0.5 and above, over 80% of the generated petroleum has been expelled in the cases of both regions. These observations of high expulsion efficiencies are in good agreement with investigations on other organic rich oil-prone source rocks as presented by Cooles *et al.* (1986) and Mackenzie *et al.* (1987).

The results of numerical modelling are presented below for the two sampling locations in the west and southeast regions whose TRs are 0.46 and 0.52, respectively (depicted as TR = 0.5 in Fig. 8), and whose petroleum expulsion efficiency was 82% and 86%.

Results of numerical modelling

One-dimensional numerical simulation of geological and geochemical porcesses were carried out utilizing the Pre-drilling Intelligence (PDI) basin modelling system. This system allows a genuine 'forward modelling' simulating all relevant geological and geochemical processes along the geological time axis. Rather detailed temperature and pressure histories are the most important

results of this numerical simulation. The temperature history was calibrated using organic geochemical maturity parameters and the present day temperature regime. These temperature and pressure histories were used to reconstruct quantity and the timing of petroleum generation and expulsion processes.

The results of numerical modelling are presented for two sampling locations, one in the west region and one in the SE region. In the west region the Posidonia Shale attained a maximum burial depth of about 2300 m and was then uplifted to a depth of about 1850 m during the basin inversion period. In the SE-area, a maximum burial depth of only 1800 m was reached and basin inversion caused a strong, two phase uplift to a rather shallow depth of less than 400 m. In both sampling locations the maturity of Posidonia Shale was between 0.7% and 0.75% R_o.

The temporal development of the temperature for the Posidonia Shale in the two sampling locations and the results of the numerical simulation of petroleum generation are presented in Fig. 9 a & b along a time axis between 100–80 Ma. In the west region the more gradual increase of temperature with a maximum value of about 125°C is a result of relatively low heating rates of less than 5°C/Ma. In the SE region there was a stronger temperature influence, with a maximum of about 135°C being attained. A stepwise increase of the heating rates to high values is observed. This temperature profile and the increase in heating rates to maximum values of 23°C/Ma for a short time span is caused by the igneous intrusions of the nearby deep-seated Vlotho Massif.

Based on these temperature histories, petroleum generation was reconstructed by kinetic modelling. In the west region the transformation of the initial petroleum potential of the Posidonia Shale progressed gradually and reached finally a transformation ratio of 0.48 (Fig. 9). The process took place over a time span of 8 Ma with a low increase in generation rates. In the SE region the temperature profile resulted in a very steep increase of the transformation ratio in a final value of 0.53 with high rates of kerogen degradation. Furthermore the whole process took place in less than half the time as compared to the west region. The significant difference between the two locations is the timing of the reconstructed petroleum generation history. In both locations the amount of gas produced by the secondary cracking of oil was minor. It should, however, be noted that gas generation in the SE region is slightly more advanced than in the west (Fig. 10 a & b).

(a)

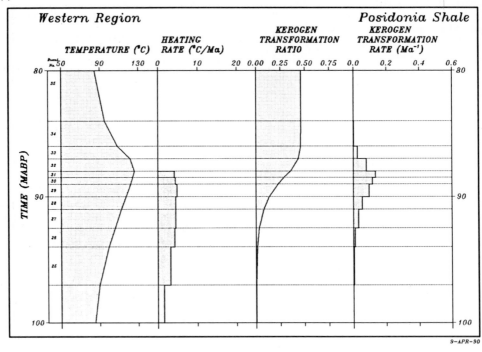

(b)

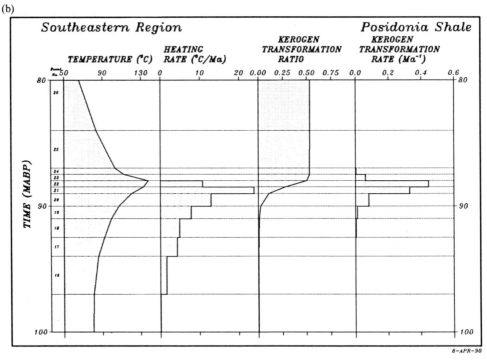

Fig. 9. Reconstruction of temperature development and petroleum generation history indicated by the kerogen transformation ratio between 100–80 Ma (*a*) for the western area and (*b*) for the south-eastern area.

(a)

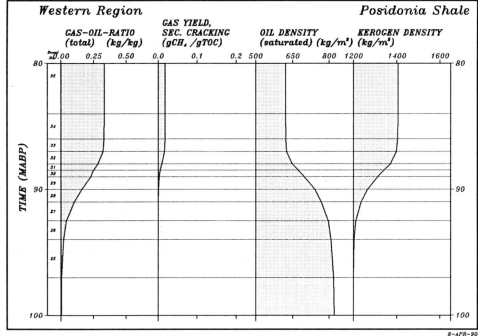

(b)

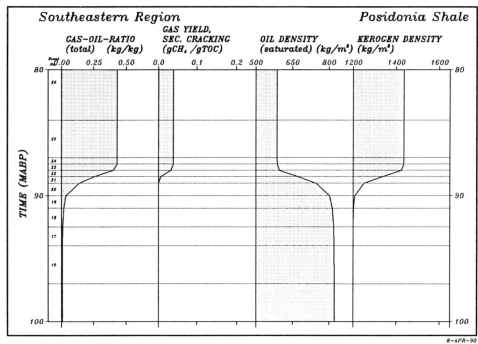

Fig. 10. Variation of the petroleum properties generated from Posidonia Shale and gas generation due to secondary cracking of oil in the two regions between 100–80 Ma (*a*) for the western area and (*b*) for the southeastern area.

(a)

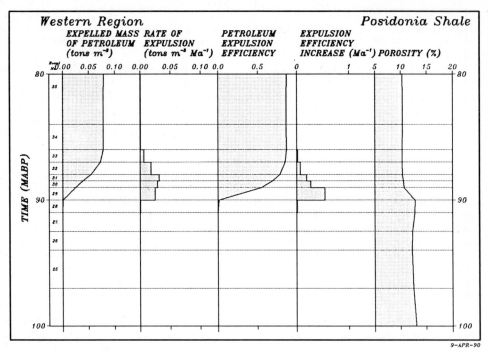

(b)

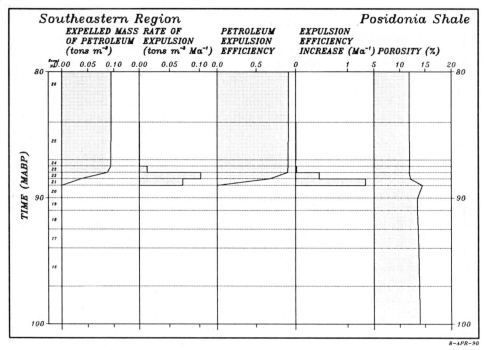

Fig. 11. Reconstruction of the petroleum expulsion for the two locations indicated by the expelled mass of petroleum in t/m³ and the petroleum expulsion efficiency between 100–80 Ma. Also presented the porosity development of Posidonia Shale during this time (*a*) for the western area and (*b*) for the southeastern area.

The variations in physical petroleum properties in the Posidonia Shale which are due to compositional changes are presented along the same time axis in Fig. 10 a & b. In the west region the total GOR of the generated petroleum plus gas from oil cracking reaches a final value of 0.33 kg/kg, the oil density decreases to a density of 621 kg/m³ and the kerogen density increases to a value of 1405 kg/m³. Because of the slightly more advanced gas generation due to oil cracking the final values of these parameters are somewhat higher in the SE-region. The GOR reaches 0.42 kg/kg, the oil density decreases to 595 kg/m³ and final kerogen density is about 1435 kg/m³. As described earlier for the case of petroleum generation, the most significant outcome of these results is the different timing of the changes of physical properties in the two regions. This resulted in differences in the pressure development due to volume expansion of organic matter. Consequently much higher pore pressures were produced in the SE region. It is important to note that during the whole time span of petroleum expulsion in both locations the calculated petroleum pressure exceeded the gas saturation pressure determined by Glaso's equations. Therefore only one liquid petroleum phase had to be taken into account for migration purposes.

The results of the numerical modelling of petroleum expulsion from Posidonia Shale are expressed as masses of expelled petroleum in tons/m³ rock and in terms of the efficiency of the expulsion process whereby values fall between 0 (no expulsion) and 1 (complete expulsion). Both sampling locations resulted in comparable final values for the masses of expelled petroleum. In the west region the total amount of expelled petroleum was about 0.08 tons/m³ whereas in the SE region 0.095 tons/m³ has been expelled (Fig. 11 a & b). The petroleum expulsion efficiency predicted from numerical simulation ranged between 0.85 and 0.90 in both locations and were in good agreement with the results of geochemical mass balance calculations, shown in Fig. 8. Again, the determined timing and rates of expulsion were significantly different between the two locations. In the west region a more gradual increase is documented and the expulsion process lasted four million years. Caused by the moderate generation of petroleum and the resulting moderate pressure devlopment the mechanism for the expulsion process, determined by the model, was bulk transport along an existing original pore system. In the SE region the whole expulsion process was condensed to a time of only 1.5 Ma with extremely high peak rates of expulsion. Here, the model calcu-

lations indicate pore pressures exceeding the rock strength, which resulted in rock fracturing and hence a modelled transport mechanism for petroleum along newly generated fractures as well as bulk transport through the pore system.

The modelled porosity development of Posidonia Shale shows in both locations an initial phase of slight porosity increase due to the increasing solid–liquid conversion of organic matter. The onset of significant petroleum expulsion causes a concomitant decrease in porosity (Fig. 11a & b).

In both locations the smallest pore throats with 5 nm radius were defined to refer to 70% of the pore volume. Because of the low free poor volume in pores with this radius, not occupied by irreducible structured water, a high oil saturation and strong pore pressure build-up is reached early. Pressure conditions sufficient for fracture formation in the SE region were only determined for pores (*sensu strictu*) of 5 and 50 nm with a pore throat size of 5 nm. The conditions for petroleum transport along the smallest pore throat class into pore classes with larger entry radii was closely interrelated with the timing and amount of generated petroleum volume. Once transport took place, either through pores or along newly-formed fractures, the petroleum saturation and pressure build-up in the four defined pore classes were changed strongly.

The sensitivity of the model with respect to variations in the organic carbon content and the pore radii distribution was tested. A decrease in initial organic carbon content to 4% or 1% resulted in significantly lower petroleum expulsion efficiencies, an absence of fracturing conditions in the SE region and a loss of distinctive difference in timing between the two locations studied. Variations in the pore class distribution showed clearly that the proportion of pore volume belonging to the smallest pore throats of 5 nm radius was very strongly influencing the expulsion conditions. Pore systems defined by pore classes with pore throats of 50 nm and above never reached the pressure conditions necessary for fracturing.

Summary and conclusions

Petroleum expulsion is a very complex process influenced by a large number of different factors. The model presented for the numerical simulation of this process takes into account:

(1) the depositional environment and the organic facies of the source rock;

(2) the continuous mass and volume balancing of kerogen degradation and petroleum generation under defined PVT conditions;

(3) the formation of microfractures predominantly parallel the bedding plane of the source rock due to pressure build-up from volume expansion of organic matter during petroleum generation and compaction disequilibrium;

(4) the pressure driven transport of a separate petroleum phase along a heterogeneous pore system and newly-generated microfractures.

The most important factors controlling the process of petroleum expulsion can only be recognised by coupling the numerical simulation of expulsion with integrated numerical modelling of geological and geochemical history of the source rock. The expulsion model presented here can be used in combination with a one, two or even three dimensional simulation approach to determine the timing, intensity and mechanism of petroleum expulsion. The compositional changes of petroleum products, the pore size distribution and the definition of fracture criterion play an important role and are therefore taken into account. The expulsion model works on the scale of pores and defines petroleum transport processes inside the source rock system between different pore classes and expulsion of petroleum out of the system. It allows the prediction of the temporal development of expulsion with well-defined petroleum quantities and composition in terms of GOR. Hence the model is useful in

predicting timing and physical behaviour for the subsequent processes of secondary migration.

The application of the expulsion model to case histories in two regions of the Lower Saxony Basin resulted in the reconstruction of timing and mechanisms for petroleum expulsion as a function of heating rates. In summary the predicted influence of igneous intrusions on petroleum generation and expulsion processes were found to be quantitatively in good agreement with present day geological and geochemical observations. Sensitivity runs showed the dependency of the modelled expulsion process on organic matter content and the pore radii distribution of the source rock.

Further case studies for different source rocks are necessary to learn more about the crucial factors determining petroleum expulsion in nature. On the other hand, better experimental data on source rock properties are required. Particular those concerning source rock fracturing behaviour, wettability and petroleum flow conditions in a multiphase system. Nevertheless, the expulsion model presented here showed very reasonable results when applied in different regions of a sedimentary basin in different geological settings.

We would like to thank U. Bayer, B. Horsfield and U. Mann for their valuable contributions. Thorough reviews of the manuscript by P. Ungerer and A. Pepper are gratefully acknowledged. Part of the samples were provided by a consortium of German oil companies (B.E.B. Erdöl & Erdgas, C. Deilmann AG, DST, Mobil AG, Texaco, Wintershall AG).

References

ALLAN, J. 1975. *Natural and artificial diagenesis of coal macerals.* PhD thesis, Univ. of Newcastle upon Tyne.

AMOTT, E. 1959. Observations relating to the wettability of porous rock. *AIME Petroleum Transactions,* **216**, 159–162.

ANDERSON, W. G. 1987. Wettability literature survey—Part 4: Effects of wettability on capillary pressure. *Journal of Petroleum Technology,* Oct. 1987, 1283–1300.

BARKER, C. 1972. Aquathermal pressuring—role of temperature in development of abnormal-pressure zones. *American Association of Petroleum Geologists Bulletin,* **56**, 2068–2071.

BETZ, D., FÜHRER, F., GREINER, G. & PLEIN, E. 1987. Evolution of the Lower Saxony Basin. *Tectonophysics,* **137**, 127–170.

BURST, L. F. 1969. Diagenesis of Gulf Coast clayey sediments and its possible relation to petroleum migration. *American Association of Petroleum Geologists Bulletin,* **53**, 73–93.

CLEMENTZ, D. M. 1982. Alteration of rcok properties by adsorption of heavy ends: Implications for enhanced oil recovery. *Society of Petroleum Engineers/Department of Environment,* **10683**, 131–134.

COMER, J. B. & HINCH, H. H. 1987. Recognizing and quantifying expulsion of oil from the Woodford Formation and age-equivalent rocks in Oklahoma and Arkansas. *American Association of Petroleum Geologists Bulletin,* **71**, 844–858.

COOLES, G. P., MACKENZIE, A. S. & QUIGLEY, T. M. 1986. Calculation of petroleum masses generated and expelled from source rocks. *In: Advances in Organic Geochemistry 1985, Organic Geochemistry,* **10**, 235–245.

COSTIN, L. S. 1981. Static and dynamic fracture behaviour of oil shale. *In:* FREIMAN, S. W. (ed.), *Fracture Mechanics of Ceramic Rocks, and Concrete.* American Society for Testing Materials, Philadelphia, STP **745**, 169–184.

DAINES, S. 1982. Aquathermal pressuring and geopressure evaluation. *American Association of Petroleum Geologists Bulletin,* **66**, 931–939.

DAKE, L. P. 1978. *Fundamentals of reservoir engineering.* Elsevier, Amsterdam.

DICKEY, P. A. 1975. Possible primary migration of oil from source rock in oil phase *American Association of Petroleum Geologists Bulletin,* **59**, 337–345.

DU ROUCHET, J. 1981. Stress fields, a key to oil migration. *American Association of Petroleum Geologists Bulletin,* **65**, 74–85.

DURAND, B. 1983. Present trends in organic geochemistry in research on migration of hydrocarbons. *In:* BJOROY ET AL (eds), *Advances in Organic Geochemistry 1981.* John Wiley & Sons, 17–128.

—— 1988. Understanding of HC migration in sedimentary basins (present state of knowledge). *In: Advances in Organic Geochemistry 1987, Organic Geochemistry* **13**, 445–459.

DUTTA, N. C. 1986. Shale compaction, burial diagenesis and geopressures: A dynamic model, solution and some results. *In:* BURRUS, J. (ed.) *Thermal modeling in sedimentary basins.* Technip, Paris, 149–172.

DÜPPENBECKER, S. J. 1991. *Genese und Expulsion von Kohlenwasserstoffen in zwei Regionen des Niedersächsischen Beckens unter besonderer Berücksichtigung der Aufheizraten.* Dissertation RWTH Aachen.

—— & HORSFIELD, B. (1990). Compositional information for kinetic modelling and petroleum type prediction. *In: Advances in Organic Geochemistry 1989,* (in press).

ENGLAND, W. A., MACKENZIE, A. S., MANN, D. M. & QUIGLEY, T. M. 1987. The movement and entrapment of petroleum fluids in the subsurface. *Journal of the Geological Society, London,* **144**, 327–347.

FATT, I. & KLIKHOFF, W. A. 1959. Effect of fractional wettability on multiphase flow through porous media. *AIME Petroleum Transactions,* **216**, 426–432.

GLASO, O. 1980. Generalized pressure–volume temperature conditions. *Journal of Petroleum Technology,* **32**, 785–795.

GOFF, J. C. 1983. Hydrocarbon generation and migration from Jurassic source rocks in the East Shetland Basin and Viking Graben of the North Sea. *Journal of the Geological Society, London,* **140**, 445–474.

HIRASAKI, G. J. 1988. Wettability: Fundamentals and surface forces. *Society of Petroleum Engineers/ Department of Environment,* **17367**, 513–528.

HUBBERT, M. K. & RUBEY, W. W. 1959. Role of fluid pressures in mechanics of overthrust faulting. *Geological Society of America Bulletin,* **70**, 115–206.

JOCHUM, J. 1988. *Untersuchungen zur Bildung und Karbonatmineralisation von Klüften im Posidonienschiefer (Lias Epsilon) der Hilsmulde.* Master Thesis, RWTH Aachen.

LARTER, S. 1988. Some pragmatic perspectives in source rock geochemistry. *Marine and Petroleum Geology,* **5**, 194–204.

LEYTHAEUSER, D., LITTKE, R., RADKE, M. & SCHAFER, R. G. 1988. Geochemical effects of petroleum migration and expulsion from Toarcian source rocks in the Hils syncline area, NW-Germany. *In: Advances in Organic Geochemistry 1987; Organic Geochemistry,* **13**, 489–502.

LINDGREEN, H. 1987. Molecular sieving and primary migration in Upper Jurassic and Cambrian claystone source rocks. *In:* BROOKS, J. & GLENNIE, K. (eds) *Petroleum Geology of North West Europe,* Graham & Trotman, 357–364.

LITTKE, R. & RULLKÖTTER, J. 1987. Mikroskopische und makroskopische Unterschiede zwischen Profilen unreifen und reifen Posidonienschiefers aus der Hilsmulde. *Facies,* **17**, 171–180.

——, BAKER, D. R. & LEYTHAEUSER, D. 1988. Microscopic and sedimentologic evidence for the generation and migration of hydrocarbons in Toarcian source rocks of different maturities. *In: Advances in Organic Geochemistry, Organic Geochemistry,* **13**, 549–559.

MCAULIFFE, C. D. 1980. Oil and gas migration: chemical and physical constrains. *American Association of Petroleum Geologists Bulletin,* **63**, 761–781.

MCCAIN, W. D. 1973. *The properties of petroleum fluids.* The Petroleum Publishing Company, Tulsa.

MACKENZIE, A. S., PRICE, I., LEYTHAEUSER, D., MÜLLER, P., RADKE, M. & SCHAEFER, R. G. 1987. The expulsion of petroleum from Kimmeridge clay source rocks in the area of the Brae Oilfield, UK continental shelf. *In:* BROOKS, J. & GLENNIE, K. (eds) *Petroleum Geology of North West Europe.* Graham & Trotman, 864–877.

MAGARA, K. 1978. Significance of the expulsion of water in oil-phase primary migration. *Canadian Petroleum Geology Bulletin,* **25**, 195–207.

MANN, U., DÜPPENBECKER, S. J., LANGEN, A., ROPERTZ, B. & WELTE, D. H. 1990. Pore network evolution of the lower Toarcian Posidonia Shale during petroleum generation and expulsion—A multidisciplinary approach. *In:* FESTBAND W. ZIMMERLE (ed.) *Zentralblatt für Geologie und Paläontologie, Part 1,* (in press).

MEISSNER, F. F. 1978. Petroleum geology of the Bakken Formation Williston Basin, North Dakota and Montana. *Williston Basin Symposium, The Montana Geological Society, 24th Annual Conference,* 207–227.

MOMPER, J. A. 1978. *Oil migration limitations suggested by geological and geochemical considerations.* American Association of Petroleum Geologists Course Notes No. **8**.

POWERS, M. C. 1967. Fluid release mechanisms in compacting marine mudrocks and their importance to oil exploration. *American Association of Petroleum Geologists Bulletin,* **51**, 1204–1254.

RIEKE, H. H. & CHILINGARIAN, G. V. 1974. *Compaction of argillaceous sediments.* Developments in Sedimentology, **16**, Elsevier, Amsterdam.

RULLKÖTTER, J., LEYTHAEUSER, D., HORSFIELD, B., LITTKE, R., MANN, U., MÜLLER, P. J., RADKE, M., SCHAEFER, R. G., SCHENK, H.-J., SCHWOCHAU, K., WITTE, E. G. & WELTE, D. H. 1988.

Organic matter maturation under the influence of a deep intrusive heat source: A natural experiment for quantitation of hydrocarbon generation and expulsion from a petroleum source rock (Toarcian Shale, northern Germany). *In: Advances in Organic Geochemistry, Organic Geochemistry*, **13**, 847–856.

SCHMIDT, R. A. 1976. Fracture mechanics of oil shale—Unconfined fracture toughness, stress corrosion cracking and tension test results. *In: Proceedings of the 18th US Symposium on Rock Mechanics, Keystone, Colorado*, 2A2-1–2A2-6.

—— & ROSSMANITH, H. P. 1983. Basics of rock fracture mechanics. *In:* ROSSMANITH, H. P. (ed.) *Rock Fracture Mechanics* CISM Courses and Lectures, No **275**, 1–29.

SCHOWALTER, T. T. 1979. Mechanics of secondary hydrocarbon migration and entrapment. *American Association of Petroleum Geologists Bulletin*, **63**, 723–760.

SNARSKY, A. N. 1962. Die primäre Migration des Erdöls. *Freiberger Forschungshefte*, **C123**, 63–73.

TALUKDAR, S., GALLANGO, O. & CHIN-A-LIEN, M. 1986. Generation and migration of hydrocarbons in the Maracaibo Basin, Venezuela: An integrated basin study. *In: Advances in Organic Geochemistry 1985, Organic Geochemistry*, **10**, 261–279.

TALUKDAR, S., GALLANGO, O., VALLEJOS, C. & RUGGIERO, A. 1987. Observations on primary migration of oil in the La Luna source rocks of Maracaibo Basin, Venezuela. *In:* DOLIGEZ, B. (ed.) *Migration of Hydrocarbons in Sedimentary Basins*. Edition Technip, Paris, 59–77.

TISSOT, B. & PELET, R. 1971. Nouvelles données sur les méchanismes de genèse et de migration du pétrole, simulation mathématique et application á la prospection. *8th World Petroleum Congress, Moscow*, **4**, Wiley, Chichester, 35–46.

UNGERER, P., BEHAR, E. & DISCAMPS, D. 1983. Tentative calculation of the overall volume expansion of organic matter during hydrocarbon genesis from geochemistry data. Implications for primary migration. *In: Advances in Organic Geochemistry 1981*, John Wiley & Sons, 129–135.

——, ——, VILLALBA, M., HEUM, O. R. & AUDIBERT, A. 1988. Kinetic modelling of oil cracking. *In: Advances in Organic Geochemistry 1987, Organic Geochemistry*, **13**, 857–868.

——, DOLIGEZ, B., CHENET, P. Y., BURRUS, J., BESSIS, F., LAFARGUE, E., GIROIR, G., HEUM, O. & EGGEN, S. 1987. A 2-D model of basin scale petroleum migration by two-phase fluid flow. Application to some case studies. *In:* DOLIGEZ, B. (ed.) *Migration of hydrocarbons in sedimentary basins*. 415–455, Edition Technip, Paris.

WELTE, D. H., SCHAEFER, R. G. & YALCIN, M. N. 1988. Gas generation from source rocks: Aspects of a quantitative treatment. *In: Origins of Methane in the Earth, Chemical Geology*, **71**, 105–116.

YARIV, S. 1976. Organophilic pores as proposed primary migration media for hydrocarbons in argillaceous rocks. *Clay Science*, **5**, 19–29.

Modelling 1D compaction-driven flow in sedimentary basins: a comparison of the Scotian Shelf, North Sea and Gulf Coast

B. S. MUDFORD,[1] F. M. GRADSTEIN,[2] T. J. KATSUBE[3] & M. E. BEST[2]

[1] Greenstone Geophysical Research, 5523 Sebastian Place, Halifax, Nova Scotia, Canada B3K 2K5

Present address: Unocal Science and Technology Division, 376 South Valencia Avenue, PO Box 76, Brea, CA 92621, USA

[2] Atlantic Geoscience Centre, Geological Survey of Canada, Bedford Institute of Oceanography, Dartmouth, Nova Scotia, Canada B2Y 4A2

[3] Mineral Resources Division, Geological Survey of Canada, 601 Booth Street, Ottawa, Ontario, Canada K1A 0E8

Abstract: A one-dimensional model of fluid pressure evolution in a compacting sedimentary sequence has been developed and used to investigate some of the physical processes which occur during sediment deposition and burial. It is well known that the permeability assigned to low permeability lithologies, such as shales, has a large effect upon the fluid pressures predicted by basin models. Shales are common in many sedimentary basins, however, the behaviour of their permeability as a function of pressure is poorly constrained. This leads to considerable uncertainty in the predictions of many basin models. Recent measurements of the permeability of overpressured Jurassic shales from the Scotian Basin, off Nova Scotia have yielded values around 10^{-21} m^2 (approximately 1 nanodarcy) for a range of effective pressures between 10 and 60 MPa. We have used these results in a number of single phase simulations of one dimensional, compaction-driven, fluid pressure evolution in the Scotian Basin, the North Sea, and the United States Gulf Coast. In all these areas chronostratigraphic data can be used to determine Neogene sedimentation rates, which strongly affect the present-day levels of overpressure. In the Scotian Basin predicted present-day pressures fall below those observed. This is in contrast to the North Sea and Gulf Coast where rapid Neogene sedimentation, together with Cenozoic sediments containing a high percentage of shale, causes predicted levels of compaction-driven overpressuring similar to those observed. The poor agreement between predicted and observed pressures in the Scotian Basin indicates that mechanisms other than compaction disequilibrium, for instance lateral fluid migration, mineral diagenesis, or ice loading, should be considered as causes of overpressuring.

A quantitative model of the evolution of sedimentary basins is an important tool in studies of fluid flow, and the development of fluid pressures in sedimentary sequences. In many cases fluid pressures in excess of hydrostatic pressure (known as overpressure) develop and have a large impact on the migration of fluids. Quantitative models help in understanding the evolution of pressures through time, and allow us to investigate the relative importance of the mechanisms which contribute to the observed fluid pressures. Numerical simulation also places constraints on the timing of overpressuring, and the permeabilities necessary for the development and maintenance of overpressures on geologic timescales. A number of quantitative investigations of pressure evolution, at varying levels of complexity, have been undertaken.

One-dimensional studies of compaction disequilibrium have been carried out by Gibson (1958), Bredehoeft & Hanshaw (1968), Smith (1971), Sharp & Domenico (1976), Bishop (1979), Keith & Rimstidt (1985), and Thorne & Watts (1989). The influence of clay diagenesis on overpressure generation in a one dimensional compacting sedimentary sequence has been modelled by Hanshaw & Bredehoeft (1968), Bethke (1986), and Dutta (1986, 1988). More complex two dimensional studies of single phase fluid flow in compacting sedimentary basins have been presented by Bethke (1985), Bethke et al. (1988), and Bredehoeft et al. (1988). Shi & Wang (1988) have presented a two-dimensional study of the evolution of overpressures in an accretionary prism.

The simulation of overpressure generation in a sedimentary sequence with hydrocarbon sources

From England, W. A. & Fleet, A. J. (eds), *Petroleum Migration*
Geological Society, Special Publication No. 59, pp. 65 85.

must include the effects of two phase flow. Simplified analytical studies of two-phase flow in compacting sedimentary columns have been carried out by Rosenblat (1985) and Barker (1988). Welte & Yukler (1981) developed a three-dimensional model of sedimentary basin evolution which included hydrocarbon generation by correlating Lopatin's temperature–time index to vitrinite reflectance and then to extractable hydrocarbons. Nakayama (1987) and Jabour & Nakayama (1988) have developed a one-dimensional model of a sedimentary section in which they calculate the present-day quantities of oil and gas expelled from the source rock. Their hydrocarbon generation model is based on that proposed by Lerche *et al.* (1984), which relates measured vitrinite reflectance to palaeoheat flux, and then allows a time–temperature integral to be calculated. A two dimensional model of basin evolution, which includes hydrocarbon generation, maturation and migration has been developed by a group at Institut Francais du Petrole (Ungerer *et al.* 1984, 1990; Doligez *et al.* 1986). This model has been applied to basins in the North Sea (Ungerer *et al.* 1984, 1985; Doligez *et al.* 1986, 1987); to the Gulf of Lion passive margin in France (Ungerer *et al.* 1984); and to the Mahakam Delta in Indonesia (Ungerer *et al.* 1984). Wei & Lerche (1988) have developed a two-dimensional model of sedimentary basin formation in which hydrocarbon generation and faulting have been included. They have used this model to analyse the development of the Pinedale anticline in Wyoming. A one-dimensional version of this model has been applied to wells in the North Sea (Cao & Lerche 1987) and the Navarin Basin, Alaska (Cao & Lerche 1989).

The usefulness of any basin model as a predictive tool is directly related to the quality of the input data. A quantitative model of a sedimentary column requires a large number of parameters to describe the fluids and sediments. Unfortunately, many of these parameters are very poorly constrained. An important use of basin models is to evaluate the sensitivity of the model to the parameters, and determine which paramaters have the greatest impact on the results from the model. Further data can then be generated for the more important parameters. Analyses of this type (for example Smith 1971; Dutta 1988), has focused attention on the physical properties of shales. These properties have a large impact on the pressure evolution and are poorly constrained. Results of recent measurements of shale permeabilities are presented in this paper, and are used to constrain a number of one-dimensional simulations of fluid flow in compacting sedimentary sequences.

The results reported in this paper are from numerical simulations of one-dimensional, single phase fluid flow in the Scotian Basin, off Nova Scotia (Fig. 1), the Central North Sea (Fig. 2), and an idealized United States Gulf Coast situation. The aim of this paper is to compare the response of the fluid pressures in these three areas to compaction-driven fluid flow under the constraints imposed by measured shale permeabilities and high resolution chronostratigraphic data. The overpressuring mechanisms included in the model for this study are compaction disequilibrium and aquathermal pressuring. Hydrocarbon generation has not been included.

Model

A one-dimensional, two-phase model has been developed to investigate the evolution of fluid pressures through geological time in a compacting sedimentary sequence (Mudford 1990). In the present paper the results of single-phase simulations only are presented, hence this outline of the model will focus on the single-phase version. The mathematical details are described in Appendix A. The model combines the equations of fluid and solid mass conversion for a sedimentary column, together with Darcy's Law and equations inter-relating permeability, k, porosity, ϕ, and effective stress, P_e, ($P_e = P_t - P_w$, where P_t represents the overburden pressure, and P_w denotes the pore fluid pressure). This set of equations is a closed, nonlinear system that is solved for the fluid pressure and porosity as functions of depth and time.

The model has been developed with the specific aim of quantitatively studying the evolution of fluid pressures through geological time. Thermal maturation studies have not been carried out. To do so would require the solution of an additional equation describing heat flow through the sediments. This equation would be coupled to the pressure and porosity equations presented in Appendix A. Fluid pressures predicted by the model are relatively insensitive to variations in the geothermal gradient over a wide range of geologically sensible values. For this reason the geothermal gradient is assumed to be constant during each model run. The model can be split into two separate parts. In the first of these the sediments are backstripped using a variation of the method first presented by Sclater & Christie (1980), and further detailed by Stam *et al.* (1987). No geodynamic modelling is undertaken in the backstripping module. The sediments are simply decompacted along a given porosity–depth profile, which can change from

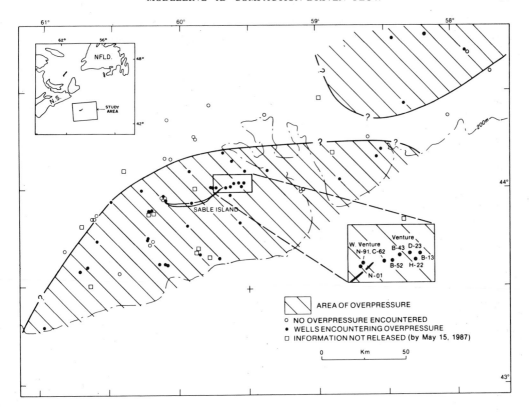

Fig. 1. Map showing the areal extent of overpressuring on the Scotian Shelf (after Wade & MacLean, 1989) and the position of the Venture field with the Venture wells, B-13, B-43, B-52, D-23, and H-22.

one sedimentary unit to another. The effects of variable water depth through time are included.

In the second part of the model the sediments are deposited and compacted, and the fluid pressure distribution down the sediment column is calculated. A finite difference solution technique is used with centred differences in space and backward differences in time. The minimum timesteps used in the model range from 0.5 ka to 50 ka depending on the rapidity with which the pressure profiles change. In general the porosity and the thickness of the sediment column are both functions of the pressure. This leads to a non-linear system of equations which must be solved by an iterative technique at each timestep. The porosity is not a rapidly varying function of pressure and no more than five iterations are usually required for convergence. An unevenly spaced vertical grid is used to allow thin, low permeability shale layers to be represented.

Data

The geological data needed for the backstripping module are a lithological column, chronostratigraphic and palaeobathymetric information, and porosity versus depth curves for each lithology. In the forward model we assume porosity is an exponential function of effective pressure. In addition, permeability is taken to be a power law function of porosity. Given the functional dependence of porosity, permeability can also be written as an exponential function of effective pressure. It is straightforward to convert expressions for permeability which are in terms of porosity (or effective pressure) to expressions which are in terms of effective pressure (or porosity). The fluid viscosity is determined from an empirical expression given by Mercer *et al.* (1975). Parameter values and equations common to all modelling runs are given in Tables 1 and 2 respectively.

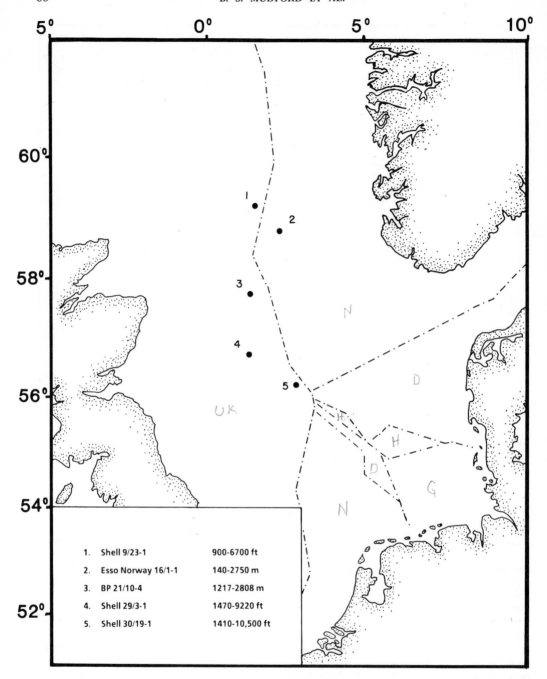

Fig. 2. Map of the Central North Sea showing the location of the wells used in this study (after Gradstein *et al.* 1988).

Porosity data

In the Venture field, off Nova Scotia (Fig. 1), we have used porosities measured on cores together with porosities determined using sonic and poro-

sity logs. The porosity functions used in modelling the Venture field are given in Table 3. Default porosity functions exist for a number of sedimentary basins around the world (Baldwin & Butler 1985). For the central North Sea wells

Table 1. *Parameter values common to all modelling runs*

Parameter	Value	Parameter definition
g	9.81 m s^{-1}	Acceleration due to gravity
a_w	$7 \times 10^{-4} \text{ K}^{-1}$	Brine isobaric thermal expansion coefficient
β_w	$6 \times 10^{-10} \text{ Pa}^{-1}$	Brine isothermal compressibility
$\bar{\rho}_b$	2350 kg m^{-3}	Average bulk density
ρ_w	1030 kg m^{-3}	Brine density

Table 2. *Parameter equations common to all modelling runs*

Permeability

$k = a\phi^m$	for sandstone	(Ref.1)	$a = 2.2 \times 10^{-11} \text{ m}^2$
			$m = 5.3$
	for shallow shale	(Ref.2)	$a = 4 \times 10^{-15} \text{ m}^2$
			$m = 8$
$k = k_0 \exp(-\kappa P_e)$	for deep shale	(Ref.3)	$k_0 = 2.12 \times 10^{-20} \text{ m}^2$
			$\kappa = 8.29 \times 10^{-8} \text{ Pa}^{-1}$

Brine viscosity (Ref.4)
$\mu = (5380 + 3800\Theta - 260\Theta^2)$ $\Theta = (T - 150)/100$ T expressed in °C

References: (1) Mudford and Best (1989); (2) Smith (1971); (3) this paper; (4) Mercer *et al.* (1975).

Table 3. *Venture field model input data for all lithologies*

Porosity versus depth (decompaction)	$\phi = \phi_0 \exp(-bz)$ $\phi_0 = 0.64, b = 4.1 \times 10^{-4} \text{ m}^{-1}$
Porosity versus effective pressure (compaction)	$\phi = \phi_0 \exp(-\beta P_e)$ $\beta = 3.3 \times 10^{-8} \text{ Pa}^{-1}$
Temperature gradient	$0.024°\text{C m}^{-1}$

Table 4. *North Sea model input data*

Porosity versus depth (decompaction)		$\phi 1 = \phi_0 \exp(-bz)$
	Sandstone	$\phi_0 = 0.29, b = 2.6 \times 10^{-4} \text{ m}^{-1}$
	Shale	$\phi_0 = 0.50, b = 4.7 \times 10^{-4} \text{ m}^{-1}$
Porosity versus effective pressure (compaction)		$\phi = \phi_0 \exp(-\beta P_e)$
	Sandstone	$\beta = 2.0 \times 10^{-8} \text{ Pa}^{-1}$
	Shale	$\beta = 3.7 \times 10^{-8} \text{ Pa}^{-1}$
Temperature gradient		$0.03°\text{C m}^{-1}$

modelled in this study we have used default porosity functions for the North Sea presented by Gradstein *et al.* (1989). The location of these wells is shown in Fig. 2, and the porosity functions used are given in Table 4. In a normally pressured environment, porosity versus depth functions can be readily transformed to porosity versus effective pressure using the expression

$$P_e = (\bar{\rho}_b - \rho_w)gz \qquad (1)$$

where ρ_b denotes the average bulk density between the sediment surface and depth z, ρ_w represents the fluid density, and g represents acceleration due to gravity. Reasonable estimates of the average bulk sediment density usually fall in the range of 2300 kg m^{-3} to 2350 kg m^{-3} for depths at which the porosity has fallen to approximately 25% or less.

In an overpressured environment data from the normally pressured part of the sedimentary section is used to generate a porosity versus effective stress curve. To get good agreement between observed and predicted sediment thicknesses in overpressured areas several complete modelling runs may need to be made, with each run using the final porosity–depth profile from the previous run to backstrip the sediments. This is especially critical in areas such as the United States Gulf Coast where the sediments are significantly undercompacted.

Biostratigraphic and palaeobathymeric data

The Central North Sea and Scotian Shelf wells analysed in this study, using the foraminiferal record, are listed in Table 5. These wells form part of a much larger data set involving over 85 wells, analysed by one of us (F.M.G.) for circum-North Atlantic stratigraphy and palaeobathymetry in Cenozoic time.

Table 5. *Cenozoic interval of wells studied for biostratigraphy and palaeobathymetry and used in fluid pressure modelling*

Central North Sea	
Esso Norway 16/1-1	140–2750 m
BP 21/10-4	1217–2808 m
Shell 29/3-1	1470–9220 ft
Shell 30/19-1	1410–10500 ft
Shell 9/23-1	900–6700 ft

The age versus depth interpretations for the central North Sea wells make use of a probabilistic zonation, based on the last occurrences of over 160 taxa in 31 deep exploration wells. These wells have a diversified, mostly deep water agglutinated foraminiferal record in the Palaeogene, and a shallower water, calcareous record in the Neogene. There is no sample record for the section of post-Pliocene age, which is a slight drawback to this particular study, although its thickness and approximate lithology are known. The Central North Sea zonation is illustrated in Fig. 3; zonal resolution approaches 1 interval zone per 5 million years or less, with the succession of individual events per zone providing slightly higher resolution. Details on the (ranking and scaling) methods that generated this particular zonation are in Gradstein *et al.* (1985). For the chronostratigraphic interpretation of the Pliocene planktonic foraminiferal record, use was made of the North Atlantic temperate to subpolar zonation of Weaver & Clement (1986).

The physical logs from the particular North Sea wells analysed indicate massive shale depo-

sition for most of the Tertiary, following the break-up of the Danian carbonate platform facies. The Quaternary record is sandy. From the biostratigraphic record in the wells, it is apparent that over half of the Cenozoic sediments were deposited in the last 10 Ma. Esso Norway 16/1-1 has a late Miocene hiatus and in all wells this interval of time may have seen some erosion of an originally thicker unit. In general, the following Cenozoic stratigraphic levels may be distinguished: Pleistocene, middle Pliocene, lower Pliocene, upper Miocene, middle Miocene, upper Oligocene–lower Miocene, lower Oligocene (Rupelian), upper Eocene–lower Oligocene, middle to upper Eocene, middle Eocene, lower Eocene (Ypresian), Palaeocene–Eocene boundary, upper Palaeocene (Selandian), lower Palaeocene (Danian) and upper Maastrichtian. The resolution is sufficient to deduce changes in sedimentation rate over less than 5 Ma.

Table 6. *Micropalaeontological criteria for palaeobathymetry in the Central North Sea and Scotian Shelf*

(1) **Non-marine, terrestrial**; spores and pollen, no foraminifers

(2) **Shallow neritic, less than 100 m**; foraminiferal assemblages of low generic and species diversity and rare planktonics; most upper Neogene section with *Cibicidoides, Elphidium; Cassidulina, Bulimina, Melonis, Quinqueloculina*, gastropods and bryozoans.

(3) **Deep neritic, 100–200 m**; foraminiferal assemblages of varying diversity, planktonics locally common; In Neogene beds occur *Uvigerina, Pullenia, Gyroidina, Alabamina, Sphaeroidina, Fursenkoina, Sigmoilopsis, Spiroplectammina, Ceratobulimina;* Palaeogene benthic assemblages reflect Midway type-fauna.

(4) **Upper bathyal (upper slope), 200–500 m**, or possibly **750 m deep**; foraminiferal generic and species diversity is high; assemblages may be dominated by coarse, often large-sized agglutinated taxa, incl. cyclammnids and tubular forms; in more basinward wells planktonics may occur frequently and radiolarians.

(5) **Middle bathyal (middle slope), 750–1000 m or slightly deeper**; foraminiferal diversity is high, calcareous benthics include *Cibicidoides wuellerstorfi, Melonis pompilioides, Uvigerina rustica* (all Neogene), and *Pleurostomella, Osangularia, Stilostomella* and *Nuttalides* (all Palaeogene). Agglutinated assemblages are diverse and contain both coarse and finer grained, smaller taxa particularly *Cystammina, Rzehakina* and *Labrospira*.

Fig. 3. Probabilistic zonation using ranking and scaling (RASC) on the last occurrences of 160 taxa in 31 Central North Sea wells. Each event as shown occurs in at least 7 wells, except for rare ones (● ●) (after Gradstein *et al.* 1988).

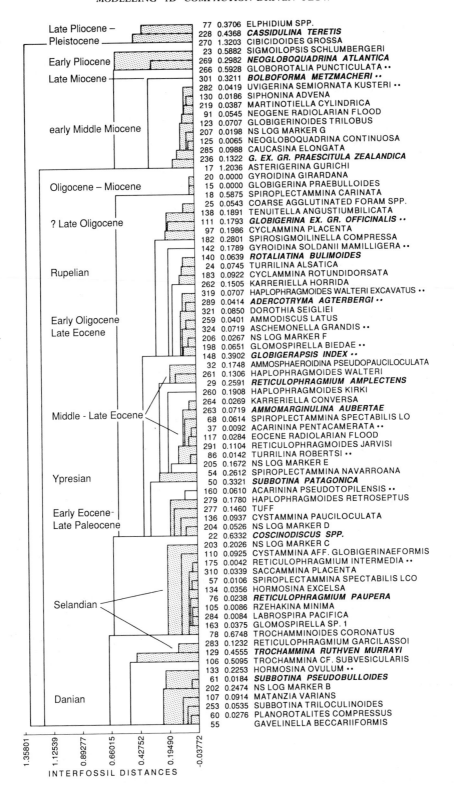

77	0.3706	ELPHIDIUM SPP.
228	0.4368	*CASSIDULINA TERETIS*
270	1.3203	CIBICIDOIDES GROSSA
23	0.5882	SIGMOILOPSIS SCHLUMBERGERI
269	0.2982	*NEOGLOBOQUADRINA ATLANTICA*
266	0.5928	GLOBOROTALIA PUNCTICULATA ··
301	0.3211	*BOLBOFORMA METZMACHERI* ··
282	0.0419	UVIGERINA SEMIORNATA KUSTERI ··
130	0.0186	SIPHONINA ADVENA
219	0.0387	MARTINOTIELLA CYLINDRICA
91	0.0545	NEOGENE RADIOLARIAN FLOOD
123	0.0707	GLOBIGERINOIDES TRILOBUS
207	0.0198	NS LOG MARKER G
125	0.0065	NEOGLOBOQUADRINA CONTINUOSA
285	0.0988	CAUCASINA ELONGATA
236	0.1322	*G. EX. GR. PRAESCITULA ZEALANDICA*
17	1.2036	ASTERIGERINA GURICHI
20	0.0000	GYROIDINA GIRARDANA
15	0.0000	GLOBIGERINA PRAEBULLOIDES
18	0.5875	SPIROPLECTAMMINA CARINATA
25	0.0543	COARSE AGGLUTINATED FORAM SPP.
138	0.1891	TENUITELLA ANGUSTIUMBILICATA
111	0.1793	*GLOBIGERINA EX. GR. OFFICINALIS* ··
97	0.1986	CYCLAMMINA PLACENTA
182	0.2801	SPIROSIGMOILINELLA COMPRESSA
142	0.1789	GYROIDINA SOLDANII MAMILLIGERA ··
140	0.0639	*ROTALIATINA BULIMOIDES*
24	0.0745	TURRILINA ALSATICA
183	0.0922	CYCLAMMINA ROTUNDIDORSATA
262	0.1505	KARRERIELLA HORRIDA
319	0.0707	HAPLOPHRAGMOIDES WALTERI EXCAVATUS ··
289	0.0414	*ADERCOTRYMA AGTERBERGI* ··
321	0.0850	DOROTHIA SEIGLIEI
259	0.0401	AMMODISCUS LATUS
324	0.0719	ASCHEMONELLA GRANDIS ··
206	0.0267	NS LOG MARKER F
198	0.0651	GLOMOSPIRELLA BIEDAE ··
148	0.3902	*GLOBIGERAPSIS INDEX* ··
32	0.1748	AMMOSPHAEROIDINA PSEUDOPAUCILOCULATA
261	0.1306	HAPLOPHRAGMOIDES WALTERI
29	0.2591	*RETICULOPHRAGMIUM AMPLECTENS*
260	0.1908	HAPLOPHRAGMOIDES KIRKI
264	0.0269	KARRERIELLA CONVERSA
263	0.0719	*AMMOMARGINULINA AUBERTAE*
68	0.0614	SPIROPLECTAMMINA SPECTABILIS LO
37	0.0092	ACARININA PENTACAMERATA ··
117	0.0284	EOCENE RADIOLARIAN FLOOD
291	0.1104	RETICULOPHRAGMOIDES JARVISI
86	0.0142	TURRILINA ROBERTSI ··
205	0.1672	NS LOG MARKER E
54	0.2612	SPIROPLECTAMMINA NAVARROANA
50	0.3321	*SUBBOTINA PATAGONICA*
160	0.0610	ACARININA PSEUDOTOPILENSIS ··
279	0.1780	HAPLOPHRAGMOIDES RETROSEPTUS
277	0.1460	TUFF
136	0.0937	CYSTAMMINA PAUCILOCULATA
204	0.0526	NS LOG MARKER D
22	0.6332	*COSCINODISCUS SPP.*
203	0.2026	NS LOG MARKER C
110	0.0925	CYSTAMMINA AFF. GLOBIGERINAEFORMIS
175	0.0042	RETICULOPHRAGMIUM INTERMEDIA ··
310	0.0339	SACCAMMINA PLACENTA
57	0.0106	SPIROPLECTAMMINA SPECTABILIS LCO
134	0.0356	HORMOSINA EXCELSA
76	0.0238	*RETICULOPHRAGMIUM PAUPERA*
105	0.0086	RZEHAKINA MINIMA
284	0.0084	LABROSPIRA PACIFICA
163	0.0375	GLOMOSPIRELLA SP. 1
78	0.6748	TROCHAMMINOIDES CORONATUS
283	0.1232	RETICULOPHRAGMIUM GARCILASSOI
129	0.4555	*TROCHAMMINA RUTHVEN MURRAYI*
106	0.5095	TROCHAMMINA CF. SUBVESICULARIS
133	0.2253	HORMOSINA OVULUM ··
61	0.0184	*SUBBOTINA PSEUDOBULLOIDES*
202	0.2474	NS LOG MARKER B
107	0.0914	MATANZIA VARIANS
253	0.0535	SUBBOTINA TRILOCULINOIDES
60	0.0276	PLANOROTALITES COMPRESSUS
55		GAVELINELLA BECCARIIFORMIS

Late Pliocene – Pleistocene

Early Pliocene

Late Miocene

early Middle Miocene

Oligocene – Miocene

? Late Oligocene

Rupelian

Early Oligocene Late Eocene

Middle - Late Eocene

Ypresian

Early Eocene- Late Paleocene

Selandian

Danian

1.35801 1.12539 0.89277 0.66015 0.42752 0.19490 -0.03772

INTERFOSSIL DISTANCES

A critical parameter for subsidence and burial analysis is palaeo-waterdepth. Five basic categories are recognized in the North Sea wells, classified according to both waterdepth and distance from shore: (1) non-marine, terrestrial; (2) shallow neritic, less than 100 m; (3) deep neritic, 100–200 m (4) upper bathyal (upper slope), 200–500 m or possibly 750 m deep; (5) middle bathyal (middle slope), 750–1000 m or slightly deeper. The palaeontological criteria for this classification are listed in Table 6.

All North Sea well sites were bathyal in the Palaeogene and neritic in the Neogene, largely becoming inner neritic in the late Neogene.

The Venture area, offshore Nova Scotia is a Gulf Stream influenced site, with a rich planktonic foramineral record throughout the upper Cretaceous and most of the Cenozoic. The sample record, as in the North Sea, comprises ditch cuttings complemented with side wall cores. For the chronostratigraphic interpretation the results of Gradstein & Agterberg (1982) and Weaver & Clement (1986) were used. Appendix B lists the biostratigraphic results from 310 m down to 1300 m, late Pliocene through Maastrichtian. A fossil record break at 450 m indicates a latest Miocene or earliest Pliocene hiatus, part of the early Miocene is missing at 550 m, a possible late Oligocene hiatus occurs at around 800 m and an early Oligocene one at 850 m. Water depth varied from upper bathyal in the Palaeogene through neritic in the Neogene, becoming less than 50 m in the late Pliocene.

Permeability data

The permeabilities assigned to low permeability lithologies in a sedimentary column can have a major impact on the pressure evolution. Over geological timescales flow in clean sandstones (with $k > 10^{-15}\,\text{m}^2$) effectively occurs instantaneously, consequently, the impact of variations in sandstone permeability on fluid pressure development is slight. Measurements of sandstone permeability are routinely carried out on cores and plots of permeability versus porosity yield reasonable fits to Cozeny Karmen formulae (England *et al.* 1987; Mudford & Best 1989). We use the results of such a determination in our model.

Shales generally have much lower permeabilities than sandstones ($k < 10^{-20}\,\text{m}^2$) and consequently have a much greater effect on the pressure regime over geological timescales. Unfortunately, there is very little data available on the physical properties of shales (Bredehoeft & Hanshaw 1968; Magara 1971; Brace 1980;

Morrow *et al.* 1984). Many of the earlier measurements were on shales with relatively high porosities (>20%) Bredehoeft & Hanshaw 1968) and it is difficult to extrapolate these results to deeper, less porous shales. In many models the shale permeability–porosity relation used has been either an equation appropriate for fine sand (see for example Smith 1971; Doligez *et al.* 1986; England *et al.* 1987; Mudford & Best 1989) which is given in Table 2, or an equation has been fitted to the few available data points for clays and shales (Dutta 1988). However, as has been remarked upon by a number of authors (Smith 1971; Dutta 1988), the actual permeability–porosity relationship for shales has yet to be determined.

The absolute permeabilities of two shale samples from the overpressured zone in the Venture gas field have been measured (Katsube *et al.* 1991). The shale samples come from depths below sealevel of 4693 m in Venture B-13 and 5556 m in Venture B-52. In Table 7 the relevant data for each sample is presented together with the calculated regression curves. The correlation coefficients for both these curves are greater than 0.95. A transient pulse technique (Walls *et al.* 1982) was used to determine the absolute permeabilities of the two samples. The measurements were carried out at a number of different confining pressures, ranging from 10 MPa to 60 MPa, and a fixed pore pressure of 10 MPa. The measured permeabilities at in situ pressure conditions were found to be $6.7 \times 10^{-22}\,\text{m}^2$ for the Venture B-13 sample and $4.0 \times 10^{-22}\,\text{m}^2$ for the Venture B-52 sample (note that $10^{-21}\,\text{m}^2 = 1.01$ nanodarcy). Measurements of the physical properties of these shale samples are still continuing. One of the aims of these investigations is to ascertain the reasons for the different functional relationships between k and P_e for the two samples. The pore-size distributions of the shales are being investigated together with their mineral composition, using the results of X-ray diffraction analysis.

For shales in the deeper part of the sedimentary section we use the exponential relation between k and P_e obtained for the sample from Venture B-52, while for shallower shales we use the relation for fine sand (Table 2). Our experimentally determined permeability–effective pressure relations lead to unacceptably low permeabilities in the shallow parts of the sedimentary section. This probably occurs because mineral diagenetic reactions during burial contribute to the decrease in shale permeability with depth. Hence, an extrapolation from deep in the sedimentary section does not account for the fact that near-surface shales are too shallow for

Table 7. *Shale permeability measurements (Katsube* et al. *1990)*

Well	Venture B-13	Venture B-52
Depth below sea level (m)	4693.4	5556.05
Porosity (%)	4.8	5.7
Permeability–effective pressure relationship	$k = a(P_e/P_0)^v$	$k = k_0 \exp(-\kappa P_e)$
	$a = 1.59 \times 10^{-19}\,\mathrm{m}^2$	$k_0 = 2.12 \times 10^{-20}\,\mathrm{m}^2$
	$P_0 = 0.101\,\mathrm{MPa}$	$\kappa = 8.29 \times 10^{-8}\,\mathrm{Pa}^{-1}$
	$v = 0.932$	
In situ P_e(MPa)	36	20
In situ k(m)	6.7×10^{-22}	4.0×10^{-21}

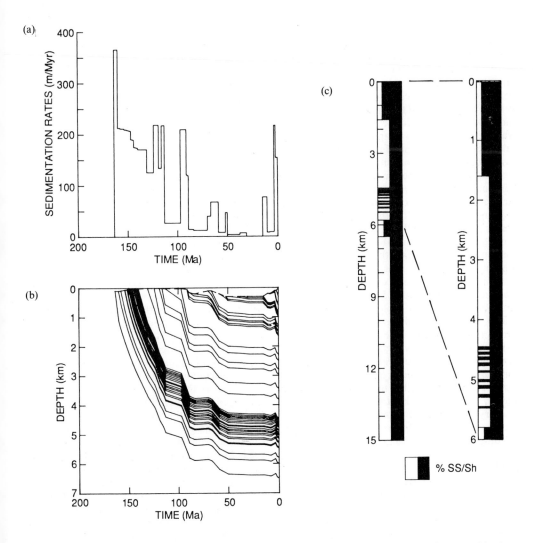

Fig. 4.(a) Corrected sedimentation rates; (b) corrected burial history, and (c) lithological column used in the composite Venture field model.

much cementation to have occurred. Clearly, the use of the same permeability–effective pressure relation for deep shales in both the North Sea and the Scotian Basin, where the porosity–effective pressures relations are different, amounts to using different permeability–porosity parametrizations in the two areas. In all likelihood, if more data were available for shallower shales we would need to use different parametrizations at shallower depths in these two areas as well.

Model Results

The Scotian Basin

Much of the data used in our model for this area has been presented in Mudford & Best (1989). We have concentrated on the Venture gas field (Fig. 1) because of the large amount of data available for these wells and the fact that the wells have been drilled to depths between 5200 m and 6000 m below sealevel. The top of the overpressure in this area is at approximately 4500 m below sealevel and none of the wells reached the base of the overpressure zone. The input data for the model, which is specific to the Venture field, is given in Table 3. The chronostratigraphic data for the Cenozoic has come from Venture D-23 (Appendix B), and for the Cretaceous from Venture B-43 (Ascoli 1982, 1983). In Fig. 4 the corrected sedimentation rates and burial history are shown together with the lithologic column used in the modelling runs. The total thickness of sediment in the vicinity of the Venture field is interpreted to be about 15 km (Wade & MacLean 1989).

The predicted fluid pressure and porosity curves at a number of different times are shown in Fig. 5. The relatively high sedimentation rates in the Early Cretaceous lead to the build up of some overpressures (Fig. 5a). These overpressures remain at much the same level, or decline slightly throughout the Late Cretaceous and Tertiary (Fig. 5b). The present-day level of overpressuring predicted by our model falls below the observed pressures seen in the Venture field. The reasons for the discrepancy between predicted and observed present-day pressures will be discussed below. The porosity–depth curves depict both the porosity curve that is used to backstrip the sediments (dashed line), and the porosity curve calculated in the forward part of the model (solid line). The mismatch between these two curves in the upper part of the section does not imply undercompaction, but is due to the fact that the average bulk density varies with depth. An average bulk density value of 2350 kg m^{-3} was used in equation (1) to convert porosity from a function of effective pressure to a function of depth. This value is too large for shallow sediments which have high porosities and low bulk densities.

A number of simulations, using the composite Venture field sedimentation rates and model lithological column (Fig. 4), have been carried out to investigate the effect on predicted, present-day overpressures of variations in the permeability. The results of these runs indicate that, even if the permeability of much of the sedimentary section at depths below 4500 m falls to values approaching those of shale (10^{-20} to 10^{-21} m^2), pressure values similar to the observed pressures cannot be generated by the model using the calculated sedimentation rates. One alternative is to invoke hydrocarbon generation from in situ organic carbon. Unfortunately, the measured values of total organic carbon (TOC) in Venture field sediments are no greater than 2%. This level of TOC is insufficient to produce a significant increase in the overpressures by hydrocarbon generation.

The results of the modelling runs for the Venture field indicate that the overpressuring mechanisms currently included in the model (compaction, aquathermal pressuring, and in situ hydrocarbon generation) do not lead to the currently observed levels of overpressuring in the Venture field. Mineral diagenetic reactions that cause dynamic permeability reductions, may contribute to the overpressuring. This has been proposed by a number of people (I. Harris, pers. comm. 1988; Jansa & Urrea 1990). However, calculations of fluid volumes necessary for mineral diagenetic reactions usually require far more fluid to pass through the pore space than is indicated by the results of fluid flow calculations (Giles 1987). Another mechanism that can have a significant effect on overpressures is ice loading during glaciations. This loading will happen on timescales of the order of 1 ka and will have a similar effect to that of rapid sedimentation. Ice thicknesses over the Scotian Shelf are inferred to have been no greater than 500 m, and probably somewhat thinner (Quinlan & Beaumont 1981, 1982; Mosher *et al.* 1989; Gipp & Piper 1989). The consequences of ice loading for subsurface fluid pressures in the Venture field are currently under investigation. Hydrocarbons and water may have migrated laterally into traps in the Venture field, and this two-dimensional process could lead to an increase in fluid pressures. However, this process cannot be investigated in a one-dimensional model.

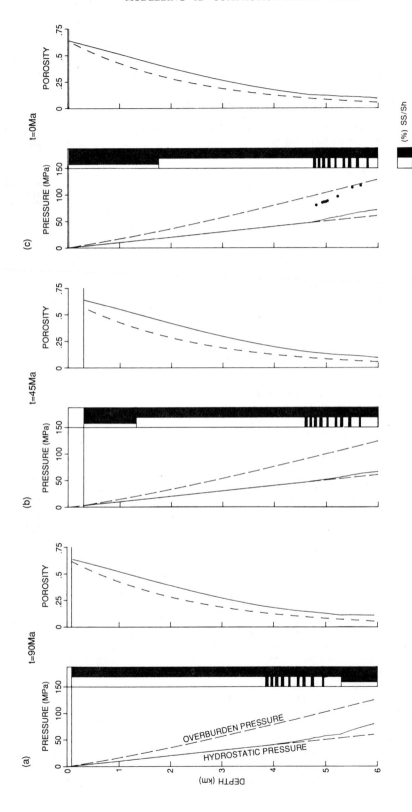

Fig. 5. Plots of predicted fluid pressure and porosity as a function of depth for the Venture field (at (**a**) t = 90 Ma, (**b**) t = 45 Ma, and (**c**) t = 0 Ma. The dashed porosity–depth curve is the porosity used to backstrip the sediments and the solid curve is the predicted porosity calculated during sediment deposition. The points plotted on curve c are Drill Stem Test data from Venture H-22.

The North Sea

Detailed stratigraphic and lithologic data for the Cenozoic are available for the wells in Table 5. The results of modelling runs for two of these wells (Esso 16/1-1 and Shell 30/19-1) (cf. Fig. 2) are typical of the results obtained for all 5 wells used in the study. The corrected sedimentation rates and model lithological columns for Esso 16/1-1 and Shell 30/19-1 are shown in Figs 6 and 7 respectively. Modelling runs have been carried out with the assumptions that the total depth to basement is around 5 km and that sediment deposition began 208 Ma, at the beginning of the Jurassic. In addition, it has been assumed that the composition of Cretaceous and older sediments is 50% sandstone and 50% shale. The details of exactly when sediment deposition began and the total depth of sediment deposited do not have much impact on the predicted present-day pressure profiles. This is because the effect, on the present-day pressure, of sedimentation rate and lithological variation early in the history of the sedimentary basin will decay away before the present day. The parameter values specific to the North Sea modelling runs are given in Table 4. We had no data on permeability or porosity for North Sea sediments, but instead used an appropriate porosity–depth profile given in Gradstein *et al.* (1989). This was converted to a porosity–effective pressure relation using equation (1), assuming that the average bulk density was 2350 kg m^{-3}. The permeability–porosity relationships used for sandstone and shallow shale in the North Sea were the same as those used for the Venture field, while the relationship found between permeability and effective pressure for deep shales from Venture B-52 was also used in the North Sea.

The predicted pressure and porosity profiles at approximately 5 Ma, 2 Ma, and the present-day are shown in Figs 8 and 9 for wells 16/1-1 and 30/19-1 respectively. The present-day overpressuring in both wells seems to be largely caused by the late Cenozoic increase in sedimentation rate. We have some mudweight data for well 16/1-1 which is plotted in Fig. 8c. The predicted fluid pressures in the overpressured sediments are about 15% less than the equivalent pressures calculated from mudweights. The data points on the 0 Ma plot of pressure versus depth in well 30/19-1 (Fig. 9c) are mudweights from a well in the Norwegian sector of the northern North Sea. For the purposes of comparison, the Cenozoic section from the northern North Sea was stretched to equal that in well 30/19-1. The predicted level of overpressuring is of a similar

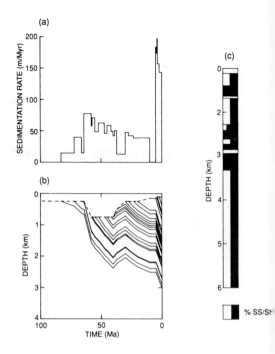

Fig. 6.(a) Corrected sedimentation rates, (b) corrected burial history, and (c) lithological column for Esso 16/1-1.

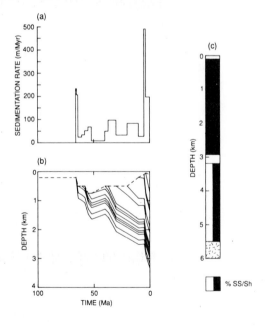

Fig. 7.(a) Corrected sedimentation rates, (b) corrected burial history, and (c) lithological column for Shell 30/19-1.

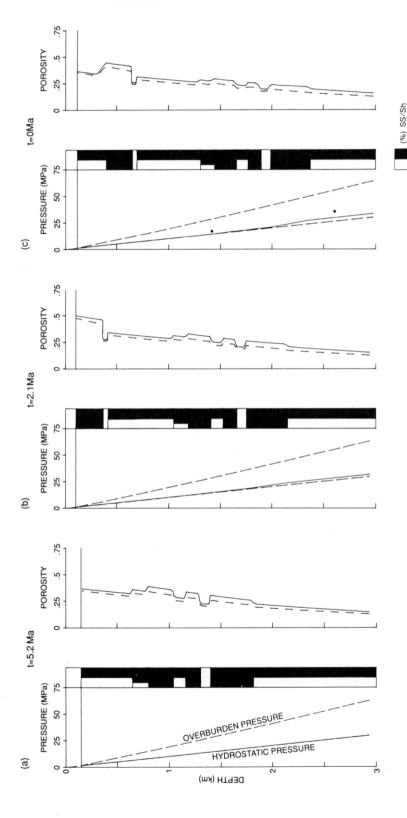

Fig. 8. Plots of predicted fluid pressure and porosity as a function of depth for Esso 16/1-1 at (**a**) t = 5.2 Ma, (**b**) t = 2.1 Ma, and (**c**) t = 0 Ma. The dashed porosity–depth curve is the porosity used to backstrip the sediments and the solid curve is the predicted porosity calculated during sediment deposition. The points plotted on curve c are mudweights used in this well.

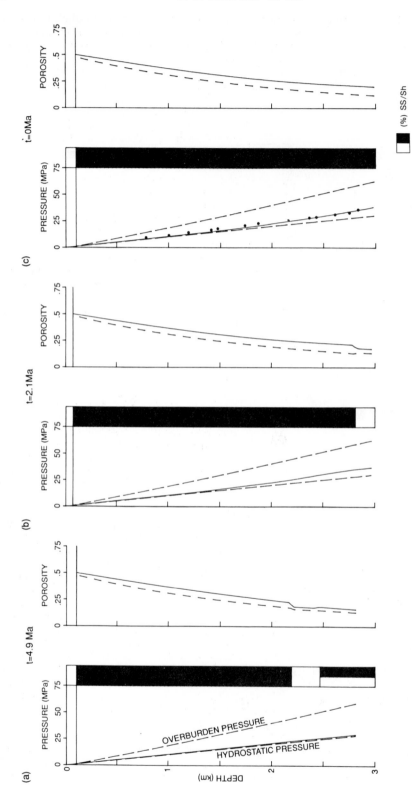

Fig. 9. Plots of predicted fluid pressure and porosity as a function of depth for Shell 30/19-1 at (**a**) $t = 4.9$ Ma, (**b**) $t = 2.1$ Ma, and (**c**) $t = 0$ Ma. The dashed porosity–depth curve is the porosity used to backstrip the sediments and the solid curve is the predicted porosity calculated during sediment deposition. The points plotted on curve c are mudweights from a well in the Norwegian Sector of the northern North Sea, in which the Cenozoic section has been stretched to the same thickness as in 30/19-1.

order of magnitude to the observed overpressures. The overpressures predicted in the central North Sea are insufficient to cause significant undercompaction as is clear from the present-day predicted porosity profiles. Possible effects of ice loading, if relevant, will be the subject of a separate investigation.

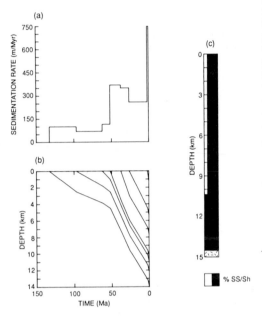

Fig. 10.(a) Corrected sedimentation rates, (b) corrected burial history, and (c) lithological column typical of the Texas Gulf Coast. The sediment thicknesses and age data are taken from Sharp & Domenico (1976).

The Gulf Coast

Sedimentation rates throughout the Cenozoic along the United States Gulf Coast have been much higher than during the same period in either the Scotian Basin or the North Sea. For this reason a simulation run was undertaken using typical Gulf Coast sedimentation rates so that comparisions could be made with the pressure and porosity profiles predicted in areas of lower sedimentation rate. The porosity and permeability parameters used in the Gulf Coast model are the same as those used in the North Sea (Table 4). The age–depth data used to backstrip the sediment column is taken from Sharp & Domenico (1976), and is representative of the Texas Gulf Coast. The model lithological column has a sandstone to shale ratio of 1:3 for the

Cenozoic section and is composed of 100% shale below that. The corrected sedimentation rate and burial history curves are shown in Fig. 10, together with the model lithological column.

The predicted pressure and porosity profiles at a number of geologic times are shown in Fig. 11. The levels of overpressuring and undercompaction predicted for the Gulf Coast are clearly much greater than those predicted for the Scotian Basin and North Sea. The data points plotted on Fig. 11d are typical mudweights which have been converted to an equivalent pressure. The pore fluid pressures predicted by the model fall some 15% below the mudweights. There are significant amounts of undercompaction between depths of 2.5 km and 5.5 km with porosities remaining between 15% and 20% throughout this depth interval. Clearly, considerably more data than we have used would be required to produce a realistic model of pressure and porosity evolution on the Gulf Coast. However, these modelling results are in general agreement with the gross details of the published porosity and pressure data (Magara 1971).

Conclusions

In this paper we have presented the results of modelling the fluid pressure and porosity evolution over geological timescales in the Scotian Basin, the North Sea, and the United States Gulf Coast. Recently acquired data on the permeability of low porosity shales has been used in our model. Measured shale permeabilities at in situ pressure conditions in the overpressure zone of the Venture gas field are approximately 10^{-21} m². Corrected sedimentation rates have been calculated by backstripping the sediments using detailed chronostratigraphic data from the Scotian Basin and North Sea. Age–depth estimates appropriate to the Texas Gulf Coast have been used to determine the backstripped sedimentation rates for this area.

Sedimentation histories and rates have been similar in the Scotian Basin and North Sea since the beginning of the Tertiary. However, the Cenozoic sediments in the North Sea contain much more shale than do those in the Scotian Basin. This difference in lithologic composition leads to compaction-driven overpressuring in North Sea sediments older than 25 Ma (Oligocene). The levels of overpressuring predicted for the North Sea wells are similar to equivalent pressures calculated from mudweight data. The overpressuring is caused by an increase in sedimentation rate in the Neogene. The overpres-

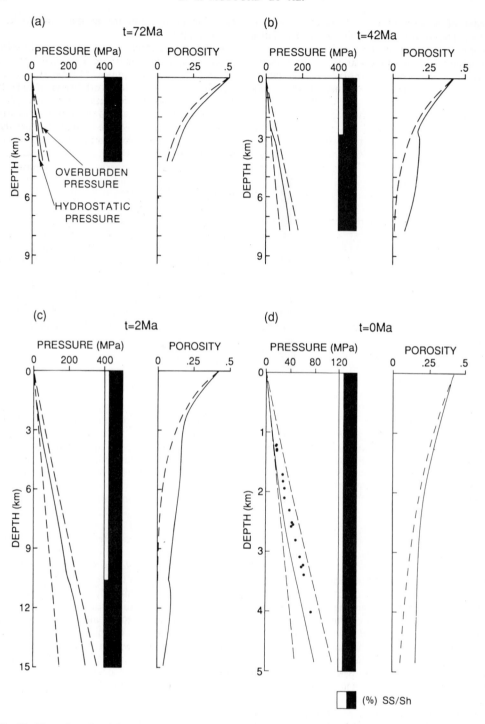

Fig. 11. Plots of predicted fluid pressure and porosity as a function of depth for typical Texas Gulf Coast parameters, (**a**) t = 72 Ma, (**b**) t = 42 Ma, (**c**) t = 2 Ma, and (**d**) t = 0 Ma. The dashed porosity–depth curve is the porosity used to backstrip the sediments and the solid curve is the predicted porosity calculated during sediment deposition. The points on curve d are typical mudweights from the Gulf Coast. Note the change of depth scale on curve d.

sured sediments are predicted to be only slightly undercompacted, due to the relatively small amount of overpressuring.

In the Scotian Basin low levels of compaction-driven overpressuring are produced in sediments of earliest Cretaceous and Jurassic age. This overpressuring is initiated by the relatively high sedimentation rates during the period ending about 90 Ma. The level of overpressuring falls as the sedimentation rate drops during the Late Cretaceous and Tertiary, but never entirely dissipates. The undercompaction predicted by the model is slight because there are only small amounts of overpressuring. The reasons for the differences in overpressure predicted for the Scotian Basin and the North Sea can be ascribed to the differences in lithology between the two areas. Larger amounts of shale deposited during the Cenozoic in the Scotian Basin would almost certainly have led to overpressuring higher up in the section, as is the case in the North Sea.

Results have also been presented for a model run using typical United States Gulf Coast sedimentation rates. These sedimentation rates are much higher throughout the Cenozoic than those in the Scotian Basin and North Sea. Significant levels of overpressuring are predicted for the Gulf Coast model. These overpressures have a magnitude similar to the equivalent pressures calculated from mudweights (Fig. 11), and to Gulf Coast pressures given by Magara (1971). The lithology in the Cenozoic section of the Gulf Coast is dominated by shale. Higher sedimentation rates on the Gulf Coast have led to greater levels of overpressuring. In addition, significant amounts of undercompaction ($\phi = 16\%$ at a depth of 5.5 km) are predicted for the Gulf Coast.

The poor agreement between the predicted and observed pressures on the Scotian Shelf is clearly due to our neglect of an overpressuring mechanism, which is important in this area. One possible mechanism is the dynamic formation of low permeability barriers by mineral diagenetic reactions. However, the lack of sensitivity of the pressure profiles to the amount of shale in the Early Cretaceous and Jurassic sediments, together with the slow vertical fluid velocities argues against mineral diagenesis as the major factor in the observed overpressures. Another possible overpressuring mechanism is ice loading during Pleistocene glaciations. This has a similar effect to that produced by a sudden increase in the sedimentation rate. The effects of both of these mechanisms in the Scotian Basin are presently being investigated. The migration of water and gas, from deeply buried source rocks, up-dip to the overpressure zone may also provide a plausible explanation for at least part of the overpressuring.

In summary, compaction-driven flow in sedimentary columns dominated by shales appears to explain a large component of the overpressuring observed in both the North Sea and the Gulf Coast. Although the Scotian Basin has experienced similar sedimentation rates to those in the North Sea, there is insufficient shale to produce compaction-driven overpressuring of the magnitude observed in the area.

We are grateful to several colleagues who supplied us with supporting well data, ranging from mud weights to depths of (Pleistocene) ice groundings or Quaternary isopach literature, including D. Piper, J. Syvitski and F. Thomas (Dartmouth), D. Fillon (New Orleans), D. Smith (Sunbury), G. Jones (Brea), I. L. Kristiansen (Bergen), and C. Griffiths (Trondheim). Critical reviews by S. Bell, C. Bethke, R. Courtney, and D. Mann helped to improve the paper. The work of B.S.M. was funded by the Office of Energy Research and Development, project number 61109. Geological Survey of Canada contribution number 49989.

Appendix A

Mathematical description

The one dimensional, single phase equations describing fluid pressure and porosity evolution through geological time in a compacting sedimentary column can be derived from the equations of fluid and solid mass conservation together with Darcy's Law. The equation for the water pressure can be written.

$$\phi\beta_w\frac{dP_w}{dt} = \frac{\partial}{\partial z}\left[\lambda(\frac{\partial P_w}{\partial z} + g\rho_w)\right] - \frac{1}{1-\phi}\frac{d\phi}{dt} + \phi a_w\frac{dT}{dt} \quad (A1)$$

where ϕ represents porosity; P_w denotes the water pressure; t represents the time; z is the height above the basement; λ is the fluid mobility, $\lambda = k/\mu$ where k represents permeability, and μ denotes the fluid viscosity; β_w denotes the fluid isothermal compressibility; ρ_w denotes the water density; g represents acceleration due to gravity; a_w denotes the water isobaric thermal expansion coefficient; and T represents the temperature. The porosity is assumed to be a known function of the effective pressure, $P_e (= P_t - P_w$, where P_t represents the overburden pressure, so that the thickness of the sediment at time t, $L(t)$, is given implicitly by the equation

$$\int_0^{L(t)}[1 - \phi(P_e)]dz = \int_0^t[1 - \phi_0(t')][\dot{\Gamma}(t')dt' \quad (A2)$$

where $\dot{\Gamma}(t')$ is the backstripped sedimentation rate through time, and $\phi_0(t')$ is the porosity at the surface of the sediments, which may vary with lithology.

The boundary and initial conditions are

$$\frac{\partial P_w}{\partial z} = -\rho_w g \text{ at } z = 0, \tag{A3}$$

$$P_w = P_a \text{ at } z = L(t), \tag{A4}$$

$$P_w = P_a \text{ at } t = 0. \tag{A5}$$

where P_a represents the fluid pressure at the top of the sediment column, and $L = 0$ at $t = 0$.

In equation (A1) the terms on the right hand side containing $d\phi/dt$, and dT/dt model the effects of changes in sedimentation and temperature respectively on the water pressure. The first term on the right hand side of equation (2) comes from Darcy's Law and represents the pressure changes caused by fluid flow in response to non-hydrostatic pressure gradients. The use of equation (A1) allows us to calculate the pressure profiles due to compaction and aquathermal effects.

Each sedimentary layer within the sediment column can have a mixed composition. The porosity of a mixed layer, assumed to be the harmonic mean of the porosities of the components, is given by the expression

$$1/\phi = \sum_{i=1}^{N} f_i/\phi_i \tag{A6}$$

where ϕ is the porosity of a layer composed of N components, each making up a fraction f_i of the layer and each having porosity ϕ_i. The permeability of a mixed layer, k, is given by the equation

$$k = \prod_{i=1}^{N} (k_i)^{f_i} \tag{A7}$$

where k_i denotes the permeability of each component. The permeability of each component is assumed to be a known function of either porosity or effective pressure.

Appendix B

Cenozoic Diostratigraphy for Mobil Texaco PEX Venture D-23, Scotian Shelf

Age in Ma	Depth in m	Relative age	Water depth in m
1.6–2	310–380	late Pliocene undefined	< 50

Neogloboquadrina aff. pachyderma, N. pachyderma, Globigerinoides trilobus, Globorotalia siakensis (reworked), *G. acrostoma* (reworked), *Spiroplectammina carinata* (reworked), *Uvigerina canariensis, Fursenkoina gracilis.*

2.2	380	late Pliocene	100–200

Nonionella pizarrense, Neogloboquadrina dutertrei, N. atlantica, Globigerina bulloides, Globorotalia inflata, Globoquadrina altispira.

2.2–3	410–420	'middle' Pliocene	100–200

Neogloboquadrina humerosa, N. atlantica, N. pachyderma, Globorotalia aff. margaritae.

3.5	430–440	N19/20, early Pliocene	100–200

Pulleniatina primalis, Globigerinita glutinata, G. uvula, Globorotalia cibaoensis group, *G. scitula, G. crassaformis, G. margaritae, Globoquadrina dehiscens* (rare), *Neogloboquadrina dutertrei, N. atlantica.*

break in fossil record at 450 m; possible hiatus.

4.5	460–470	late Miocene	200–300

Globoquadrina dehiscens–praedehiscens, G. dehiscens (common), *Neogloboquadrina acostaensis, Globorotalia acrostoma.* 490–400 m is sandy on the log and appears barren.

12	530–750	middle Miocene	200–300

Globorotalia praebulloides, G. mayeri, Siphonina advena, Marginulopsis bachei, Ceratobulimina contraria, Sigmoilopsis tenuis, Spiroplectammina carinata, Melonis barleanum, hispid and costate uvigerinids, including *U. peregrina* and *U. macrocarinata, Globoquadrina venezuelana, G. baroemoensis.* Interval 550 m is barren and may represent a hiatus. *Globorotalia angulisuturalis* at 580 m, and *Globigerina gortanii* at 750 m are reworked from upper Oligocene strata.

| 18 | 757–800 | early Miocene–late Oligocene | 300–500 |

Globoquadrina venezuelana, Catapsydrax unicavus, coarse agglutinated taxa.

| 25 | near 800 | possible hiatus | 300–500 |

break in fossil record, and much diagenetically induced tests dissolution.

| 30 | 840–850 | middle Oligocene | 300–500 |

Globoquadrina venezuelana, Globigerina gortanii, G. munda, G. decoraperta, G. ouachitaensis, Globorotalia opima opima.

| 33 | near 860 | possible hiatus | 300–500 |

| 38 | 870 | late Eocene | 300–500 |

Cibicidoides granulosus, Subbotina linaperta, Turborotalia pomeroli, Globigerinatheka index.

| | approx 890 | possible middle–late Eocene hiatus | 300–500 |

| 50 | 900–910 | early part of Middle Eocene | 300–500 |

Subbotina boweri, Acarinina aragonensis, A. bulbrooki, A. pentacamerata, A. senni, abundant radiolarians and diatoms.

| 52 | 940–970 | early Eocene | 300–500 |

Acarinina pentacamerata, Subbotina inequispira, Spiroplectammina spectabilis, radiolarians.

| 59 | 972–1035 | late Palaeocene, P4 | 300–500 |

Planorotalites chapmani, P. pseudomenardii, Morozovella aequa, Gavelinella beccariiformis, Acarinina mckannai, Spiroplectammina spectabilis.

| 67 | 1300–1430 | Maastrichtian | 100–200 |

Globotruncana stuartii, G. stuartiformis, G. arca, G. fornicata, g. linneanum group, *Globotruncanella havanensis, Stensoeina pommerana, Loxostomum gemmum.*

References

Ascoli, P. 1982. *Report on the biostratigraphy (foraminifera and ostracoda) and depositional environments of the Mobil–Texaco–Pex Venture B-43 well, Scotian Shelf, from the top of Cretaceous (1320 m) to 5380 m.* Geological Survey of Canada, Report No. EPGS-PAL.21-82PA.

—— 1983. *Report on the biostratigraphy (foraminifera and ostracoda) and depositional environments of the Mobil–Texaco–Pex Venture B-43 well, Scotian Shelf, from 5400 to 5871 m.* Geological Survey of Canada, Report No. EPGS-PAL.12-83PA.

Baldwin, B & Butler, C. O. 1985. Compaction curves. *American Association of Petroleum Geologists Bulletin,* **69**, 622–626.

Barker, C. 1988. Generation of anomalous internal pressures in source rocks and its role in driving petroleum migration. *Revue de l'Institut Francais de Petrole,* **43**, 349–355.

Bethke, C. 1985. A numerical model of compaction-driven groundwater flow and heat transfer and its application to the paleohydrology of intracratonic sedimentary basins. *Journal of Geophysical Research,* **90**, 6817–6828.

—— 1986. Inverse hydrologic analysis of the distribution and origin of Gulf Coast-type geopressured zones. *Journal of Geophysical Research,* **91**, 6535–6545.

——, Harrison, W. J., Upson, C. & Altaner, S. P. 1988. Supercomputer analysis of sedimentary basins. *Science,* **239**, 261–267.

Bishop, R. S. 1979. Calculated compaction states of thick abnormally pressured shales. *American Association of Petroleum Geologists Bulletin,* **63**, 918–933.

Brace, W. 1980. Permeability of crystalline and argillaceous rocks. *International Journal of Rock*

Mechanics and Mining Sciences and Geomechanics Abstracts, **17**, 241–251.

BREDEHOEFT, J. & HANSHAW, B. 1968. On the maintenance of anomalous fluid pressures: I. Thick sedimentary sequences. *Geological Society of America Bulletin*, **79**, 1097–1106.

——, DJEVANSHIR, R. D. & BELITZ, K. R. 1988. Lateral fluid flow in a compacting sand–shale sequence: south Caspian Basin. *American Association of Petroleum Geologists Bulletin*, **72**, 416–424.

CAO, S. & LERCHE, I. 1987. Geohistory, thermal history and hydrocarbon generation history of the northern North Sea Basin. *Energy Exploration and Exploitation*, **5**, 315–355.

—— & —— 1989. Geohistory, thermal history and hydrocarbon generation history of Navarin Basin COST no. 1 well, Bering Sea, Alaska. *Journal of Petroleum Geology*, **12**, 325–352.

DOLIGEZ, B., BESSIS, F., BURRUS, J., UNGERER, P. & CHENET, P. 1986. Integrated numerical simulation of the sedimentation heat transfer, hydrocarbon formation and fluid migration in a sedimentary basin: the THEMIS model. *In*: BURRUS, J. (ed.) *Thermal modeling in sedimentary basins, proceedings of 1st IFP exploration conference, Carcans, France, June 3–7, 1985*. Technip, Paris, 173–195.

——, UNGERER, P., CHENET, P., BURRUS, J., BESSIS, F. & BESSEREAU, G. 1987. Numerical modeling of sedimentation, heat transfer, hydrocarbon formation and fluids migration in the Viking Graben, North Sea. *In*: BROOKS, J. & GLENNIE, K. W. (eds) *Petroleum Geology of northwest Europe*. Graham & Trotman, London, 1039–1048.

DUTTA, N. 1986. Shale compaction, burial diagenesis, and geopressures: a dynamic model, solution and some results. *In*: BURRUS, J. (ed.) *Thermal modeling in sedimentary basins, proceedings of 1st IFP exploration conference, Carcans, France, June 3–7, 1985*. Technip, Paris, 149–172.

—— 1988. Fluid flow in low permeable porous media. *Revue de l'Institut Francais du Petrole*, **43**, 165–180.

ENGLAND, W., MACKENZIE, A., MANN, D. & QUIGLEY, T. 1987. The movement and entrapment of petroleum fluids in the subsurface. *Journal of the Geological Society, London*, **144**, 327–347.

GIBSON, R. 1958. The progress of consolidation in a clay layer increasing in thickness with time. *Géotechnique*, **8**, 171–182.

GILES, M. 1987. Mass transfer and problems of secondary porosity creation in deeply buried hydrocarbon reservoirs. *Marine and Petroleum Geology*, **4**, 188–204.

GIPP, M. R. & PIPER, D. J. 1989. Chronology of Late Wisconsin glaciation, Emerald Basin, Scotian Shelf. *Canadian Journal of Earth Sciences*, **26**, 333–335.

GRADSTEIN, F. M. & AGTERBERG, F. P. 1982. Models of Cenozoic foraminiferal stratigraphy–northwestern Atlantic margin. *In*: CUBITT, J. & REYMENT, R. (eds) *Quantitative Stratigraphic Correlation*, J. Wiley & Sons, 119–181.

——, ——, BROWER, J. & SCHWARZACHER, S. 1985 *Quantitative Stratigraphy*. Reidel Publ. Co. and Unesco.

——, FEARON, J. & HUANG, Z. 1989. *BURSUB and DEPOR 3.50 – two FORTRAN 77 programs for porosity and subsidence analysis*. Geological Survey of Canada – Open File, no. 1283.

——, KAMINSKI, M. & BERGGREN, W. 1988. Cenozoic foraminiferal biostratigraphy of the central North Sea. *Abhandlungen Geologischen Bundes Anstalt*, **41**, 97–108.

HANSHAW, B. & BREDEHOEFT, J. 1968. On the maintenance of anomalous fluid pressure: II. Source of layer at depth. *Geological Society of America Bulletin*, **79**, 1107–1122.

JABOUR, H. & NAKAYAMA, K. 1988. Basin modeling of Tadla Basin, Moroco, for hydrocarbon potential. *American Association of Petroleum Geologists Bulletin*, **72**, 1059–1073.

JANSA, L. F. & URREA, V. H. N. 1990. Geology and diagenetic history of overpressured sandstone reservoirs in the Venture gas field, offshore Nova Scotia, Canada. *American Association of Petroleum Geologists Bulletin*, **74**, 1640–1658.

KATSUBE, T. J., MUDFORD, B. S. & BEST, M. E. 1991. Petrophysical Characteristics of shales from the Scotian Shelf. *Geophysics*, in press.

KEITH, L. & RIMSTIDT, J. 1985. A numerical compaction model of overpressuring in shales. *Mathematical Geology*, **17**, 115–136.

LERCHE, I., YARZAB, R. & KENDALL, C. 1984. Determination of paleoheat flux from vitrinite reflectance data. *American Association of Petroleum Geologists Bulletin*, **68**, 1704–1717.

MAGARA, K. 1971. Permeability considerations in generation of abnormal pressures. *Society of Petroleum Engineers Journal*, **11**, 236–242.

MERCER, J., PINDER, F. & DONALDSON, I. 1975. A galerkin-finite element analysis of the hydrothermal system at Wairakei, New Zealand. *Journal of Geophysical Research*, **80**, 2608–2621.

MORROW, C., SHI, L. & BYERLEE, J. 1984. Permeability of fault gouge under confining pressure and shear stress. *Journal of Geophysical Research*, **89**, 3193–3200.

MOSHER, D. C., PIPER, D. J., VILKS, G. V., AKSU, A. & FADER, G. B. 1989. Evidence for Wisconsinan glaciations in the Verrill Canyon area, Scotian Slope. *Quaternary Research*, **31**, 27–40.

MUDFORD, B. S. 1990. A one dimensional, two phase model of overpressure generation in the Venture Gas Field, offshore Nova Scotia. *Bulletin of Canadian Petroleum Geology*, **38**, 246–258.

MUDFORD, B. S. & BEST, M. E. 1989. The Venture gas field, offshore Nova Scotia: a case study of overpressuring in a region of low sedimentation rate. *American Association of Petroleum Geologists Bulletin*, **73**, 1383–1396.

NAKAYAMA, K. 1987. Hydrocarbon-expulsion model and its application to Niigata area, Japan. *American Association of Petroleum Geologists Bulletin*, **71**, 810–821.

QUINLAN, G. & BEAUMONT, C. 1981. A comparison of

observed and theoretical postglacial relative sea level in Atlantic Canada. *Canadian Journal of Earth Sciences*, **18**, 1146–1163.

—— & —— 1982. The deglaciation of Atlantic Canada as reconstructed from the postglacial relative sea-level record. *Canadian Journal of Earth Sciences*, **19**, 2232–2246.

ROSENBLAT, S. 1985. A mathematical model of compaction-induced migration. *Journal of Geophysical Research*, **90**, 779–789.

SCLATER, J. G. & CHRISTIE, P. 1980. Continental stretching: an explanation of the post-mid-Cretaceous subsidence of the central North Sea Basin. *Journal of Geophysical Research*, **85**, 3711–3739.

SHARP, Jr., J. & DOMENICO, P. 1976. Energy transport in thick sequences of compacting sediment. *Geological Society of America Bulletin*, **87**, 390–400.

SHI, Y. & WANG, C. 1988. Generation of high pore pressures in accretionary prisms: inferences from the Barbados subduction complex. *Journal of Geophysical Research*, **93**, 8893–8910.

SMITH, J. 1971. The dynamics of shale compaction and evolution of pore fluid pressures. *Mathematical Geology*, **3**, 239–263.

STAM, B., GRADSTEIN, F., LLOYD, P. & GILLIS, D. 1987. Algorithms for porosity and subsidence history. *Computers and Geoscience*, **13**, 317–349.

THORNE, J. A. & WATTS, A. B. 1989. Quantitative analysis of North Sea subsidence. *American Association of Petroleum Geologists Bulletin*, **73**, 88–116.

UNGERER, P., BESSIS, F., CHENET, R., DURAND, B., NOGARET, E., CHIARELLI, A., OUDIN, J. & PERRIN, J. 1984. Geological and geochemical models in oil exploration; principles and practical examples. *In*: DEMAISON, G. & MURRIS, R. J. (eds) *Petroleum Geochemistry and Basin Evaluation*. American Association of Petroleum Geologists Memoir, **35**, 53–77.

——, BURRUS, J., DOLIGEZ, B., CHENET, P. & BESSIS, F. 1990. Basin evaluation by integrated two-dimensional modelling of heat transfer, fluid flow, hydrocarbon generation and migration. *American Association of Petroleum Geologists Bulletin*, **74**, 309–335.

——, CHIARELLI, A. & OUDIN, J. 1985. Modelling of petroleum genesis and migration with a bidimensional computer model in the Frigg sector, Viking graben. *In: Petroleum geochemistry in exploration of the Norwegian Shelf*. Norwegian Petroleum Society, 121–129.

WADE, J. & MACLEAN, B. 1989. The geology of the south-eastern margin of Canada, part 2: aspects of the geology of the Scotian Basin from recent seismic and well data. *In*: KEEN, M. & WILLIAMS, G. (eds) *Geology of the continental margin off eastern Canada*. Geological Survey of Canada no. 2 (also Geological Society of America, The Geology of North America, vol. 1), in press.

WALLS, J., NUR, A. & BOURBIE, T. 1982. Effects of pressure and partial water saturation on gas permeability in tight sands: experimental results. *Journal of Petroleum Technology*, 930–936.

WEAVER, P. & CLEMENT, B. 1986. Synchroneity of Pliocene planktonic Foraminiferal datum in the North Atlantic. *Marine Micropaleontology*, **10**, 295–307.

WEI, Z. & LERCHE, I. 1988. Quantitative dynamic geology of the Pinedale anticline, Wyoming, U.S.A.: an application of a two dimensional simulation model. *Applied Geochemistry*, **3**, 423–440.

WELTE, D. & YUKLER, M. 1981. Petroleum origin and accumulation in basin evolution—a quantitative model. *American Association of Petroleum Geologists Bulletin*, **65**, 1387–1396.

PART II:

SECONDARY MIGRATION

Are numerical models useful in reconstructing the migration of hydrocarbons?
A discussion based on the Northern Viking Graben

J. BURRUS, A. KUHFUSS, B. DOLIGEZ & P. UNGERER

Institut Français du Pétrole, BP 311, 92506 Rueil-Malmaison Cedex, France

Abstract: The 2D-finite-difference code TEMISPACK is used to discuss the history of hydrocarbon (HC) generation, expulsion and migration along a 160 km long transect across the Northern Viking Graben. In the first section, HC generation is discussed based on hypotheses on the crustal heat flow; a kinetic model is used to calculate the maturity. Observations are shown to be consistent with a 67 mW m^{-2} heat flow, possibly 60 mW m^{-2} in the graben axis. The precision of the calibrated heat flow is not more than 10%, which results in an uncertainty of the timing of HC generation of 5–10 Ma. The development of overpressures, observed today in the deep Jurassic reservoirs is shown to be essentially controlled by compaction, with a minor contribution due to gas generation; the model also shows that faults related to the Jurassic rifting are most certainly impermeable. In the second section, the previous overpressures and maturity reconstructions form the framework in which expulsion and migration of HC are modelled, based on the two-phase (water, hydrocarbon) Darcy law. The computation explains the filling of the tilted blocks located in the Tampen Spur zone as well as the presence of gas at Troll. A sensitivity analysis shows the importance of poorly constrained parameters such as the relative permeability curves, capillary pressures, or the properties of the petroleum fluids.

Significant advances in the understanding of the migration of hydrocarbons (HC) have been achieved during the last decade. It is now recognized that some expulsion processes contemplated 15 years ago are irrelevant (see review by Tissot & Welte 1984; Leythaeuser *et al.* 1987; England *et al.* 1987; Durand, 1987). The transport of hydrocarbons as colloidal solutions (disproved by the absence of enrichment of HC in N, S, O compounds), or as individual HC bubbles (in contradiction with the high capillary pressure generally prevailing in source-rocks) are no longer considered. Other mechanisms, initially thought to be dominant, appear possible at a local scale only. For instance, molecular solution of HC in mobile water is limited by the very low solubility of HC in water (except toluene and benzene); in addition the compaction-related flow of water expelled out from a shale at depth around 3–4 km (i.e. in the oil window) is very small, as most porosity is lost in the first two kilometres of burial. Molecular diffusion can also play a significant role locally for light hydrocarbons. Most authors however agree that the dominant mechanism for both primary and secondary migration is a separate phase flow, driven by the hydrodynamic potential (i.e. by overpressure gradients), gravity forces, and capillary pressures. This recognition led several groups in the mid-80s to try to reconstruct the expulsion and migration

phenomena by numerical modelling, i.e. by coupling the equations which describe the compaction history and subsequent overpressure development, the thermal history and subsequent HC generation history. This approach, illustrated by authors such as Welte & Yukler (1981), Durand *et al.* (1983), Ungerer *et al.* (1984), Doligez *et al.* (1986), Nakayama & Lerche (1987) presents new advantages, but also raises new questions.

Among the advantages claimed by modellers, are: (1) the possibility of estimating quantitatively the reserves in place; (2) the possibility of addressing migration problems in a dynamic geological framework, with an explicit description of time-dependent effects, and of coupling effects; (3) the possibility of testing various and conflicting hypotheses against data. Against the use of models, many explorationists feel (e.g. Mackenzie & Quigley 1988): (1) that present knowledge enables the demonstration of qualitative concepts but not quantitative ones; (2) too many physical parameters are in general ignored; (3) knowledge of basic geological data is too limited to enable quantitative reconstruction.

The purpose of this paper is to discuss the numerical approach to reconstructing migration phenomena occurring at a regional scale. The discussion is based on a numerical modelling study carried out in 1987–1988 with Elf-Aqui-

From England, W. A. & Fleet, A. J. (eds), *Petroleum Migration*
Geological Society, Special Publication No. 59, pp. 89–109.

taine in the Northern Viking Graben of the North Sea. The model used in this study is IFP's TEMISPACK finite-difference software. The first section briefly outlines the geological con-

text; the second section describes the model used; in the third section, numerical results are presented, and their significance is discussed in the fourth section.

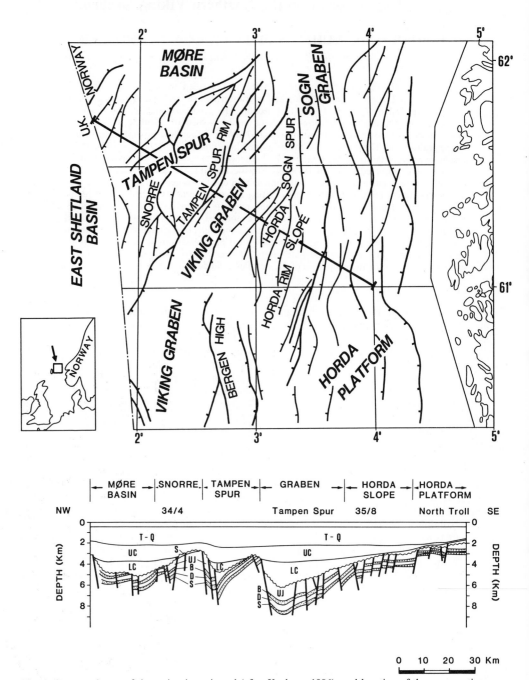

Fig. 1. Structural map of the region investigated (after Karlsson 1986), and location of the cross section investigated in this study. (S, Statfjord; D, Dunlin; B, Brent; UJ, Upper Jurassic; LC, Lower Cretaceous; UC, Upper Cretaceous; T-Q, Tertiary Quaternary).

The geological framework; an overview:

The 160 km long section modelled (Fig. 1) is located across the Northern Viking Graben, in the Norwegian Sector of the North Sea.

This region includes from east to west: the Horda Platform, the Viking Graben and the Tampen Spur area. The structural style illustrated by the cross section in Fig. 1 is typical. The Permo–Jurassic rifting is responsible for the major faults that bound the Graben to the West, for the numerous minor faults that affect the Horda platform, and for the tilted blocks in the Tampen Spur–Snorre region. The Jurassic synrift sequence is bounded by the Cimmerian unconformity, overlain by Cretaceous and Tertiary sequences deposited during the post-Jurassic flexural sagging of the margin (Goff 1983; Field 1985; Cornford et al. 1986). The following lithologies are considered (see Fig. 1).

The Palaeozoic, encountered to the North and South, has not been recognized in the area.

The Triassic is continental/fluviatile, and neither a good reservoir, nor a good source rock.

The Jurassic includes five major formations (see the review by Field 1985; Karlsson 1986): (1) the Statfjord Fm. (Lower Jurassic) is a fluvio-deltaic gas-producing group, with a poor source-rock potential, but is a good reservoir; (2) the Dunlin Fm. mixes silts, shales and sandstones; the shales have a small potential due to the presence of some coals (average TOC 3%); (3) the Brent Fm. (Middle Jurassic), associated with a northward-prograding delta, is essentially sandy, containing interbedded coals (average TOC 5–10%); it could have generated oil (Durand et al. 1983) and is a good reservoir; (4) the Heather Fm. (Upper Jurassic) is a transgressive shale, with an average TOC of 2–4%; (5) the Draupne Fm. (or Kimmeridge Clay) forms the top of the transgression; it corresponds to an excellent anoxic Type II source rock (TOC 5%) and is the principal source rock in the region.

The Cretaceous covers the extensional structures, and is essentially shaly.

The Tertiary is again shaly, but with a higher sand content, in particular on the eastern and western flanks of the Graben. Neither the Cretaceous nor the Tertiary produce hydrocarbons in the region.

The area investigated is a major petroleum province (Fig. 2). To the west, the tilted blocks of the Tampen Spur area (Statfjord, Snorre, Tampen Spur, Gulfaks) contain medium to light overpressured oil accumulations (30–40°API), located at short distances (5–15 km) from the kitchens, found in syncline positions The 'tilted block' play is indeed considered as the most prolific play in the region. To the east of the Graben, the Horda platform contains the giant gas accumulation found at Troll, with its heavy, biodegraded oil rim, and some gas and condensate accumulations such as in Block 35/8, or in Huldra (Fig. 2).

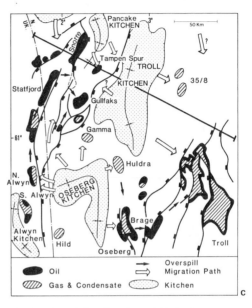

Fig. 2. Map of the oil and gas fields in the area investigated, with tentative migration paths between kitchens and structures (after Thomas et al. 1985).

Numerous authors have addressed some aspects of petroleum exploration in the region.

The maturity of the source-rocks and the thermal regime have been discussed in numerous studies (Evans & Coleman 1974; Carstens & Finstad 1981; Eggen 1985; Thomas et al. 1985; Leadholm et al. 1985). There is no consensus on the present heat flow distribution (or average thermal gradient); reported heat flows range from 50 to 75 mW m^{-2} (or 25–45°C km^{-1}). The existence of a thermal asymmetry (with the eastern flank warmer than the western) is also controversial. Modelling observed maturity was used to constrain the palaeothermal regime. Not surprisingly, no better consensus on palaeo-heat flows is achieved; whether the Jurassic rifting has had an effect (Mackenzie & McKenzie 1983) or not (Goff 1983) on maturation is controversial. In addition, many studies are based on heterogeneous or discredited methods (such as TTI, Ro/TTI calibration; see comments in Tissot et al. 1987), making it difficult to compare conclusions.

Another debatable question relates to the permeability level of major faults, and to the distribution of overpressures. Previous studies (Herring 1973; Carstens 1978; Lindberg *et al.* 1980; Chiarelli & Duffaud 1980; Carstens & Dypvik 1981; Buhrig 1989) have shown that the Upper Jurassic reservoirs are overpressured ($\Delta P \geqslant 13$–16 MPa) in the Tampen Spur region, that the Horda slope has intermediate pressures ($\Delta P < 11$ MPa), while the Horda platform is normally pressured. Many authors agree that compaction disequilibrium is the principal cause of the overpressures observed, but the exact contribution of the other causes (clay diagenesis, and HC generation below 4000 m) is disputed. Due to the existence of lateral pressure gradients, it has been suggested that most major faults are impermeable (Karlsson 1986; Cornford *et al.* 1986); most migration would occur along the dip slope of the tilted blocks, and not along the faulted scarp. In contradiction with this view, good drainage capacity has been attributed to the faults, in order to explain palaeothermal anomalies attributed to hot water flow (Thomas 1986). The pattern of expulsion and migration has been addressed qualitatively and by geochemical mass balance, but many questions remain unsolved. The Draupne Fm. is considered to have fed most accumulations. The contribution of other source-rocks (Brent) has not been established. To the west of the Graben, the expulsion efficiency has been estimated to be above 25% for the principal source-rock (Goff 1983), up to 60–90% for the richest mature sources (Cooles *et al.* 1986; Mackenzie *et al.* 1987); the entrapment ratio would be as high as 20–30% (Goff 1983). In contrast, to the eastern platform, the entrapment efficiency would be much lower (Field 1985); this is interpreted by the poor drainage across the numerous faults which affect the gently dipping Horda Platform. Therefore, the presence of the huge gas field of Troll, at the top of the platform, appears anomalous to many. Conflicting explanations have been proposed as to the origin of this gas, localized either to the northwest in the 'Troll Kitchen' (Fig. 2), or to the southwest (spilling from Huldra, Oseberg and Brage), or the north (Sogn Graben). In summary, three different aspects of petroleum geology are questioned:

palaeo-heat flow, importance of rifting in the maturity development;

origin of overpressures, and permeability of major faults;

factors controlling the expulsion and migration of HC.

Conflicting answers have been given to these questions essentially based on the qualitative interpretation of observations. The following section presents numerical reconstructions pertaining to these questions.

TEMISPACK model, and modelling results

The model

The model used is briefly described below. More detailed presentations can be found elsewhere (Doligez *et al.* 1986; Ungerer *et al.* 1987*a*, *b*; Burrus & Audebert 1990). Equations are grouped in Appendix 1.

TEMISPACK is a finite difference model which simulates numerically the following processes on a 2D vertical grid.

(a) Geometrical evolution as a result of sedimentation, erosion, compaction (assumed to result in a vertical displacement of grain particles).

(b) Computation of geopressures at every time step, obtained by coupling a water flow equation (Darcy's law) and a deformation equation (Terzaghi's effective stress, related to porosity). In addition to compaction disequilibrium, HC generation is also considered as a source of overpressures; the development of overpressures is bounded by hydraulic fracturing (permeabilities are increased where overpressures become too high with respect to geostatic stress).

(c) Computation of palaeotemperatures, as a consequence of conductive and convective heat transfer, and of the geodynamic evolution.

(d) Computation of the maturity of each source rock, according to a first-order kinetic model, calibrated on experimental Rock–Eval pyrolysis for each source rock (described by its initial potential and by a distribution of activation energies and a pre-exponential factor).

(e) Computation of the rate of HC expulsion and migration, obtained by coupling the geopressure reconstruction and the maturity module; the displacement of the water + HC fluid phases is described by a two-phase Darcy's law. No expulsion efficiency factor, or drainage efficiency factor is introduced. These efficiencies are outputs of the model, and depend on parameters such as the fluid properties (density, viscosity), the capillary pressures, and the relative permeabilities used (see discussion below).

Presentation of model results

The calibration of the thermal history, and of the permeability structure of the faults (in par-

Table 1. *Petrophysical parameters for various lithologies*

	Compaction				Heat flow			
	Surface Porosity (%)	S_o $m^2 m^{-3}$	K at $\Phi = 10\%$ (Darcy)	K_h/K_v	λ W/m/K	ρ kg m^{-3}	C J/kg/K	Q μW m^{-3}
Shales (Dunlin, Heather Draupne, Cretaceous)	65	2×10^8	10^{-9}	500	1.9	2750	840	1.5
Silts (Triassic, Heather East; Tertiary)	60	1.5×10^8	10^{-8}	500	2.9	2675	950	0.7
Sandstones (Brent, Statfjord, Heather East; Eastern and Western Tertiary)	42	10^5	2×10^{-2}	1	4.2	2650	1080	0
Basement	—	—	—	—	3.15	2800	1390	

ticular of the major faults) are presented first: lithological information and petrophysical parameters used can be found in Table 1. The computation of migration follows the discussion on thermal history and permeability structure.

Calibration of heat flow against maturity. Figure 3 shows computed present-day temperatures and maturity profiles obtained for two distinct heat flow assignments (55 and 67 mW m^{-2}). Present temperatures (Fig. 3a) agree well with the 67 mW m^{-2} curve, while maturity profiles, calculated with kinetic parameters (found in Table 2) adjusted for the Draupne Fm. (Type II) and for the Brent Fm. (Type III) with the constant 67 mW m^{-2} assumption, are found in agreement with the position of the top, peak and bottom of the oil window for both flanks of the Graben (Fig. 3b), while the 61 mW m^{-2} assumption is preferred in the Graben axis (Fig. 3c). Figure 3 clearly indicates that the 55 mW m^{-2} heat flow hypothesis does not adequately account for the present temperatures and maturity. The preferred heat flow has thus been used to model migration phenomena (see below). Figure 4 represents the distribution of computed present-day isotherms reconstructed by TEMISPACK while Fig. 5 shows the source rock maturity history using the preferred heat flow hypothesis, calculated for source rocks located at km 40 of the section (Western syncline, Fig. 4). Other past heat flows could result in the same present-day state but with a different maturity history. The result of a sensitivity

analysis is therefore indicated in Fig. 6. Several maturation histories of the Draupne and Brent source rocks, have been computed with different assumptions for the heat flow in the Graben axis (km 65 of Fig. 4). The curves obtained with constant heat flows of 55, 61 and 67 mW m^{-2} or with variable heat flow (enhanced during the Jurassic due to rifting, following the uniform extension model with $\beta = 1.7$; McKenzie 1978) all show that the deep graben is at present in the gas window. In addition, introducing the rifting hypothesis does not significantly modify the maturation history; the difference with the 67 mW m^{-2} curve is around 5 Ma. This suggests that although the 67 mW m^{-2} might not be a unique solution, alternative solutions probably result in a rather similar maturation history.

Calibration of rock and fault permeability against overpressures. The Kozeny–Carman formula (see Appendix 1) is used to represent the permeability of rocks as a function of porosity; the 'specific surface area' of the matrix and the permeability anisotropy are therefore the input data used to describe the permeabilities. In this study, specific surface areas and anisotropies were taken, for the various lithological classes, from Table 1. These values had already been tested successfully in previous modelling studies in the North Sea (Doligez *et al.* 1987; Ungerer *et al.* 1987*b*). With these parameters, the permeability of sandy reservoirs with 15% porosity is of the order of several hundred mD, and the permeability of compacted ($\Phi = 10\%$) shales is

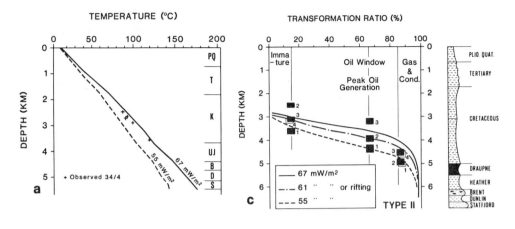

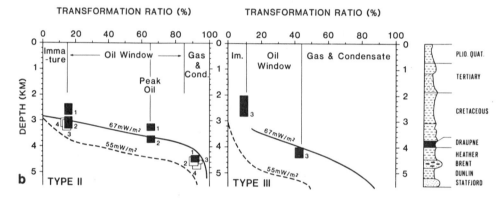

Fig. 3. Calibration of the basement heat flow: **(a)** on corrected BHT (Block 34/4); **(b)** on the maturity profile observed on the flanks of the Graben (1, Goff, 1983; 2, Field, 1985; 3, Brosse & Huc, 1986; 4, Karlsson, 1986); **(c)** on the maturity profile observed in the Graben axis (1, Field 1985; 2, Cornford *et al.* 1986; 3, Goff, 1983; 4, Brosse & Huc 1986). The 67 mW m^{-2} curve is preferred on the flanks, while the 61 mW m^{-2} heat flow is preferred in the graben axis.

Table 2. *Geochemical parameters for the Draupne and Brent Formations*

Kinetic parameters
Type II

E_i	46	48	50	52	54	56	58	(kcal/mole)			
X_{io}	10.5	29.1	149.7	321	65.7	19.2	7.3	(mg HC/g.C_{org})			

Type III

E_i	50	52	54	56	58	60	62	64	66	68	70	72
X_{io}	1.1	3.2	12.5	75.3	62.7	19.0	10.8	6.7	5.2	4.2	2.7	2.4

Initial potential (in kg HC/ton tock)

Draupne	9	(up to 25 on deep Horda Slope)
Heather	4	
Brent	3	(up to 6 in synclines)
Dunlin	6	

After Espitalié *et al.* (1988)

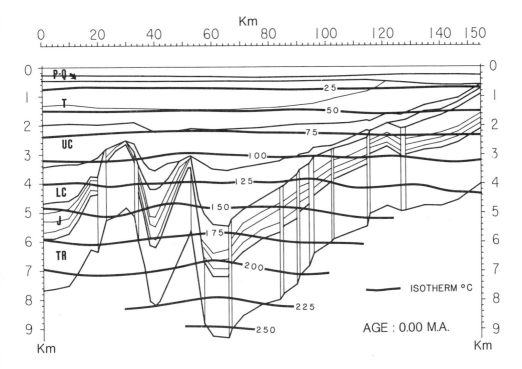

Fig. 4. 2D reconstruction of the temperatures along the section with the preferred heat flow indicated on Fig. 3. Note the representation of the Jurassic faults as vertical shear columns.

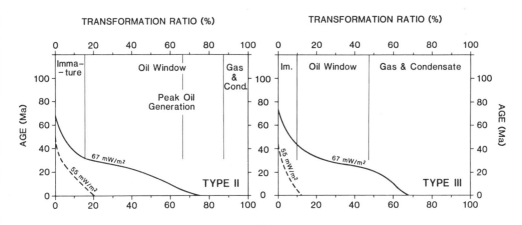

Fig. 5. Theoretical history of maturity for source-rocks located in the synclines between Snorre and Tampen Spur (Type II and Type III have different kinetic parameters, see Table 2).

around 10^{-9} D. These values are in agreement with measurements on reservoir cores in the region, and the literature on experimental measurements for shales (Morrow *et al.* 1984; Lin *et al.* 1986).

Computed overpressures can be checked against the overpressures observed in reservoirs (DST and mud density, corrected for HC col-

umns), in the vicinity of the cross section. The faults are described in TEMISPACK as vertical shear zones, with finite lateral width (around 2 km, see Fig. 4). Both lateral and vertical permeabilities are assigned to these faulted zones, permitting displacement of fluid across and parallel to the fault planes. All faults are sealed by the Cretaceous shales in the model.

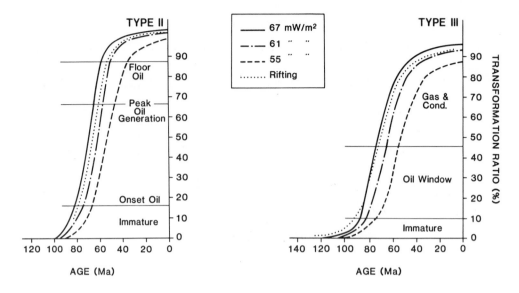

Fig. 6. Theoretical history of maturity for the graben axis: a sensitivity analysis based on various assumptions on the heat flow.

Figure 7 represents the calculated overpressures (assuming no HC generation) due to compaction disequilibrium obtained with two extreme hypotheses for fault permeabilities: all faults impermeable (like shale dykes), and all faults permeable (like sandstone dykes). Figure 7a shows that with the permeable hypothesis, the Brent and Stratfjord sandstones are found to communicate laterally from the Tampen Spur structure (km 50), up to the Horda platform, and are not overpressured anywhere along the section. In contrast, with the impermeable hypothesis (Fig. 7b), the Jurassic reservoirs being laterally confined by the faults, the overpressures cannot relax as easily towards the Horda platform, and the deep reservoirs are found to be overpressured. The model also shows, in both cases, that water flow is oriented parallel to stratification in reservoirs, and perpendicular to strata in shales (with much lower velocity). This is a classical result in hydrology. The computed pressure profiles are compared with observations in Fig. 8; computations do not differ for the most western block (Snorre, Fig. 8a), indicating that this block is hydraulically independent from the section located further to the East; the computed pressures are in good agreement with the measured ones (10 versus 13 MPa). On the Horda slope (Fig. 8b, km 98), the agreement of calculated overpressure with observations is good only with the impermeable fault hypothesis (discrepancy less than 2 MPa);

the previous permeable hypothesis is totally disproved by observations. Both profiles were computed in 2D neglecting the generation of HC (therefore underestimating the pressures in the mature source rocks). The contribution of HC generation to overpressure development is illustrated in Fig. 8c. Theoretical overpressures profiles are compared for three cases: (1) source rock contribution neglected; (2) source rock contribution assumed to be an oil charge (moderate volume increase); (3) same, but gas charge (high volume increase). As expected, the overpressures obtained with the gas hypothesis are higher than with the oil hypothesis, which are in turn higher than without HC generation. It is however noticeable that the difference between these profiles is very small (less than 5–7 MPa), disproving the claim that HC generation is the dominant cause of overpressure development in the region. In the following migration studies, all faults were assumed to be impermeable (except at Troll, where a detailed investigation, not shown here, showed better agreement for geopressures if faults were assumed to be permeable at Troll).

Migration simulation: coupling of previous steps; additional uncertainties. In HC migration reconstruction, it is necessary to state if water only, HC only, or a combination of both fluids are displaced. This distinction is introduced using relative permeability curves. It is also necessary to account for the pressure difference

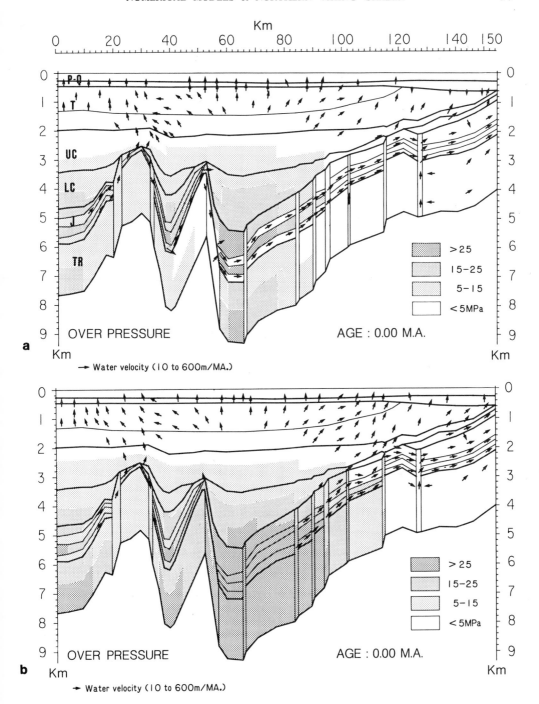

Fig. 7. Computed overpressures calculated at Present for two scenarios on the permeability of faults: (**a**) faults assumed to be permeable; (**b**) faults assumed to be impermeable. Expulsion of water appears essentially restricted to the two first kilometres (arrows: Darcy velocity).

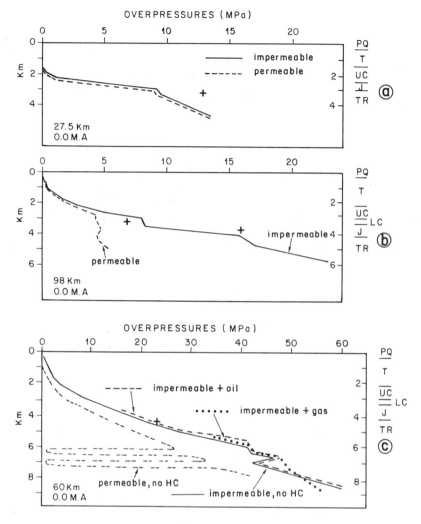

Fig. 8. Comparison of observed overpressures (crosses) with computed overpressure profiles, (**a**) at Snorre (permeable and impermeable faults); (**b**) at Block 35/8 (see Fig. 1) with permeable and impermeable faults; (**c**) in the Graben axis, with impermeable faults (neglecting the generation of HC in the source rocks, and assuming HC are generated, oil and gas), and with permeable faults (neglecting HC generation).

between the HC phase and the water phase (i.e. for capillary forces). Finally, buoyancy forces are described through the density of the HC phase, which in turn needs to be evaluated. The uncertainty in these parameters can be significant.

Relative permeability curves are usually determined experimentally on reservoir cores. The traditional shape described by reservoir engineers is shown in Fig. 9a; when the HC saturation is below a certain saturation threshold, only the water is found to be mobile; for intermediate saturations, both water and HC

are mobile; above an irreducible saturation, the water is immobile. Although these curves are known to be dependent on P–T conditions (see Fig. 9a), they are relatively well-constrained for rocks with reservoir characteristics ($K10$–100 mD). Unfortunately, no such experimental determinations are available for shales ($K = 10^{-6}$–10^{-9} D). However, numerous practical observations (existence of saturation threshold in shale source rocks, existence of residual saturation after expulsion) are in agreement with the consequences of using the relative permeability concept to model expulsion. In addition,

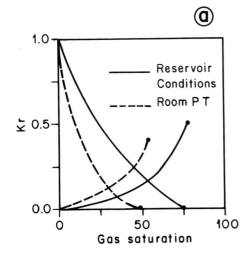

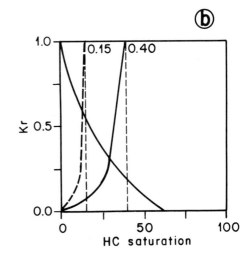

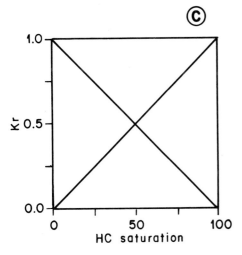

Fig. 9. Types of relative permeability curves. (**a**) A typical relative permeability curve measured in the laboratory at room and reservoir conditions; (**b**) theoretical curves used in the TEMISPACK modelling with a saturation plateau of 15% (shales) and 40% (sandstones); (**c**) theoretical curves assuming HC and water are transported as two independent, isolated phases (no HC/water interactions).

some theoretical work on rock–fluid interactions (e.g. Kalaydjian 1987) supports the idea that the concept might be generalized to shales. Still, in the absence of calibrated experimental curves for shales, three different sets of relative permeability curves were tested (Fig. 9b): (1) an exponential curve with a low saturation plateau in both shales and sandstones (15%); (2) an exponential curve with a low saturation plateau in shales (15%), and a high plateau (40%) in sandstones (more in agreement with experiments in sandstones); (3) linear curves without saturation thresholds (these curves correspond to the independent transport of the HC phase and of the water phase; they describe exactly the trans-

port of HC along a saturated film located at the top of the carriers).

Capillary pressures are commonly determined in reservoir samples, and are known to vary with saturation between 10^{-3} and 10^{-1} bar. At the scale of the model, these values are considered as constant. No direct P_c measurements are available for shales. The only possibility is to infer the capillary pressure P_c theoretically from the interfacial tension and the pore radius r:

$$P_c = \frac{2\gamma}{r} \quad \text{(water wet rock)}$$

Interfacial tensions decrease with depth and are currently taken to lie between 25 and 10 dyne cm^{-1} for oil–water contacts, and between 50 and 25 dyne cm^{-1} for gas–water contacts (Schowalter 1979). Pore radius r can be indirectly derived from the theoretical permeability K of a capillary tube model of porosity ϕ: $r = \sqrt{8} \, K/\phi$.

With K of the order of 10^{-6}–10^{-9} D for porosity around 10%, this relation gives r in the range 10^{-8}–10^{-7} m (this range of pore size distribution is confirmed by recent mercury injection studies, this volume). With $\gamma = 12$ dyne cm^{-1} (average in the oil window), P_c would be of the order of 10–100 bars for oil–water interfaces, and with $\gamma = 25$ dyne cm^{-1} (average in the gas window), P_c would be of the order of 25–250 bars for gas–water interfaces. In order to account for this variability, P_c values of 10 and 100 bars are tested for the shales.

Density of HC is a function of P, T and of the composition of HC. Taking data from England *et al.* (1987) and local thermal gradient and pressures, the average density of liquid HC is found to be around 500 kg m^{-3} in the oil window (3000–4000 m), the corresponding mean gaseous density is 400 kg m^{-3}. At a depth of 2000 m, the density of liquid HC is around 600 kg m^{-3}, whereas the density of gaseous HC is around 230 kg m^{-3}; below the oil window, the densities of both liquid and gaseous HC are close (around 450 kg m^{-3}). The extreme values are hence 200 and 600 kg m^{-3}. In order to test the influence of these variations (and because density is assumed to be constant in the model), the following HC densities are tested: 200 kg m^{-3}, 350 kg m^{-3} and 500 kg m^{-3}.

Model results. Figures 10 and 11 represent the calculated distribution of the HC saturations (and direction of HC flow) at the end of the Cretaceous (65 Ma), at the end of the Eocene (38 Ma) and at the present day. These simulations are based on the same relative permeability curves (saturation plateau of 15% in shales, 40% in drains), the same capillary pressures (10^{-2} bar in carriers, 10 bars in shales); they differ only in the density and viscosity of HC considered. Figure 10 considers a medium oil ($\rho = 500$ kg m^{-3} and $\mu = 10^{-2}$ to 10^{-3} Pa s), whereas Fig. 11 represents a compressed gas ($\rho = 350$ kg m^{-3} and $\mu = 10^{-4}$–10^{-5} Pa s). These simulations show similarities: in both cases, expulsion out of the three generative kitchens has already started at 65 Ma; the structures present to the West begin being filled (and begin leaking through the Cretaceous cover) at 38 Ma; residual saturations in the Brent and Statfjord Fms are low (<5%), except

when the HC are accumulating against the impermeable faults, in particular across the Horda Slope (Fig. 11, at 39 Ma); the Tampen Spur block is initially sourced from the deep graben, across the master fault, then by the syncline along the tilted flank of the block. The Snorre block is essentially fed from the West. In both cases, the HC do not reach the Troll structure. However, additional results (not shown) show that a 70% saturation is found at Troll at the present day if the density of gas is reduced to 200 kg m^{-3}. Some significant differences are observed between Fig. 10 and Fig. 11. Gas invades a larger domain in the Cretaceous than does the oil. Present-day residual saturations in the source rocks (in particular in Draupne Fm., but also in the Heather Fm., labelled UJ) are much higher for oil (15–25%) than for gas (5–10%). Figure 12 represents a simulation similar to the one illustrated by Fig. 11 (compressed gas), except for the shape of the relative permeability curves (which are assumed to be linear). The basic dynamic of the migration processes obtained for both hypotheses is very similar (timing of expulsion and filling). There are however rather important differences: (1) with the linear k_r (Fig. 12), the importance of losses through the Cretaceous cover is much reduced; the efficiency of expulsion is diminished (the Draupne Fm. is more saturated with gas in Fig. 12 than in Fig. 11); in other words, migration through shales is more difficult; (2) the efficiency of migration through the carriers is improved (smaller residual saturations in the Brent): an important consequence is that the structure of Troll is now filled as early as the Eocene.

The last simulation tests the influence of capillary pressures in the shales. Figure 13 shows the computed saturations calculated for the same hypotheses as in Fig. 11 (compressed gas with a density of 350 kg m^{-3}, and exponential relative permeabilities with saturation plateaux of 15% for shale and 40% for carriers), except that the capillary pressure in the shales was assumed to be 100 bars instead of 10 bars. This change significantly modifies the model output, as the new model predicts that the three structures of Troll, Snorre and Tampen Spur should contain significant accumulations (none of which was found in Fig. 11). Also, the efficiency of expulsion of HC out of the source rocks is slightly higher (saturations are lower in the Jurassic shales in Fig. 13 than in Fig. 11).

Discussion

Principal results are discussed below.

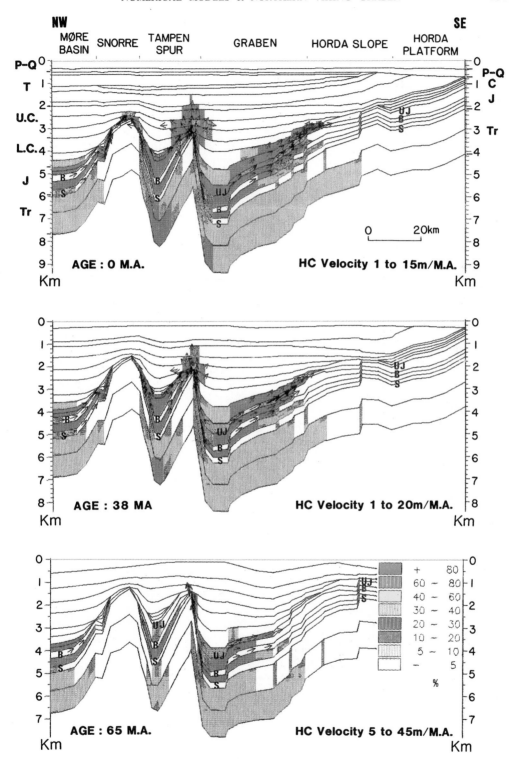

Fig. 10. Computed saturations (in %) of HC displayed at 65, 38 and 0 Ma with the assumptions of impermeable faults and heat flow calibrated as on Fig. 3; capillary pressure of shales 10 bars; density of HC 500 kg m^{-3}; relative permeabilities of Fig. 9b (light oil hypothesis). (S, D, B, UJ: see Fig. 1).

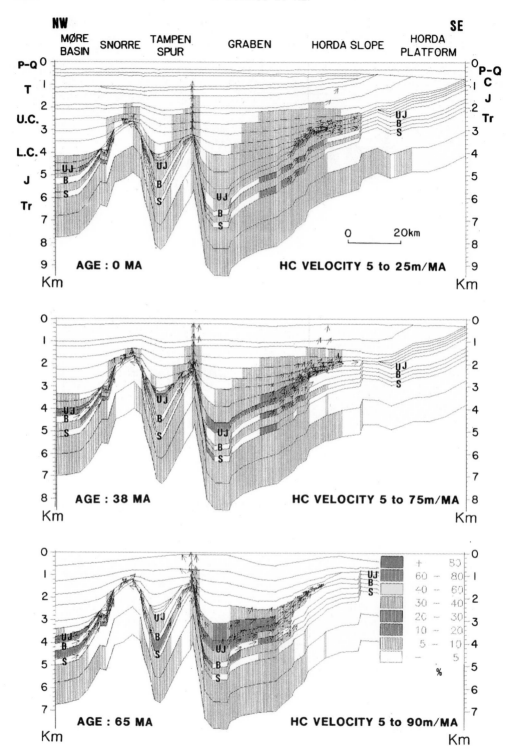

Fig. 11. Computated saturation (in %) of HC as in Fig. 10, except the density of the HC: 350 kg m^{-3} instead of 500 kg m^{-3} (compressed gas hypothesis).

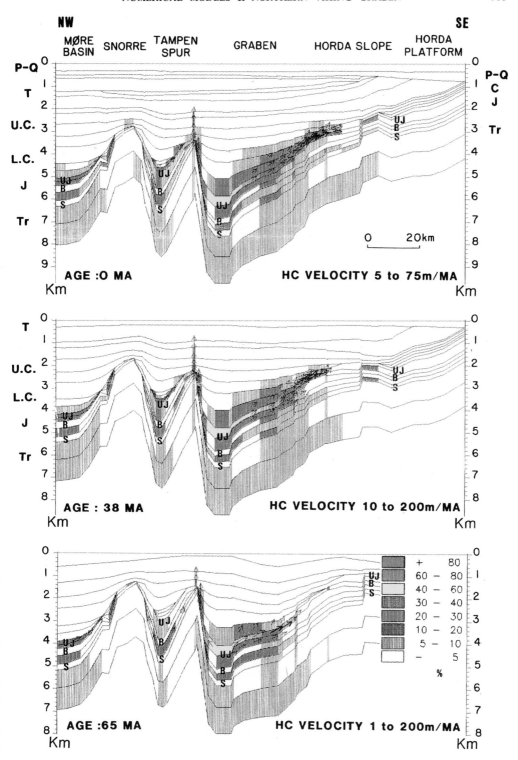

Fig. 12. Computed saturations of HC as in Fig. 11, except the shape of the relative permeabilities, taken linear as on Fig. 9c (compressed gas hypothesis).

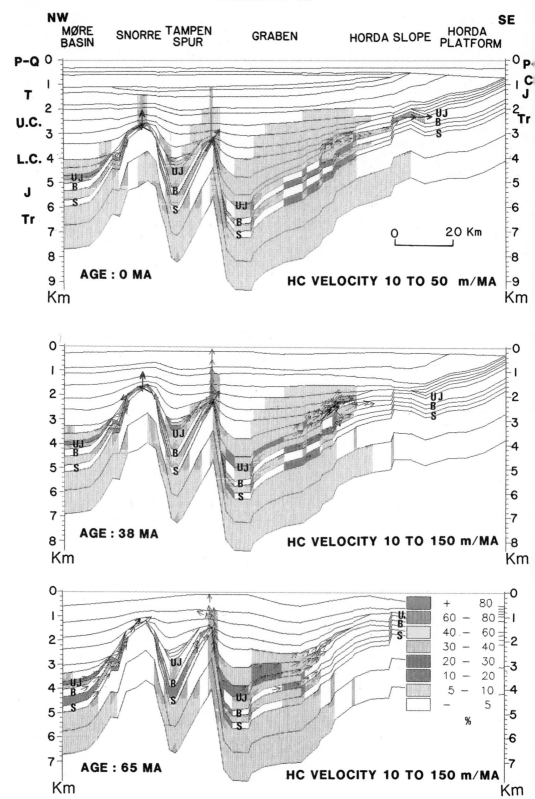

Fig. 13. Computed saturations of HC as in Fig. 11 (350 kg m^{-3} and relative permeabilities of Fig. 9b) except the capillary pressure of shales (100 bars instead of 10 bars).

A constant and uniform heat flow (around $67\,mW\,m^{-2}$, possibly 10% lower in the graben axis) permits a fairly good reconstruction of the observed maturity of both Type II and Type III source-rocks, and appears to agree with the few subsurface temperatures available. Our study confirms previous conclusions that the Jurassic rifting has had little effect (in terms of enhanced heat flow) on the maturity history. Our study also indicates that the uncertainty commonly found in the literature as to the position of the base, peak and top of the oil window is so great that the past heat flow cannot be discussed more precisely to within 10%; the effect of this uncertainty on the timing of HC generation is ± 5–$10\,Ma$. Within this uncertainty, our timing of HC generation in the Graben axis for the Draupne (onset of oil generation around 85 ± 5 Ma, peak oil generation around 67–70 Ma, and top gas window reached at $55 \pm 5\,Ma$) and for the kitchens of the Tampen Spur area (35 Ma, 30 Ma, 0–5 Ma respectively) appears about 10–20 Ma later than previously published (e.g. Leadholm *et al.* 1985). This difference is not very important; it is however believed to reflect the conceptual difference between our kinetic approach (bases on activation energies centered around 55 kcal/mole) and the TTI approach generally found in the literature (implicitly based on activation energies of 20 kcal/mole; Tissot *et al.* 1987).

Compaction disequilibrium appears to be the principal factor controlling the development of overpressures in the region; gas generation appears to play only a minor role in the overpressuring of the Jurassic. This conclusion is based on a sensitivity analysis comparing the overpressures in the absence of HC, and if oil or gas generation is considered.

We did not consider the effect of clay diagenesis (smectite–illite transformation) or of aquathermal pressuring (thermal expansion of water). Our study therefore does not prove that these processes are non-existent. Simply, as the observed overpressures could be reproduced with realistic permeability values (as low as $10^{-9}\,D$ in shales and as high as $10^{-1}\,D$ in sandstones), it can be concluded that clay diagenesis and aquathermal pressuring probably play a minor role. The study also shows that the observed overpressures are incompatible with permeable faults (except in the Troll region); the assumption that faults have always had a low permeability (like shales) is consistent with the observed overpressures. This does not prove that faults have always been impermeable. For instance it is possible to imagine that faults were temporarily more permeable, as a result of favourable orientation of regional stress. Our study however indicates that fluid losses through the faults during such hypothetical events have never been considerable (because overpressures should then be much lower than observed).

According to our results, HC generation and overpressure reconstructions cannot be simply 'extrapolated' to derive detailed scenarios of HC migration. The sensitivity analysis shows that key parameters such as petroleum properties (density, viscosity), magnitude of capillary barriers in seals, differential permeability towards water and HC play a significant role at a regional scale. Many of these parameters are difficult to evaluate. It is anticipated that the present knowledge on parameters such as petroleum fluid properties or petrophysical characterization of shales (based on very high pressure techniques) will be considerably improved in the near future. In the mean time, careful sensitivity analysis of the less well constrained parameters provides useful conclusions. Our study, for instance, explains that a structure like Troll is likely to have been sourced from the deep graben along the section studied and suggests that only the most mobile HC (gas) could cross the rather confined Horda slope, up to the Troll structure; heavier HC (oil) are most likely to be stopped by the impervious faults during the eastward migration; this leads to the conclusion that the oil found in the rim at Troll has been transported after dissolution in gas. Similarly the model predicts an age of filling of the structures during Palaeocene/Eocene for the tilted blocks to the West of the Graben which is in good agreement with dating of diagenetic events related to reservoir invasion in the region (Thomas 1986; Jourdan *et al.* 1987); this dating does not change significantly if the shape of the relative permeability curves are changed.

The efficiency of HC expulsion is dynamically related in the model to HC generation (because of the dependency of k_r on saturations), to pressure build-up and to capillary forces. It is interesting to note that depending on the type of relation considered between k_r and HC saturations (see Fig. 9), the residual HC saturation of the Draupne (now in the gas window) is found to lie between 15 and 50% (Fig. 11 and 12); it drops to 5–10% if very high capillary pressures (100 bars instead of 10 bars) are considered (Fig. 13). Better controls on the residual saturations in source-rocks (either by direct observations on cores, or indirectly by geochemical mass balance) should therefore help to calibrate the parameters (P_c, k_r).

It is interesting to note that all simulations predict significant amounts of HC to be lost by tertiary migration through the Cretaceous shales. Although gas chimneys have not been described in the area, this suggests that any attempt to discuss the global entrapment/ generation ratio or expulsion efficiency based on the differences in volumes of HC accumulated, lost in carriers, and generated, needs also to account for the quantities lost in the cover. This factor has not received enough attention in the past.

Conclusion

The reconstruction of hydrocarbon migration at the basin scale by numerical modelling involves two distinct steps. In the first step, the dynamic geological framework is reconstructed; this step involves the computation of the burial history, of temperatures, of HC generation, of overpressure development. Currently, available concepts permit us to address these histories with confidence. One of the most important advantages of numerical modelling, during this step, is that several conflicting hypotheses can be tested, to compare model output with observations, and to select one or more scenarios that appear consistent with reality. In this study, the 2D TEMISPACK model shows that a uniform heat flow of $67\,\mathrm{mW}\ \mathrm{m}^{-2}$ (possibly reduced to $61\,\mathrm{mW}\ \mathrm{m}^{-2}$ in the Graben axis) can account for the maturity observed in two source rocks of distinct types (Type II Draupne, Type III Brent). The model showed that, due to the scatter of maturity observations, the precision of the reconstruction to within is 10% for the heat flow, and around 10 Ma for the maturity history (which does not appear to be sensitive to the Jurassic rifting). The model also shows that compaction disequilibrium is much more important than HC generation in the development of the geopressures observed in the Jurassic reservoirs; the observed overpressures are correctly accounted for only if the faults are assumed to have a low shale-type permeability; with the opposite assumption, the reservoirs in the deep graben and even in the tilted blocks to the West of the graben would be close to hydrostatic pressure, which is inconsistent with observations.

The reconstruction of this dynamic framework is however not sufficient to address migration phenomena. Additional description of capillary pressures, of relative permeabilities, and petroleum fluid properties (density, viscosity) is introduced in the second step. Many of these parameters are not precisely known; their influence is investigated through a sensitivity analysis. Our computations, based on

this input, confirm some of the previous qualitative interpretations: expulsion from the overpressured Draupne Fm. into the less pressured Brent Fm., followed by migration along the tilted flanks of the blocks in the Tampen Spur area; rather poor drainage through the Horda Slope; filling of Troll by the deep graben possible for gaseous HC only; timing of filling of the tilted structures around the early Tertiary. Our simulations also show that, in the absence of sufficient knowledge about the residual saturation in the source rocks, the losses in the carriers, and the losses in the Cretaceous shales (predicted to be significant), the efficiency of expulsion and of entrapment cannot be 'predicted' by the modelling procedure. Therefore, volumetric estimates of the quantities accumulated in structures cannot be achieved by modelling alone (even in the case where the maturity history and pressure history are correctly reconstructed). At the present time, however, the modelling approach enables the discussion of migration scenarios based on sensitivity analysis of the less well constrained parameters, within a dynamic geological framework. In addition studies of the type shown enable one to identify the parameters which need to be specified more accurately: petrophysical characterization of shales, and predicting petroleum fluid properties (density, viscosity, composition) rank among the most important. In this sense, numerical models appear useful in reconstructing HC migration phenomena.

We thank the SNEA-P company who provided the data and many ideas used in this study, partly funded by the Fonds de Soutien des Hydrocarbures.

Appendix 1 The equations in TEMISPACK

Five equations are used to describe the evolution of the sediments during compaction.

State equations

The densities of both fluids and solids are assumed constant. Fluids are represented by one water and one HC Phase.

Continuity equations

For the water:

$$\rho_{\mathrm{w}}\,\mathrm{div}(\vec{u}_{\mathrm{w}}.\phi.(1-s)) + \rho_{\mathrm{w}}\,\frac{\partial}{\partial t}\,((1-s).\phi) = 0 \quad (1)$$

For the hydrocarbons:

$$\rho_{HC} \, \mathrm{div}(\vec{u}_{HC}.\phi.s) + \rho_{HC} \frac{\partial}{\partial t} (s.\phi) = q_{HC} \quad (2)$$

Transport equations

For the water:

$$\vec{u}_w = -k_{rw}.\bar{K}.\rho_w.g/\mu_w.\mathrm{grad}(P/\rho_w g - z) \quad (3)$$

For the HC:

$$\vec{u}_{HC} = -k_{rHC}.\bar{K}.\rho_{HC}.g/\mu_{HC}.\mathrm{grad}((P+P_c)/\rho_{HC}.g - z)$$
$$(4)$$

Consolidation law

$$\sigma_{eff} = S - P \quad (5)$$

with

$$\rho_{eff} = a\phi^2 + b\phi + c + d/\phi + e/\phi^2 \quad (6)$$

where the coefficients a, b, c, d depend on the lithology, and can be calibrated in hydrostatic conditions.

Heat equation

$$\mathrm{div}(\bar{\lambda}.\mathrm{grad}\,T) + \mathrm{div}(Q_r) = \frac{\partial}{\partial t} (\rho C.T) \quad (7)$$

Kinetic generation scheme

$$dX_i = -A_i \exp(-E_i/RT).X_i dt \quad (8)$$

$$Q_i = X_{io} - X_i \quad (9)$$

$$Q = \sum_i Q_i \quad (10)$$

The transformation ratio is:

$$TR = \sum(X_{io} - X_i)/\sum X_{io} \quad (11)$$

The absolute quantities in kg of HC formed by ton source rock can be calculated if the initial potential (kg HC producible/ton rock) is known. This is necessary to simulate the migration history.

Non-linearity of parameters

Most parameters depend in a non-linear way on the porosity, lithology and sometimes P/T, and saturation.

Thermal conductivity:

$$\lambda = \lambda_w^\phi.\lambda_m^{1-\phi}. \frac{1}{1 + \alpha(T - T_o)}$$

Heat capacity

$$\rho c = \rho c_w.\phi + \rho c_m(1 - \phi)$$

Intrinsic permeability adapted from Kozeny Carman formula

$$K = 0.2\phi^3/(S_o^2(1 - \phi)^2)$$

Water viscosity (Bingham formula)

$$\mu_w = 1/(aT + (bT^2 + cT + d)^{1/2} + e)$$

HC viscosity (Andrade formula)

$$\mu_{HC} = a \exp(bT + c)$$

Relative permeability

See Fig. 9.

List of variables

A_i	: pre-exponential factor of ith primary reaction (s^{-1})
α	: coefficient in conductivity–temperature relation (K^{-1})
ϕ	: porosity (no dimension)
γ	: interfacial tension (dyne cm^{-1})
E_i	: activation energy for i^{th} primary reaction (kcal/mole)
$\bar{K}$	: permeability tensor (Darcy)
k_{ri}	: relative permeability to oil (i = HC) or water
$\bar{\lambda}$:	thermal conductivity tensor (W/m/K)
λ_i	: matrix (i = m) or water (i = w) thermal conductivity (W/m/K)
μ_i	: viscosity of water (i = w) or HC (i = HC)
P	: water pore pressure (Pa)
P_c	: capillary pressure (Pa)
Q_r	: radiogenic production (Wm^{-3})
Q_i	: relative mass of HC generated by i^{th} reaction (mg HC/g initial TOC)

Q	: relative mass HC generated (mg HC/g initial TOC)	σ_{eff}	: effective stress (Pa)
		S	: total geostatic stress (Pa)
ρ_i	: density of matrix (i = m) or water (i = w) $(kg\,m^{-3})$	T	: temperature (K)
		TR	: Transformation ratio (no dimension)
ρc_i	: heat capacity $(J\,m^{-3})$	t	: time (s)
R	: Boltzman constant (8314 J/mole.K)	$\vec{u}_i$	: Darcy velocity for water (i = w) or
r	: pore radius (m)		HC (i = HC) (m/s)
s	: HC volumetric saturation (no dimension)	X_i	partial potential for i^{th} reaction (mg HC producible/g initial TOC)
S_o	: specific surface $(m^2\,m^{-3})$		

References

BROSSE, E. & HUC, A.Y. 1986. Organic parameters as indicators of thermal evolution in the Viking Graben. *In*: BURRUS, J. (ed.) *Thermal modelling of sedimentary basins*. Technip, Paris, 517–530.

BUHRIG, C. 1989. Geopressured Jurassic reservoirs in the Viking Graben: Modelling and geological significance. *Marine and Petroleum Geology*, **6**, 31–48.

BURRUS, J. & AUDEBERT, F. 1990. Thermal and compaction processes in a rifted basin in the presence of evaporites, the Gulf of Lions case study. *American Association of Petroleum Geologist Bulletin*, **74**, 1420–1440.

CARSTENS, H. 1978. Origin of abnormal formation pressures in Central North-Sea Lower Tertiary clastics. *Log Analysis*, **19**, 24–28.

—— & DYPVIK, H. 1989. Abnormal formation pressure and shale porosity. *American Association of Petroleum Geologist Bulletin*, **75**, 344–350.

—— & FINSTAD, K. G. 1981. Geothermal gradients of the North Sea Basin, 59–62°N. *In*: ILLING, L. V. & HOBSON, G. D. (eds) *Petroleum Geology of the continental shelf of North-West Europe*. Heyden, London, 152–161.

CHIARELLI, A. & DUFFAUD, F. 1980. Pressure Origin and distribution in Jurassic of Viking Basin (United Kingdom–Norway). *American Association of Petroleum Geologist Bulletin*, **64**, 1245–1266.

COOLES, G. P., MACKENZIE, A. S. & QUIGLEY, T. M. 1986. Calculation of petroleum masses generated and expelled from source rocks. *Organic Geochemistry*, **10**, 235–245.

CORNFORD, C., NEEDHAM, C. E. J. and DE WALQUE, L. 1986. Geolochemical Habitat of North Sea oils and gases. *In*: SPENCER, A. M. (ed.) *Habitat of the hydrocarbons of the Norwegian shelf*. Graham & Trotman, London, 39–54.

DOLIGEZ, B., BESSIS, F., BURRUS, J. UNGERER, P. & CHENET, P. Y. 1986. Integrated numerical modelling of the sedimentation, heat transfer and fluid migration in a sedimentary basin: the Themis model. *In*: BURRUS, J. (ed.) *Thermal modeling of sedimentary basin*. Technip, Paris, 173–195.

——, UNGERER, P., CHENET, P. Y., BURRUS, J., BESSIS, F. & BESSEREAU, G. 1987. Numerical modelling of sedimentation, heat transfer, hydrocarbon formation and fluids migration in the Viking graben, North sea. *In*: BROOKS, J. & GLENNIE, K. (eds) *Petroleum Geology of Northwest Europe*. Graham & Trotman, London, 1039–1048.

DURAND, B. 1987. Understanding of HC migration in sedimentary basins (present state of knowledge). *Advances in Organic Geochemistry*, **13**, 445–459.

——, UNGERER, P., CHIARELLI, A. & OUDIN, J. L. 1983. Modélisation de la migration de l'huile: Application à deux exemples de bassins sédimentaires. *Proceedings of the 11th World Petroleum Congress, London, Sept. 1983*, **PD(1)**, 3–15.

EGGEN, S. 1985. Modelling of subsidence, hydrocarbon generation and heat transport in the Norwegian North Sea. *In*: DURAND, B. (ed.) *Thermal phenomena in sedimentary basins*. Technip, Paris, 271–283.

ENGLAND, W. A., MACKENZIE, A. S., MANN, D. M. & QUIGLEY, T. M. 1987. The movement and entrapment of petroleum fluids in the subsurface. *Journal of the Geological Society*, **144**, 327–347.

ESPITALIE, J., UNGERER, P., IRWIN, I. & MARQUIS, F. 1988. Primary cracking of kerogens. Experimenting and modelling C_1, C_2–C_5, C_6–C_{15} and C_{15}^+ classes of hydrocarbons. *In*: MATTAVELI, L., NOVELLI, L. (eds) *Advances in Organic Geochemistry 1987, Venice, Organic Geochemistry*, **13**, 893–899.

EVANS, T. R. & COLEMAN, N. C. 1974. North Sea geothermal gradients. *Nature*, **247**, 28–30.

FIELD J. D. 1985. Organic geochemistry in exploration of the Northern North Sea. *In*: THOMAS, B. M. (ed.) *Petroleum Geochemistry in Exploration of the Norwegian Shelf*. Norwegian Petroleum Society, Graham & Trotman, London, 39–57.

GOFF, J. C. 1983. Hydrocarbon generation and migration from Jurassic source rocks in the E. Sheltand Basin and Viking Graben of the northern North Sea. *Journal of the Geological Society, London*, **140**, 445–474.

HERRING, E. A. 1973. Estimating abnormal pressures from Log Data in the North Sea. *In*: *Proceedings of the 2nd annual Society of Petroleum Engineers, Europe Meeting*. Preprint Society of Petroleum Engineers, 4301.

JOURDAN, A., THOMAS, M., BREVART, O., ROBSON, P., SOMMER, F. & SULLIVAN, M. 1987. Diagenesis as the control of the Brent sandstone reservoir properties in the Greater Alwyn area (East Shetland Bains). *In*: BROOKS, J. & GLENNIE, K. W. (eds) *Petroleum Geology of Northwest Europe*. Graham & Trotman, London, 951–961.

KALAYDJIAN, F. 1987. A macroscopic description of multiphase flow in porous media involving space — time evolution of fluid–fluid interface. *Trans-*

port in Porous Media, **2**, 537–552.

KARLSSON, W. 1986. The Snorre, Statfjord and Gullfaks oil fields and the habitat of hydrocarbons on the Tampen Spur, offshore Norway. *In*: SPENCER, A. M. (ed.) *Habitat of the hydrocarbons on the Norwegian continental shelf*, Graham & Trotman, 181–197.

LEADHOLM, R. H., THOMAS, T. Y. & HO, S. K., Sahai 1985. Heat flow, geothermal gradients and maturation modelling on the Norwegian continental shelf using computer methods. *In*: THOMAS, B. M. (ed.) *Petroleum geochemistry in Exploration of the Norwegian Shelf*. Norwegian Petroleum Society, Graham & Trotman, London, 131–143.

LEYTHAEUSER, D., SCHAEFFER, R. G. & RADKE, M. 1987. On the primary migration of petroleum. *Proceedings of the 12th World Petroleum Congress, Special Publication* **2**, 1–10.

LIN, C., PIRIE, G. & TRIMMER, D. A. 1986. Low Permeability Rocks: laboratory measurements and three dimensional microstructure analysis. *Journal of Geophysical Research*, **91**, 2173–2181.

LINDBERG, P., RIISE, R. & FERTL, W. H. 1980. Occurrence and distribution of overpressures in the Northern North Sea area. *In: 55th Annual Society of Petroleum Engineers Fall Technical Conference, Dallas*. Preprint, Society of Petroleum Engineers, 9339.

MACKENZIE, A. S. & McKENZIE, D. P. 1983. Isomerization and aromatization of hydrocarbons in sedimentary basins formed by extension. *Geological Magazine*, **120**, 5, 417–528.

—— & QUIGLEY, T. M. 1988. Principles of Geochemical prospect appraisal. *American Association of Petroleum Geologists Bulletin*, **72**, 399–415.

——, PRICE, I., LEYTHAEUSER, D., MULLER, P., RADKE, M. and SCHAEFFER, R. G. 1987. The expulsion of petroleum from Kimmeridge clay source-rocks in the area of the Brae field, UK continental shelf. *In*: BROOKS, J. & GLENNIE, K. (eds) *Petroleum Geology of north Western Europe*. Graham & Trotman, London, 865–877.

McKENZIE, D. 1978. Some remarks on the development of sedimentary basins. *Earth and Planetary Science Letters*, **40**, 25–32.

MORROW, C. A., SHI, L. Q. & BYERLEE, J. D. 1984. Permeability of Fault Gouge under confining pressure and shear stress. *Journal of Geophysical Research*, **89**, B5, 3193–3200.

NAKAYAMA, K. & LERCHE, I. 1987. Basin analysis by model simulation: effects of geologic parameters on 1D and 2D fluid flow systems with application to an oil field. *Gulf Coast Association of Geological Society Transactions*, **37**, 175–184.

SCHOWALTER, T. T. 1979. Mechanics of hydrocarbon secondary migration entrapment. *American Association of Petroleum Geologists Bulletin*, **63**, 723–760.

THOMAS, M. 1986. Diagenetic sequences and K/Ar dating in Jurassic sandstones, Central Viking graben: effects on reservoir properties. *Clay Mineralogy*, **21**, 695–710.

THOMAS, B. M., MOELLER-PEDERSEN, P., WHITAKER, M. F. & SHAW, N. D. 1985. Organic facies and hydrocarbon distribution in the Norwegian North Sea. *In*: THOMAS, B. M. (ed.) *Petroleum Geochemistry in Exploration of the Norwegian Shelf*. Norwegian Petroleum Society, Graham & Trotman, London, 3–26.

TISSOT, B. P. & WELTE, D. H. 1984. *Petroleum formation and occurrence*, Springer Verlag, Second edition.

——, PELET, R. & UNGERER, P. 1987. Thermal history of sedimentary basins, maturation indices, and kinetics of oil and gas generation, *American Association of Petroleum Geologists Bulletin*, **71**, 1445–1446.

UNGERER, P., BESSIS, F., CHENET, P. Y., DURAND, B., NOGARET, E., CHIARELLI, A., OUDIN, J. L. and PEERIN, J. F. 1984. *In*: DEMAISON, G. (ed.) *Geological and geochemical models in oil exploration: principles and practical examples*. American Association of Petroleum Geologists Memoir, **35**, 53–77.

——, DOLIGEZ, B., CHENET, P. Y., BURRUS, J., BESSIS, F., LFARGUE, E., GIROIR, G., HEUM, O. & EGGEN, S. 1987a. A 2D model of basin scale petroleum migration by two phase fluid flow. Application to some case studies. *In*: DOLIGEZ, B. (ed.) *Migration of hydrocarbons in sedimentary basins*. Technip, Paris, 415–456.

——, BESSEREAU, G., JUNCA, J. & RABILLER, P. 1987b. Application d'un modèle de migration des hydrocarbures à l'évaluation des permis. *Proceedings of the 12th World Petroleum Congress, Houston, PO1* (3), Wiley, 19–30.

WELTE, D. H. & YUKLER, M. A. 1981. Petroleum origin and accumulation in basin evolution—a quantitative model. *American Association of Petroleum Geologists Bulletin*, **65**, 1387–1396.

Modelling of secondary migration and entrapment of a multicomponent hydrocarbon mixture using equation of state and ray-tracing modelling techniques

ØYVIND SYLTA

Continental Shelf and Petroleum Technology Research Institute (IKU) A/S, N-7034 Trondheim, Norway

Abstract: Equation of state modelling of multicomponent hydrocarbon mixtures has been incorporated into a program which models the secondary migration of oil and gas. Migration is modelled by means of a ray-tracing technique which operates on backstripped depth maps of a carrier bed. Reservoir beds are represented as rectangular grids (arrays) in the computer, ensuring the high resolution required for proper secondary migration modelling. The resulting software, named SEMI, can be calibrated against the hydrocarbon composition in drilled traps and then used to compute the composition in undrilled prospects. This is especially useful in moderately explored basins where some discoveries have already been made. A model for the loss of hydrocarbons during secondary migration is utilized and, by modelling secondary migration within a synthetic basin, it is shown that compositional differences in traps may result from differences in loss of the vapour and liquid phases. Using a simple model for primary migration, it is shown that the primary and secondary migration losses influence the hydrocarbon composition in traps differently. Thus it is hoped that the proposed methodology can be used to quantify some of the unknown effects of migration and that a decrease in the quantified uncertainty in exploration for oil and gas may result.

Secondary migration of hydrocarbons (HC) is a very difficult process to investigate because of the difficulty in performing experiments at the right scale and at the appropriate temperatures and pressures. Models that describe the compaction of sediments and the associated fluid flow have been published (Smith 1971; Welte & Yukler 1981 and others). Secondary hydrocarbon migration along permeable carrier beds was modelled by Lehner *et al.* (1988). However, when these models are extended to the three dimensional cases and gas and oil migration is included, the resulting procedures become extremely time-consuming. Therefore, it was decided to develop a migration modelling program based upon another formulation of the migration phenomena. The chosen method, ray-tracing, makes some simplifications but benefits in that it is very fast and three-dimensional volumetrics are possible. As a result, ideas on phase relationships during migration can be modelled and tested. This paper describes and discusses some of the concepts which have evolved during the development of the modelling tool, named SEMI. In particular, it discusses whether the modelling of a multicomponent hydrocarbon mixture during secondary migration is feasible and if the results can be used to increase the understanding of the migration phenomena.

Methodology

The SEMI program uses a ray-tracing procedure in order to simulate the movement of hydrocarbons in the subsurface. The vapour and liquid phases are considered to migrate along the top of a single carrier bed, i.e. within a permeable zone, and upwards along the steepest dip towards the lowest potential according to the buoyancy (Hubbert 1953) (see Fig. 1). Once the capillary pressure resistance (Berg 1975) has been overcome by building up a continuous hydrocarbon phase along the top of the carrier bed, hydrocarbon flow is possible in preferred paths within the carrier beds (Dembicki & Anderson 1989).

The ray-tracing method is used in many scientific and technical applications as a means of simulating the effects of some sort of movement. These include seismic modelling and the computer imaging of three dimensional objects. The ray-tracing solution is significantly different from the more commonly used finite difference and finite element (FD/FE) numerical schemes for the same problems in that, while the latter two schemes simulate the movement through time in a discrete, but almost 'continuous' manner, the ray-tracing simulates the movement as a series of instantaneous events. Thus, while the FD/FE approximations can

From England, W. A. & Fleet, A. J. (eds), *Petroleum Migration*
Geological Society, Special Publication No. 59, pp. 111–122.

simulate the movement of hydrocarbons in a time span of seconds as well as millions of years, the ray-tracing can only model the net effect over a long time span, typically longer than a million years, i.e. the time span should be so long that the hydrocarbons can migrate from the source rock and into a trap location entirely within the time window and assuming no basin change during this time step. However, the ray-tracing method can model the change in some attribute during migration and one important attribute that may be modelled is the hydrocarbon loss during secondary migration. Time transient effects within each time step is not considered. The FD/FE methods, when applied to fluid flow (in particular gas flow), are forced to work with very small time steps whereas the ray-tracing procedure is only valid for fairly large time steps and, as a result, the ray-tracing tends to be significantly faster than the other methods in terms of computer time. Therefore ray-tracing may be well suited to investigate the secondary migration of hydrocarbons which is a very fast process in geologic time scale (see Sylta 1987; Dembicki & Anderson 1989).

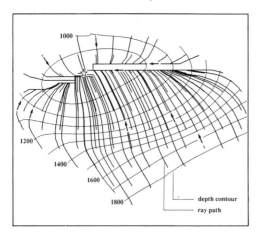

Fig. 1. Ray-tracing simulating up-dip buoyancy driven secondary migration. Inaccuracy in ray-tracing along the ridge of the dome results from the need to accommodate migration along very sharp ridges in a grid node system.

During migration, the hydrocarbons experience varying pressure and temperature but, since secondary migration is fast, any movement during drainage and spillage is always towards lower pressure and temperature even in rapidly subsiding basins. Once the hydrocarbons have

reached a trap, higher temperatures and pressures are expected due to the general subsidence of the basin as a whole. Thus, during drainage and spillage, a multicomponent mixture will tend towards the two phase region where two petroleum phases (liquid and vapour) are stable and, if both light and heavy components are present, gas will be separated out from the liquid phase. Because of the faster migration rate of vapour, due to its higher density contrast to water and its lower viscosity compared to the liquid hydrocarbons, the vapour needs to saturate a relatively smaller amount of pore space i.e. cross-sectional area, in order to migrate the same volume. This causes a difference in the amount of loss of the vapour and liquid phases. The C1 (methane) component is, for example, more abundant in the vapour phase than the C11+ components and will experience a different loss during drainage and spillage. In the SEMI program, which models the movement of a multicomponent system in the one or two-phase region, it is possible to model this difference in loss for components in the two-phase region. Changes in hydrocarbon composition are also caused by preferential leakage of vapour through the cap rock in two-phase reservoirs, secondary cracking in traps etc. These effects are not discussed in this paper.

The ray-tracing procedure requires the top of the carrier bed to be represented in the computer and this is accomplished by using rectangular grids of arbitrary size in each direction. Map backstripping is performed and the procedure is the same as for single point backstripping (Sclater & Christie 1980), i.e. compaction corrected subsidence curves for wells, except when the uppermost horizon is faulted. In the latter case the thicknesses of the faulted zones utilized in the backstripping procedure will be too thin because of structural reorientation of beds. If the palaeodepth of a North Sea Middle Jurassic carrier bed is needed at a time before the end of the Jurassic then the fault zones have to be treated with care and special algorithms in which regridding of a sparse grid is employed may be utilized.

For basins in the North Sea, the distance between grid nodes may typically be set to 500–1000 m and a total of at least 15 000 grid nodes are required. Each grid node holds the value, i.e. depth, porosity, etc., for one grid element. The ray-tracing navigates within the grid node system by finding the steepest gradient of ascent (Fig. 1). The path is straight within each grid element, but the direction is changed each time a grid element border is crossed. The loss during migration is also computed (see later) for each grid element which is passed and the loss is stored within the nodes so that later loss calculations can be compared to the losses from previous time steps. The navigation procedure is tailored to handle the most commonly observed discontinuities in geology, i.e. faults (Ritter *et al.* pers. comm. 1989), salt domes cutting the carrier bed and erosional truncations leading to the possible formation of stratigraphic traps (Sylta 1987).

Equation of state modelling (EOS) of phase behaviour involves the determination of phase properties based on a given pressure and temperature and overall composition. In this paper the EOS routines provide % liquid, % vapour, the distribution of the components in the two phases and the density of the two phases. Standard procedures are available for this task in the COPEC subroutine library which has been developed at IKU for reservoir simulation purposes. The COPEC library uses methodologies described in Van Ness & Abbott (1982) and Pedersen *et al.* (1989). The procedures are based on a minimizing of the Gibbs free energy function which employs component properties such as critical pressure, temperature etc. in the fugacity relations. The fugacities incorporate an equation of state, the Peng–Robinson (PR), the Soave–Redlich–Kwong (SRK) or the Generalized Redlich–Kwong (GRK), which may be tuned to match any observed phase behaviour data through an optimization procedure also available in COPEC. Risnes *et al.* (1981) use COPEC in a discussion of EOS in the near-critical region. The differences between some commonly used EOS are discussed for reservoir simulation applications in Firoozabadi (1988) and the largest uncertainties seem to be found in the procedures for calculating density. However, the extremely high temperatures and pressures that occur in the deepest traps are not found in reservoir modelling applications. If the EOS computations are performed frequently during sediment burial, an almost continuous change in the phase properties can be modelled.

EOS calculations are only valid if 'perfect' mixing of the different petroleum components takes place and the mixing has to be discussed separately for the drainage and the trap/spill processes. In fields in production mixing of components is observed to a large extent, but compositional gradients are also frequently observed in many fields of the North Sea and elsewhere (England & Mackenzie 1989). Vertical mixing is more likely than horizontal mixing due to the differences in distances but, as a first approximation, perfect mixing in the traps is assumed in the following.

Perfect mixing of hydrocarbons expelled from the source rock and into the carrier bed within one grid node area is assumed. During drainage it may also be assumed that the hydrocarbon mixture associated with one ray is always well mixed but with the vapour phase moving at a different velocity to the liquid phase. Compositional changes during drainage are, however, not discussed in this paper. During drainage, the focusing of rays leads to the possibility that hydrocarbons generated in different parts of the basin may mix. Gas expelled from the centre of the basin may reach the same grid node at the same time as heavier components generated along the flank of the basin. However, on closer inspection, the two rays may still be separate entities even though the preferred pathways may come very close (note that mapping resolution may be larger than 100 m). As a result, they will not be able to mix. Exceptions occur when microtraps (traps below the resolution of mapping) and rerouting caused by faulting are prominent. The model assumes that there is no mixing between rays, i.e. preferred pathways. I feel that this is the best approximation until other evidence becomes available, but further research is necessary.

This paper focuses on component distributions within traps resulting from phase equilibrium and spillage involving two phases. The effects will be elaborated upon for the synthetic model and should become clear to the reader in the last part of this paper.

Secondary migration losses

A proper handling of the multicomponent system during secondary migration (drainage and spillage) requires a realistic model (conceptual and numerical) of migration losses. Because of the differences in velocity between the liquid and vapour phases, it becomes important to simulate the resulting differences in loss which may, for example, lead to enrichment of the methane component during long distance migration. A model for secondary migration losses was developed based upon Darcy flow, which is assumed to be valid for hydrocarbon migration along preferred carrier bed pathways, and conservation of mass (Sylta 1987). The permeability governing the velocity of the migrating hydrocarbons is taken from relative permeability curves but hydrocarbon saturation during flow is kept constant because of the difficulties in assessing whether any saturation gradient exists within the migration pathways. The resulting loss function is modified from Sylta (1987) and variables are defined in Table 1:

$$dV/dl = -\frac{q}{E_{sm} \sin a} - V_{ts} \tag{1}$$

and

$$E_{sm} = \frac{k \; k_r \; \Delta\rho \; g \; S_{fl}}{\tau^2 \; u \; S_{irr}} \tag{2}$$

$q = V/\Delta t$ and $V \geqslant 0$.

dV/dl is the hydrocarbon loss during migration, i.e. the change in movable volume of hydrocarbons per unit of migrating distance. Before any petroleum can reach the trap, the migrating hydrocarbons must fill all microtraps and dead ends along the migration pathway. This volume is expresed by the variable V_{ts} in equation (1). The V_{ts} loss is related to the area through which the HC flows (units are m^3/m). It is assumed that if one path saturates enough porespace within a grid node, then pore space is already saturated for the next ray passing that node, i.e. at a later time step.

Table 1. *Nomenclature*

V	=	volume of hydrocarbons migrating
E_{sm}	=	Secondary migration "efficiency"
V_{ts}	=	volume of hydrocarbons trapped in microtraps, dead ends, etc. per unit of distance along the raypath. Units are m^3/m
Δt	=	time period considered in equation (1)
l	=	horizontal distance along raypath
q	=	rate of hydrocarbon flow (volume/time)
a	=	dip of carrier bed
k	=	absolute permeability
k_r	=	relative permeability
$\Delta\rho$	=	density contrast hydrocarbon/water
S_{fl}	=	hydrocarbon saturation during flowing
S_{irr}	=	irreducible hydrocarbon saturation
τ	=	tortuosity (actual versus apparent path length)
u	=	viscosity
g	=	constant of gravity
V_{exp}	=	volume of hydrocarbons expelled from source rock
E_{exp}	=	expulsion efficiency
V_{gen}	=	volume of hydrocarbons generated in source rock
V_{te}	=	volume of hydrocarbons remaining in source rock

E_{sm} may be considered to be the secondary migration 'efficiency' and incorporates the fluid flow variables from the Darcy law. It is highly depth dependent and in particular the permeabilities show large variations. In the SEMI program, E_{sm} is input as a grid for each time step and any known functional relationship between the parameters may therefore be utilized.

Once the microtraps and dead ends have been filled, the HC may continue along its path towards lower potential as shown in Fig. 2 where a cross section along a preferred pathway is shown. Within the area of active expulsion, the thickness of the migrating HC zone increases until it reaches the maximum value of h_1 (Fig. 2) where the expulsion zone ends. The increase in thickness is caused by the additional volume which has to pass any point along the path. Once the expulsion zone is passed, the thickness of the HC zone decreases because of a decrease in volume passing caused by loss during migration. When passing through high dip areas, the velocity increases and a decrease in the required HC thickness is observed (compare h_2 versus h_3 in Fig. 2). Focusing of raypaths increases the volume passing and although individual paths do not necessarily merge, an increase in the hydrocarbon loss results due to the increase in 'q' in equation (1). Losses caused by solution of the hydrocarbon phase into water (Barker & Fuliang pers. comm. 1989) are not considered in equation (1).

Following the time step for which equation (1) holds, the irreducible HC saturation along the path becomes S_{irr} and therefore HC passing at later time steps does not need to saturate the pore space unless the flow rate is higher than the maximum flow rate (q) from earlier time steps. Further loss is only observed when q increases (i.e. E_{sm} or dip decreases) during subsequent time steps.

Estimating secondary migration losses in a simple way for exploration purposes by means of this model is not easy. A proper loss evaluation requires an elaborate modelling of the HC migration through the basin's history since parameters such as the dip, permeability and flow rate vary through time. Furthermore the permeability change may span several orders of magnitude due to compaction effects.

Assuming equation (1) holds, it may be argued that an overpressured region will be more efficient for secondary migration because of the relatively higher carrier bed permeabilities often associated with such a basin. Higher permeabilities result in higher E_{sm} and thus lower dV/dl and less loss during migration. Also, 'hot' basins (with high geothermal gradients) are preferred because of the shallower depth at which hydrocarbon generation occurs and thus high permeabilities are present in the carrier beds. On the other hand, very fast subsiding basins with high rates of hydrocarbon generation result in a high throughput; i.e. q

LOSS DURING SECONDARY MIGRATION

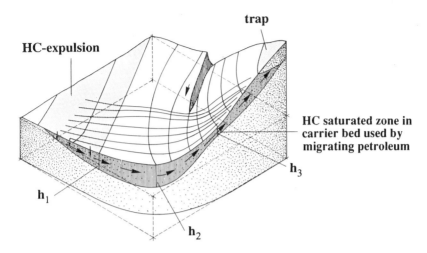

Fig. 2. Migration along preferred pathways: block diagram. Note changes in height of HC saturated zones with dip.

becomes large and an increase in the loss should be expected. Most importantly, the dip of the basin has to be considered when evaluating the efficiency of migration and a basin in which the average dip of the carrier bed is 5° will be significantly more efficient for migration than one with 1° dip. Carrier beds with a smooth upper surface may exhibit lower tortuosity and as a result be more efficient for secondary migration (see τ in E_{sm}).

A model for hydrocarbon expulsion

Hydrocarbon generation and expulsion is not modelled in the SEMI program, but are provided as input to the program. For each modelled time step, grids of expelled oil and gas volumes are computed using a program which models the hydrocarbon generation process (Ritter *et al.* 1987).

A simple model for the expulsion of hydrocarbons from the source rock is needed in order to study the effect of expulsion as opposed to that of secondary migration (Sensitivity section below). Many petroleum explorationists use a simple expulsion efficiency for routine prospect appraisal, and values from as low as 10% to as high as 90% have been proposed. Another approach used is to assume that an initial hydrocarbon saturation has to be reached

in the source rock after which all generated hydrocarbons are expelled. It is interesting to combine these concepts into a very simple mathematical expulsion model (variables are defined in Table 1):

$$V_{exp} = E_{exp} (V_{gen} - V_{te}) \qquad (3)$$

In this paper we will assume oil and gas expulsion efficiencies to be equal i.e. expulsion in a single phase is modelled, but the effect of two-phase expulsion may be investigated later as may other models, i.e. models such as the primary migration model proposed by Duppenbecker *et al.* (this volume).

A synthetic model: the Basin of Domes

Synthetic models, representing some properties of the Earth, are used extensively within seismic modelling, whereas it is unusual to use artificial basins for the basin modelling research. The use of synthetic models is proposed for several reasons; i.e. accurate definition of the model is possible; resolution can be tested in a controlled way; different algorithms can be tested easily on the same data set; ease of construction; analytical solutions may be constructed, etc. Therefore, a synthetic basin, named the 'Basin of Domes' for obvious reasons (Fig. 3), is used in

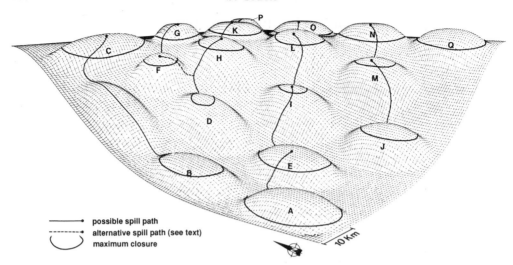

Fig. 3. Basin of Domes: perspective view. Depth ranges from *c.* 700 m to 3000 m.

this paper to ilustrate some important features of multicomponent migration modelling. The basin is defined by a very simple mathematical function representing elliptical domes superimposed on parts of an 'inverted dome' basin:

$$f(x,y) = \Sigma h/2 \ (1 + \cos \ [\pi \times ((x-x_0)^2/a^2 + (y-y_0)^2/b^2)]) \tag{4}$$

and

$$[(x - x_0)^2/a^2 + (y - y_0)^2/b_2] \leqslant 1$$

The variables are defined in Table 2 and the values for the parameters used are given in Table 3. Each of the 'cos'-functions represents a dome which can be positioned arbitrarily within the basin according to the values of x_0 and y_0. The height of each dome is given by the parameter h. The Basin of Domes measures 100 km in the east–west direction and 150 km in the north–south direction and the distances between the grid nodes are set to 1000 m in this case but this is one parameter on which sensitivity may be studied at a later stage. The basin is shown in a perspective view in Fig. 3, whereas a map view with drainage paths, etc. is shown in Fig. 4.

Table 2. *Variables used in defining the Basin of Domes*

a	= elliptical length in x direction
b	= elliptical length in y direction
x_0	= x location of centre of ellipse
y_0	= y location of centre of ellipse
h	= maximum height of dome (i.e. at x_0, y_0).

Table 3. *Basin of Domes: values assigned to parameters in equation (4)*

x_0	y_0	a	b	h
0	0	500000	750000	−1000
0	0	100000	150000	−2000
70000	40000	10000	20000	300
70000	80000	10000	20000	300
70000	120000	10000	20000	300
20000	130000	20000	10000	300
10000	20000	10000	20000	250
10000	60000	10000	20000	250
10000	100000	10000	20000	250
50000	20000	10000	20000	300
50000	60000	10000	20000	300
50000	100000	10000	20000	300
30000	40000	10000	20000	300
30000	80000	10000	20000	300
30000	110000	10000	10000	300
50000	130000	10000	10000	300
90000	20000	10000	20000	300
90000	60000	10000	20000	300
90000	100000	10000	20000	300
90000	140000	10000	20000	300

The thickness of the carrier bed is 10 m and its porosity is fixed at 20%: i.e. no compaction is considered for the 20 Ma of the basin history. The subsidence history consists of a constant burial rate for the first 10 Ma followed by linear erosion for the last 10 Ma. The basin centre and flanks experience the same subsidence and erosion rates. No overpressure is considered and the geothermal gradient is set to 30°C km^{-1}. In order to study the effect of erosion on the traps'

spill path   drainage path

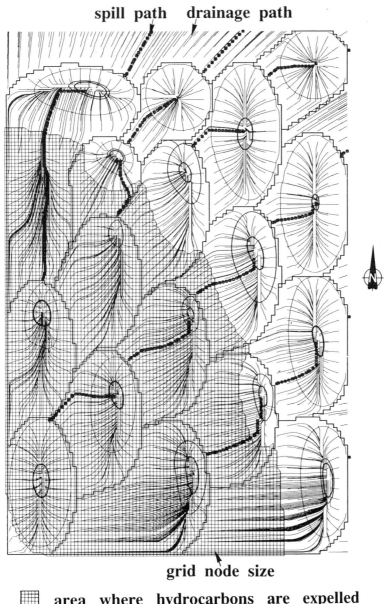

grid node size

⊞ **area where hydrocarbons are expelled
from source rock**

Fig. 4. Basin of Domes: expulsion area, drainage paths, drainage area outline, etc. (see also Fig. 5).

hydrocarbon composition when hydrocarbons migrate into the traps after the erosion has been initiated, hydrocarbon expulsion is modelled through all the 20 Ma in the area shown in Fig. 4. The composition of the expelled hydrocarbons does not change and is equal to that of trap A in Fig. 5. In this paper mixing is only considered to occur within the traps and loss is modelled during spill only, i.e. no change in hydrocarbon composition is modelled during drainage. Six components are used in the EOS modelling and these are C1 (methane), C2, C3, C5, C7 and C11+ (Fig. 5). The EOS modelling is repeated for all the traps each 2 million years but the same

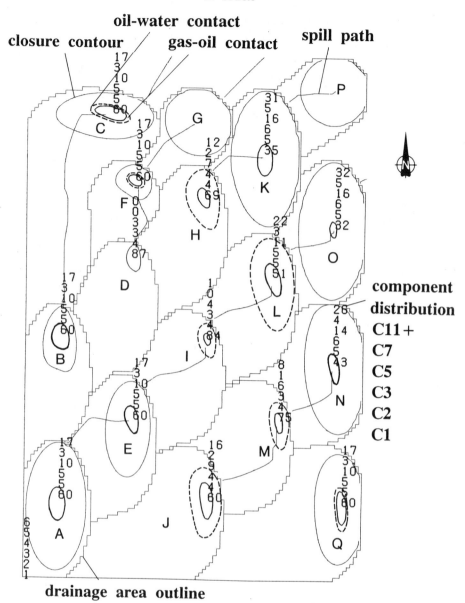

Fig. 5. Basin of Domes: component distribution (mol%) in traps. If no liquid is present, the gas–oil contact represents the gas–water interface.

results are obtained when larger time steps are used. E_{sm} is constant and no loss of the vapour phase is modelled.

The resulting drainage and spill paths are shown in Fig. 4. Drainage paths originating from each third grid node are shown and the pathways are also plotted for those grid nodes

that do not generate any hydrocarbons. Trap names run from 'A' to 'Q' (see Figs 3 and 5) and a total of 17 traps exists. The SEMI program automatically computes the outlines of the drainage area, the spill point depths, gas–oil contacts and oil–water contacts (see Figs 4 and 5). The blocky nature of the drainage outlines reflects the assumption that a grid node can only be part of one drainage area. i.e. it can only drain into one trap.

Note that in Figs 4 and 5 trap D is shown to spill into trap F, but the spill path is very close to the drainage area outline between traps F and H. The spill paths shown in these Figs. are made with a minor change in the overall definition of the basin as compared to Table 3 and in the later sensitivity analysis trap D spills into trap H instead of trap F as shown in the Figure. Similarly, trap A, if filled, also spills very close to the drainage outline between traps E and B and might very easily have spilled into trap B. One of the first uses of such a modelling tool can be to outline which areas are sensitive to mapping inaccuracies that might change the entire migration scenario in a basin.

The spill pattern resulting from using Table 3 in defining the basin is:

$B \rightarrow C$

$F \rightarrow G$

$D \rightarrow H \rightarrow K \rightarrow P$

$A \rightarrow E \rightarrow I \rightarrow L \rightarrow O$

$J \rightarrow M \rightarrow N$

Q.

Six drainage systems, which are fairly independent of each other, result. One system consists of only one trap (Q) whereas the maximum number of traps within a drainage system is five (starting in trap A and ending in trap O).

The resulting petroleum compositions (in terms of component distribution) in the modelled traps are shown in Fig. 5 for one scenario. Component distributions should add up to 100%, but the results of Fig. 5 are generated by the computer and a rounding off error is introduced. The inaccuracy is reduced if the results are printed with 0.1 mol% resolution. The drainage system AEILO (starting at A and ending in O) is focused upon because it shows a number of interesting features. Trap I spills liquids into trap L and therefore the overall methane content of I is enriched to 84 mol%. Trap L, which receives the liquids from trap I, is enriched in the heavy components (22 mol% of C11+), but it also spills liquids further into trap O and therefore trap O also shows an increase in the heavier components (32 mol% of C11+) and a corresponding decrease in the lighter components, including methane.

Sensitivity

It is important to show through a sensitivity analysis which parameters are important when simulating a multicomponent system during secondary migration. The results may be used so that further investigations can be directed towards the variables that produce the largest variations in the entrapped hydrocarbon compositions. Also, it has to be shown whether the proposed method of modelling is feasible for the study of hydrocarbon distribution within sedimentary basins. In this paper it is only possible to investigate some key variables

One possible way of 'observing' the mechanism of expulsion from source rocks is to propose different mathematical and numerical models for expulsion and secondary migration and apply them to real basins. The calibration of model results to observed trap content may then distinguish between the wrong and right models. An assumption which has to be fulfilled for this approach is that the effects of secondary migration and expulsion are distinguishable from each other. The easiest way to investigate this is to perform a sensitivity analysis, i.e. vary the input parameters for the different models and observe whether the resulting component distribution in traps changes in a different manner for the different models.

A formal procedure for this evaluation could use a statistical criterion. However, in these preliminary evaluations, it may be more realistic to apply a simple visual analysis based upon some key parameters. Plotting the methane content (C1 in mol%) versus the total trapped volume of hydrocarbons (million m^3) within the traps is proposed because the methane content can be expected to be sensitive to changes in the pressure and temperature history.

Parameters varied in the sensitivity analysis are: (1) expulsion efficiency when no delay in expulsion (E_{exp}); (2) expulsion after initial saturation is reached (V_{te}); (3) amount of erosion (500 m and 1000 m); (4) loss of liquid phase during spill (E_{sm}). In case 1, $V_{te} = 0$, $E_{sm} = 0$ and amount of erosion is 500 m. In case 2, $E_{exp} = 50\%$, $E_{sm} = 0$, amount of erosion is 500 m. In case 3, $E_{exp} = 50\%$, $V_{te} = 0$ and $E_{sm} = 0$. In case 4, $E_{exp} = 50\%$, $V_{te} = 0$ and amount of erosion is 500 m. $V_{ts} = 0$ for all cases.

The results of the simulations are shown in Fig. 6 and a number of conclusions can be made. In many traps varying the expulsion efficiency results in significant changes both in the methane content and in the volume of trapped hydrocarbons. Some traps are not sensitive to variations in input parameters since they are completely filled with gas and have reached their 'component distribution equilibrium' (CDE). When more heavy or light components migrate

MAXIMUM TRAP VOLUMES

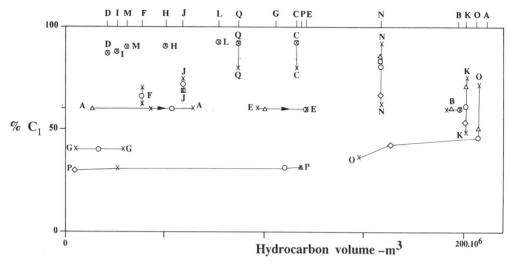

Fig. 6. Basin of domes: methane content versus hydrocarbon volume in traps.

into them they are spilled further into the next trap, which may or may not have reached its CDE.

One group of traps is the one in which nothing is received from any other traps and which is never filled to spill point (trap A etc.). The apparent constant methane content in these traps is only due to the fixed composition of the expelled hydrocarbons, and the exclusion of loss during drainage in the simulations. Thus, more work is needed before any conclusions can be made for these traps.

For this artificial basin there is a tendency that either the methane content or the trapped volume varies i.e. the lines in Fig. 6 are either horizontal or vertical, and only trap 0 shows a deviation from this characteristic. Again, the inclusion of loss during drainage may change this observation but it is a sensitivity that is expected to some degree since the spill process is very important in creating a component distribution pattern within basins where most of the migration occurs in well defined carrier beds.

If the expulsion model is changed (case 2) and the total volume expelled is the same as for $E_{exp} = 50\%$ in case 1, the hydrocarbon compositions of traps A, E, K, N, O and P

change. The sensitivity is mostly caused by some parts of the basin not having reached the necessary saturation for expulsion to occur in case 2 whereas other parts of the basin expel more for case 2 than for case 1. Thus, some traps record an increase and some traps a decrease in the volume of hydrocarbons or C1 content.

If the subsidence and erosion is increased from 500 m to 1000 m (case 3), almost no change is observed. Only for traps J and N can a minor change in the methane content (less than 2 mol%) be found and this indicates that a phase of erosion probably cannot be deduced from compositional distributions.

Loss during secondary migration as modelled in case 4, clearly leads to a change in the composition for several traps in the basin of domes. The tendency is towards a decrease in volume and a decrease in the methane content and these results are explained as a result of the delay in the filling of the reservoirs caused by the reduced input of liquid hydrocarbons when loss is modelled. This effect clearly dominates over a possible increase in the lighter components which might be expected when the heavier components are more retained during secondary migration.

Discussion and conclusions

The results presented here show that the modelling of the multicomponent hydrocarbon mixture during secondary migration is possible using ray-tracing and equation of state techniques. The model described in this paper can handle the detailed structural development found in most sedimentary basins. This opens up the possibility of using the modelled component distributions of drilled (i.e. 'observed') fields as a means of unravelling the mechanisms governing migration since different models may be tested and compared with observation.

The sensitivity of several input parameters has been tested for a simple synthetic basin, i.e. the Basin of Domes, and it is shown that the component distribution varies differently for the various parameters. The parameters are therefore inferred to be fairly independent of each other with respect to predicting the component distribution. Therefore, an optimization and calibration of the SEMI model to fields may be useful in future exploration.

Different traps behave differently to varying input parameters. Some are totally insensitive (within the limits used in this study); these traps are all filled to the spill point with gas, and have reached what might be called a state of component equilibrium in which only a major change in the feedstock component distribution will have any effect. These traps are positioned in between traps which show a sensitivity in either the total hydrocarbon volume or the component distribution. Traps in the Basin of Domes generally change from an initial volumetric-only variation to a sensitivity only in the component distribution (except for trap 0), but this effect is believed also to be a result of the simplifications made in this paper.

The expulsion efficiency influences the component distribution strongly whereas uplift and erosion do not lead to any changes in composition, except in cases where the leakage through the cap rock is enhanced. If a model for expulsion which requires an initial hydrocarbon saturation is valid, the resulting distribution in traps will be different from a model in which instantaneous expulsion occurs. Secondary migration losses, when modelled, influence the composition in traps because of different retention of the liquid and vapour phases and the sensitivity is different from the one observed when expulsion is varied.

In this paper not all parameters have been tested for their sensitivity upon the component distribution and therefore more research has to be done on parameter sensitivity. The presented model, which did not perform EOS computations during drainage, uses less than an hour of cpu on a minicomputer. When EOS modelling is performed during drainage the cpu requirements increase drastically due to the large number of calculations needed and a more efficient procedure may have to be used than the ones found in standard EOS packages. Using explicit phase behaviour models, it would be possible to represent the hydrocarbon system with as few as four pseudocomponents (Hustad 1989) and the increase in speed of the calculations required would be obtained. The multicomponent migration model has to be tested on real basins in the same way that the two-component version of SEMI has been applied (Raa 1989) and a number of case studies are required if a proper calibration and testing is to be achieved.

It is concluded that the modelling of a multicomponent system during secondary migration is feasible and has a potential for quantifying (and thereafter reducing) the uncertainty involved in drilling new prospects in mature petroleum provinces.

I thank K. O. Sandvik for the support when starting the development of this model and for believing in the concepts before they could be shown to work. The help of T. Andersen and O. S. Hustad was essential for the successful integration of the COPEC subroutines into the SEMI program and S. J. Lippard kindly helped with his comments.

References

BERG, R. R. 1975. Capillary pressure in stratigraphic traps. *American Association of Petroleum Geologists Bulletin*, **59**, 939–956.

DUPPENBECKER, S. J., MANN, U., DOHMEN, L. & WELTE, D. H. 1991. Numerical modelling of petroleum expulsion in two areas of the Lower Saxony Basin, Northern Germany. *This Volume*.

DEMBICKI, H. JR & ANDERSON, M. J. 1989. Secondary migration of oil: experiments supporting efficient movement of separate, buoyant oil phase along limited conduits. *American Association of Petroleum Geologists Bulletin*, **73**, 1018–1021.

ENGLAND, W. A. & MACKENZIE, A. S. 1989. Some aspects of the organic geochemistry of petroleum fluids. *In*: POELCHAU, H. S. & MANN, U. (eds), 'Geologic modelling—Aspects of integrated basin analysis and numerical simulation' *Geologische Rundschau*, **78**, 291–303.

FIROOZABADI, A. 1988. Reservoir-fluid phase behaviour and volumetric prediction with equations
of state. *Journal of Petroleum Technology*, April.
397–406.

HUBBERT, M. K. 1953. Entrapment of petroleum
under hydrodynamic conditions. *American
Association of Petroleum Geologists Bulletin*, **37**,
1954–2026.

HUSTAD, O. S. 1989. *An explicit phase behaviour model
for pseudocompositional reservoir simulation.*
Doctor ingeniør discertation at the Norwegian
Institute of Technology, Trondheim. Institute of
Petroleum Technology (IPT) report 1989: 3.

LEHNER, F. K., MARSAL, D., HERMANS, L. & VAN
KUYK, A. 1988. A model of secondary
hydrocarbon migration as a buoyancy-driven
separate phase flow. *Revue de l'Institut Français
du Pétrole*, **43**, 155–164.

PEDERSEN, K. S., FREDENSLAND, AA. & THOMASSEN, P.
1989. *Properties of oils and natural gases.*
Contributions in Petroleum Geology and
Engineering, **5**.

RAA, A. M. 1989. *Kalibrering av en modell for
sekundærmigrasjon (SEMI) til felt i Statfjord-
området.* (Calibration of a model for the
secondary migration (SEMI) to fields in the
Statfjord area). Thesis. Geological Institute,
Norges Tekniske Høgskole.

RISNES, R., DALEN, V. & JENSEN, J. I. 1981. Phase
equilibrium calculations in the near-critical
region. *In*: FAYERS, F. J. (ed) *Enhanced Oil
Recovery, Procedings of the 3rd European*

Symposium, Bournemouth 1981. Developments in
Petroleum Science 13. Elsevier Scientific
Publication Co. 329–350.

RITTER, U., LEITH, T. L., GRIFFITHS, C. M. & SCHOU,
L. 1987. Hydrocarbon generation and thermal
evolution in parts of the Egersund Basin,
Northern North Sea. *In*: BEAUMONT, C. &
TANKARD, A. J. (eds) *Sedimentary Basins and
Basin-Forming Mechanisms.* Canadian Society of
Petroleum Geologists, Memoir **12**, 75–85.

SCLATER, J. C. & CHRISTIE, P. A. F. 1980. Continental
stretching: an explanation of the post-mid-
Cretaceous subsidence of the Central North Sea
Basin. *Journal of Geophysical Research*, **85**, 3711–
3739.

SMITH, J. E. 1971. The dynamics of shale compaction
and evolution of pore-fluid pressure.
Mathematical Geology, **3**, 239–263.

SYLTA, Ø. 1987. *SEMI—a program for the modelling of
buoyancy driven, secondary migration of oil and
gas by means of a ray-tracing technique.*
Continental Shelf and Petroleum Technology
Reserach Institute (IKU) report no 25.2403.00/
01/87.

VAN NESS, H. C. & ABBOTT, M. M. 1982. *Classical
Thermodynamics of Nonelectrolyte Solutions. With
Applications to Phase Equilibria.* Chemical
Engineering Series. USA 1982 McGraw Hill Inc.

WELTE, D. H. & YUKLER, M. A. 1981. Petroleum
origin and accumulation in basin evaluation—a
quantitative model. *American Association of
Petroleum Geologists Bulletin* **65**, 1387-1396.

Using thermal fields to estimate basin-scale permeabilities

D. S. CHAPMAN[1], S. D. WILLETT[2] & C. CLAUSER[3]

[1] *Department of Geology & Geophysics, University of Utah, Salt Lake City, UT 84112–1183, USA*

[2] *Department of Oceanography, Dalhousie University, Halifax NS, Canada*

[3] *Niedersächisches Landesant für Bodenforschung, Stillweg 2, Hannover, Germany*

Permeabilities for characterizing fluid migration in a sedimentary basin are normally obtained from laboratory measurements on core samples (1–10 cm scale) or from well tests (10–100 m scale) but permeabilities at the basin scale (10–100 km) are more difficult to determine. However the high sensitivity of thermal fields to fluid flow offers the possibility of inferring basin scale or 'system' permeabilities by analysing the advective perturbations to temperature fields in a basin. Examples are drawn from three separate sedimentary basin studies: (a) the Uinta Basin of the western USA, (b) the Rhinegraben, and (c) the Western Canada Sedimentary Basin.

The thermal field in the Uinta Basin was determined by analysing 320 bottom hole temperatures corrected for the thermal disturbances of drilling and mud circulation. Spatially varying thermal gradients in each of four Tertiary formations were determined by a method of stochastic inversion; the three-dimensional temperature field was constructed using these gradients together with formation thickness information. The most notable feature of the thermal field in the Uinta Basin is depressed isotherms in the high elevation, north flank of the basin and raised isotherms towards the low elevation, geographic centre of the basin. This warping of the isotherms results in a temperature anomaly of $-25°C$ at the base of the Tertiary section at the north flank and $+10°C$ in the centre of the basin. The thermal field in the Uinta Basin was simulated using a finite element code for coupled heat and fluid flow. Hydrological and thermal parameters were varied in the simulations, and the resulting simulated temperature field was compared to the observed temperature field. Thermal conductivity variations within the basin and in the surrounding rocks produce a temperature anomaly with some similarities to the observed anomaly but fail to produce the magnitude of the anomalies. To account for the additional anomaly, the hypothesis of an advective perturbation to the thermal field resulting from a topographically driven regional ground water flow system was considered. A good match between the observed and modeled temperature fields is obtained if the permeability of the Duchesne River and Uinta formations is $5 \times 10^{-15} m^2$ or slightly higher: permeabilities more than an order of magnitude different from these values fail to produce the observed temperature anomaly. A comparison between the range of permeabilities needed to simulate the thermal field on a basin scale, and permeabilities compiled independently from measurements on the Duchesne River Formation is shown in Fig. 1. The mean permeability of the well scale pump tests is more than an order of magnitude higher than permeability derived from laboratory measurements on cores, probably due to a sampling bias towards sandstone aquifers in the pump tests. In contrast, the basin scale model permeability is lower than the well scale tests because it is a regional average which samples the integrated effects of sandstone interbedded with lower permeability shales. This case illustrates the danger in extrapolating core measurements or pump test results to infer permeabilities appropriate for basin scale fluid migration.

The present thermal field for the Rhinegraben (Germany) in the vicinity of Mannheim–Karlsruhe also shows evidence of an advective disturbance that can be used to estimate permeability at the kilometre and greater scale. A vertical Peclet Number analysis was made for 22 oil/gas wells of 1–3 km depth in the vicinity of the graben's western border faults. For reasonable bounds on the characteristic lengths of the vertical flow system, curvature in the temperature-depth profiles in these wells yield an average Peclet number of 0.5 (i.e. total vertical heat flow is 2/3 conductive and 1/3 advective) and bounds on the Peclet Number between 0.1 and 1.2. Numerical simulations were also made for the coupled fluid flow-heat flow in a transect across the basin, with basal heat flow and permeability of the main aquifers (Buntsandstein/Muschelkalk and fault zones) being varied systematically in the simulations. Both topographically driven flow and thermally driven free convection are

From England, W. A. & Fleet, A. J. (eds), *Petroleum Migration* Geological Society, Special Publication No. 59, pp. 123–125.

important in this setting. Local vertical Peclet Numbers, computed for each simulation, are shown in Fig. 2. The model Peclet Number depends principally on permeability and, to a lesser extent, on heat flow through the temperature dependence of fluid density and viscosity. Models that satisfy both the constraints of the observed temperature field and the previous Peclet Number analysis limit the basin scale aquifer permeability within the narrow range of 10^{-16} to $10^{-14}\,\mathrm{m}^2$.

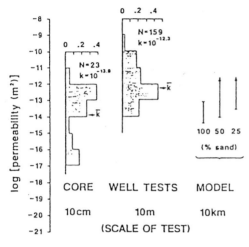

Fig. 1. A comparison of permeabilities determined for the Duchesne River Formation, Uinta Basin, USA. Left, laboratory measurements on core samples, centre, right, model permeabilities required to produce the observed temperature anomaly. Model value varies depending on the effective thickness of the permeable beds within the formation. As thickness (i.e. % sand) is decreased, the minimum permeability required to produce an equivalent thermal effect increases.

In the Western Canada Sedimentary Basin, palaeotemperature indicators along an east–west transect north of Edmonton suggest significant decreases from west to east that are difficult to explain solely by burial in a uniform geothermal gradient. Preliminary data from aromatization–isomerization reactions in hydrocarbons and apatite fission track analyses imply that maximum temperatures varied from west to east (**W**: E) as follows: Blairmore, Viking, Colorado and Cardium formations (120°C : 70°C), Lea Park Formation (110°C : < 100°C?) and Belly River Formation (80°C : 70°C). These maxima were probably obtained during the Late Cretaceous to Early Tertiary Laramide Orogeny. Palaeo-·temperatures reflect both the effects of burial and palaeogeothermal gradient. The depth of burial is constrained by a flexural model of the

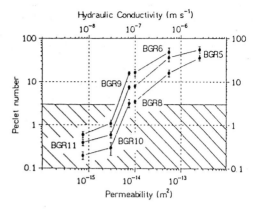

Fig. 2. Variation of vertical Peclet number in the numerical simulations with the permeability of the main aquifers and fault zones. Three simulation results shown for each permeability value correspond to different basal heat flows of 60, 80, and 100 mW/m² (bottom to top). The shaded region indicates Peclet Number range from analysis of curvature in temperature–depth profiles and is used to determine acceptable basin scale permeability solutions.

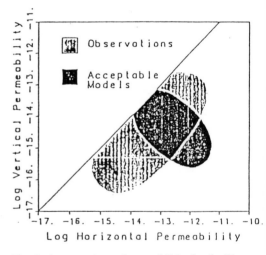

Fig. 3. A comparison of permeabilities for the Western Canada Sedimentary Basin determined by two methods: (a) basin scale or 'system' permeability found by simulating the thermal field required to explain palaeogeothermal indicators (dark shading), and (b) permeability range for all formations found by conventional hydrological methods (light shading). Horizontal permeabilities are found to be more than 100 times the vertical permeability in both cases. The 'system' permeability solution forms a subset within the formation permeabilities.

evolution of the foreland basin. If the burial predicted by this model is correct, the palaeo-temperature data imply a laterally varying palaeogeothermal gradient that increases eastward from 20–30°C/km near the edge of deformation to as high as 70°C/km on the flanks of the basin. This variation can be explained by heat advection in a gravitationally driven fluid flow system within the basin sediments. Numerical models of the coupled fluid flow/heat flow in the restored section demonstrate that the inferred thermal regime does not constrain the palaeotopography, but is particularly sensitive to permeability and permeability anisotropy. The basin scale permeability in these simulations is constrained to a range of $10^{-14}\,m^2$ to $10^{-12}\,m^2$ with horizontally enhanced anisotropy of about 100. Acceptable solutions form a subset within the formation permeabilities for this sedimentary section estimated by conventional hydrological techniques (Fig. 3).

Integration of geological data into hydrodynamic analysis of hydrocarbon movement

RICHARD W. DAVIS

Geotechnical Corporation, PO Box 1292, Laramie, WY 82070, USA

Abstract: As a general rule, when hydrodynamic analysis is used in the course of exploration for petroleum, it is used as a means of testing individual geological structures or stratigraphical variations to see if they can serve as traps for hydrocarbons. Hydrodynamic analysis can be used to accomplish much more, especially in the search for subtle and unusual traps. It can be used to evaluate whole basins, not just restricted areas of one prospect. In addition, by utilizing all the geological information available, it is possible to use the analysis as a test of the consequences of the geological model upon petroleum migration as well as entrapment. It is possible to use the geological information to improve the accuracy of potentiometric surface data. It cannot be stressed too much that the value of the hydrodynamic evaluation is only as good as the accuracy and detail of the geological model.

The importance of hydrodynamics was recognized almost from the first petroleum exploration efforts in the mid-1800s, but the first fundamental analysis of the interaction between geology, moving ground water and hydrocarbons was not published until 1940. This work was later polished, improved, and republished in 1953 (Hubbert) in a work more familiar to the petroleum fraternity. Interest in the subject has waxed and waned since in response to the pace of advances in analytical technology, but in spite of considerable advances in knowledge, hydrodynamic analysis is still generally practised only in the primitive fashion of the early 1950s.

In the years since Hubbert's publications (1940, 1953) laid the groundwork, patterns of fluid circulations in large basins and the controls on that circulation have been defined (i.e. Hitchon 1969*a*, *b* and Toth 1963). Berg's (1975) and Schowalter's (1979) discussions on threshold capillary pressures have partly lifted the veil on a troublesome area covered only in the ideal by Hubbert. Several corollary issues have also been addressed. For instance, Bredehoeft & Papadopoulos (1965) quantified the relationship between vertical fluid flow and geothermics. The work of Neuzil (1983, 1986) and many others has shown how effective shales and clays can be in preventing drainage of overpressured strata. Gies' studies (1984) in the basins of western Canada have shown us another significant source of energy besides gravity to move fluids in the subsurface. All of these workers and many others not mentioned have provided information on the interaction between fluid movements and

geology and upon methods which can be used to trace the fluid movements.

This paper will (a) discuss the range of data necessary for a hydrodynamic analysis, (b) show a few of the applications of such analyses, and (c) demonstrate the importance of a good geological model to the development of a valid hydrodynamic analysis. It is this latter fact, especially, which needs to be addressed. Too many geologists, because of a failure brought about by geological error or by a lack of geological detail tend to dismiss simple, fundamental physical principles as being in error. Fluid movements are determined by the geology and by the physical state of the fluids themselves. A good hydrodynamic analysis is absolutely dependent upon a good geological model. Hydrodynamic analysis provides a method of integrating the geological data by evaluating the interaction between different components of the geological model and migrating fluids.

This paper assumes that the hydrodynamic system is in a steady state. That is, that the system is not changing with time. This is not a serious restriction, because at any point in the geologic history of a basin changes in basin hydrology are generally occurring so slowly as to be negligible. This statement is not meant to imply that the system does not change. Only that once the hydrological framework is defined for any one point in geological history, the rate of change is slow enough that the rate of change will not negate a steady state analysis for exploration purposes.

The term 'hydrodynamic analysis' as used

From England, W. A. & Fleet, A. J. (eds), *Petroleum Migration*
Geological Society, Special Publication No. 59, pp. 127–135.

here refers to the study of the movement of fluids in the subsurface, particularly those movements which affect the accumulation of hydrocarbons. In the discussions which follow hydrocarbons are assumed to migrate as discrete, immiscible stringers.

Elements of a hydrodynamic analysis

Hubbert (1940, 1953) showed that three factors, water drive, buoyancy drive and threshold capillary pressure, were responsible for the movement and entrapment of hydrocarbons. The relationship between each of these three factors and geological data will be discussed here.

Water drive analysis

Water drive is a function of four parameters:
(a) potentiometric surface gradients which are the first differential of the hydraulic potential;
(b) aquifer thickness, which is more often than not different from stratigraphic unit thickness;
(c) aquifer permeability which is usually derived from a porosity/permeability regression relationship;
(d) fluid properties which, in turn, are a function of temperature, pressure, salinity, and the composition of the oil and gas.

The reader will note that the four parameters listed are information compiled in the course of any complete exploration program. At present, hydrodynamic analysis does not usually consider any factors beyond potentiometric surface gradients and fluid densities. Yet, as will be shown below, the potentiometric surface gradient is also a function of aquifer thickness and permeability and must correlate with them. To construct a valid potentiometric surface map requires (a) an understanding of what a potentiometric surface represents, (b) definition of areas of recharge and discharge, and (c) a scientifically consistent correlation between the potentiometric surface and the geology.

Too many geologists construct a potentiometric surface map just as they would a structure map. The result is a statistically correct, but hydraulically incorrect interpretation. The uniqueness of analysing potentiometric surface data are perhaps best exemplified by a few examples. Potentiometric surface maps reflect the dynamics of a flowing mass of fluids. Potentiometric gradients other than zero indicate moving fluids. High points on a potentiometric surface map where gradient directions are all diverging from an area represent fluid sources. Since the fluids are moving away from the high points, laws governing conservation of mass require that the high can only be maintained over geological time if a source of new fluids is present. Low points on the maps similarly require that discharge be occurring. Variations in gradients across a map involve even more complicated analyses in order to be justified. A flattening for instance, in the gradient may be due either to discharge or to an increase in porosity or aquifer thickness. It could also be due to a change in fluid properties as the fluids enter a new thermal regime.

The work of Gies (1984) has shown that apparently anomalous potentiometric surface gradients can originate from overpressuring phenomena associated with petroleum source beds. His work should serve as a caution that we may not have yet identified all the reasons for anomalous gradients. However, the effects of the gradients are the same, regardless of the source. They still represent either a change in geology or a change in volume of fluid flow. Identifying a source of anomalous gradients is necessary in order to ensure a complete understandng of the geology.

Potentiometric surfaces constructed in hydrocarbon exploration necessarily rely on widely spaced data points. Further, those data points have often been affected by drawdowns from producing fields making it necessary to discard even some of the reliable information. This makes analysis very difficult because the use of hydrodynamic analysis in exploration is presently assumed to require an accurate knowledge of original potentiometric gradients. In cases where the total area of investigation is much larger than the area of drawdowns it is usually possible to find at least four or five data points in areas where drawdowns have not occurred. If the data points are reasonably well distributed, the correlation between potentiometric gradients and transmissivity offers a means to reconstruct the original potentiometric surface. Darcy's law offers the basis for correlation in a steady state system:

$$Q = KIA$$

where: Q = volume of flow per unit time; K = hydraulic conductivity for formation water; I = the hydraulic gradient; A = size of cross sectional area normal to fluid flow.

In hydrological work this formula is frequently modified to:

$$Q = TIL$$

where: T = transmissivity = $K \times m$; m = aquifer thickness; L = horizontal length of the cross section represented by A.

This second version is somewhat easier to work with in many analyses because it shifts variables into parameters which are consistent with the way geologic data are gathered. What it shows is that changes in gradient must either be accompanied by an inversely proportionate change in aquifer permeability or aquifer thickness, or a proportionate change in volume of water flowing through the system, i.e., recharge or discharge.

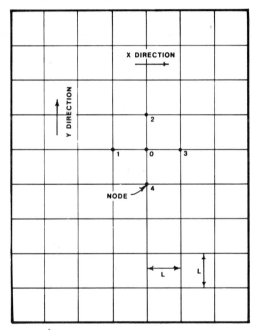

Fig. 1. Finite grid and nomenclature used in analysis of recharge.

Analysis of the recharge and discharge signified by a potentiometric surface provides some insight into how accurate the potentiometric surface map is. As a corollary, the analysis also helps in identifying those areas influenced by production drawdowns. In most instances it can be assumed that there is no vertical movement (i.e. recharge or discharge) of fluids across strata confining the aquifer. To do such an analysis, one can digitize the potentiometric surface map into nodes on a square grid separated by a distance, L (Fig. 1).

As Ferris *et al.* (1962, eq. 70) point out, the hydrological balance at each node of aquifer material is:

$$h_1 + h_2 + h_3 + h_4 - 4h_0 = (S\Delta h_{0t}/\Delta t)/T - W/T)L^2$$

where: h_n = the potentiometric heads in adjacent squares; h_0 = the head at the centre of the square; W = rate of vertical recharge to the square (negative if discharge); L^2 = map area of the node square; Δt = time since recharge began.

The term, $S(\Delta h_{0t}/\Delta t)/T$, would represent the effects of a time varying surface due to, say, a producing field. It would equal zero for purposes of hydrodynamic analysis of the original potentiometric surface where time, Δt, is assumed to be infinitely long. As noted earlier, it can usually be assumed that the recharge term, W/T, is also equal to zero. In cases where vertical geothermal gradients, closed contours around a potentiometric high or low, or other geological information indicate recharge or discharge it is necessary to consider this term. For most cases, though, when the aquifer transmissivity is properly correlated with potentiometric gradients, the equation reduces to:

$$h_1 + h_2 + h_3 + h_4 - 4h_0 = 0$$
$$\text{or: } h_1 + h_2 + h_3 + h_4 = 4h_0$$

In other words, if gradients are properly adjusted, the amount of fluid moving into each node equals the amount moving out as long as vertical discharge or recharge are zero. Any changes in gradients between node intervals must be solely a response to changes in geology. It is no great feat to program a computer to adjust the potentiometric gradients at each node to reduce the recharge or discharge toward zero. By iterating the adjusting process a number of times, the potentiometric surface can be incrementally changed until the recharge and discharge at the nodes approach acceptably close to zero. Control of the procedure is provided by (a) the outer boundary nodes, which are extended into areas of known and undisturbed heads, and (b) the internal nodes with undisturbed potentiometric heads. These nodes are not allowed to vary.

There are cases where data limitations prevent extending the model boundaries beyond the area influenced by production. In such cases it may be necessary to evaluate more than one hypothetical flow regime until one is found which meets other criteria such as the tilt of known oil fields or geochemical trends in the ground water. Even in the best of instances it is usually necessary to put as much time and effort into constructing a potentiometric surface map as went into analysing the rest of the geology.

AQUIFER THICKNESS

(FEET)

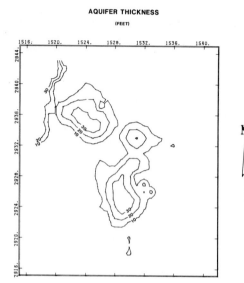

Fig. 2. Isopach (feet) of aquifer material for first text example. Map coordinates in this and all following figures are in thousands of feet.

AQUIFER POROSITY

(%)

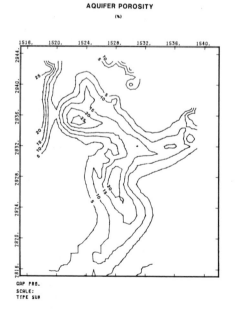

Fig. 3. Porosity (percent) of aquifer material for first text example.

Figure 2 shows the thickness and Fig. 3 shows the porosity of a geological unit which will be used for illustrative purposes. The regional structure across the map area is essentially a monoclinal dip in a southwesterly direction. Figure 4 shows the initial potentiometric surface contour map of the area. Hydraulic gradients on Fig. 4 were used to calculate recharge at each node. Figure 5 shows the recharge (negative values are discharge) across the area necessary to balance the potentiometric surface gradients shown in Fig. 4 with the aquifer parameters shown in Figs 2 and 3. There are obviously some problem areas.

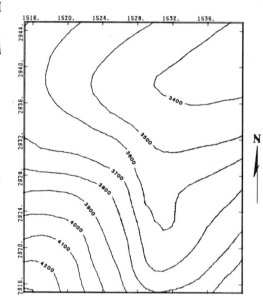

Fig. 4. Initial potentiometric surface (feet above mean sea level) map for first text example.

Using a formula directly relating porosity and permeability, the porosities in Fig. 3 were used to calculate permeabilities at each node. The formula used was:

$$K = c \times n^B$$

where: c and B are constant within each area on Fig. 6; n = porosity (%).

Table 1. *Constants used in calculating permeabilities from porosities in areas shown in Fig. 6*

Area index	Value of constant	
	c	B
A	1	1
B	2	0.1
C	3	0.01
D	4	0.001

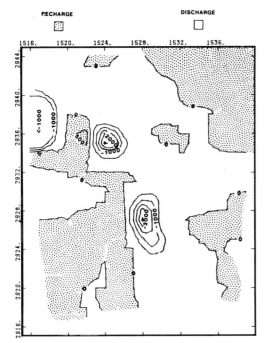

Fig. 5. Recharge/discharge map calculated from Fig. 4. Recharge (negative if discharge) is in cubic metres per day per 1000 square foot grid area.

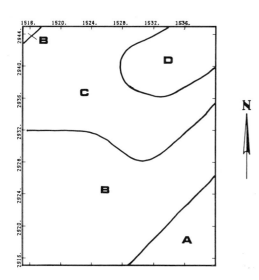

Fig. 6. Index map showing areas in which the formula constants listed in Table 1 were applied to calculate aquifer permeabilities.

The values of the constants, c and B, vary across the map area. Table 1 lists the value sets of c and B used in calculating permeability and

Fig. 6 delineates the areas used for each of the value sets.

Figure 7 shows the potentiometric surface modified to bring all recharge values nearly to zero. Careful study shows a number of significant changes in contours. In effect, the geology has been used in concert with a few wells believed to represent pre-production potentiometric heads to bring the potentiometric surface into conformance with the geological parameters.

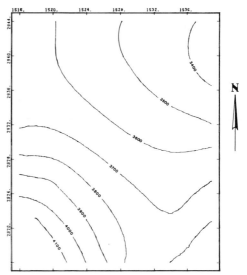

Fig. 7. Potentiometric surface map revised from Fig. 4 to match geology shown in Figs 2 and 3. Recharge and discharge values have been reduced to less than 10 cubic metres per day per 1000 square feet.

Fluid properties are a very important part of any hydrodynamic analysis. The density and viscosity of water are a function of salinity, temperature, and pressure. At depths explored for petroleum, pressure effects on water density are negligible. Likewise, the density and viscosity of the hydrocarbon are important and are a function of temperature, pressure, and fluid composition. Maps of geothermal gradients, depth to top of the aquifer, and aquifer water salinity are important geological inputs to determination of water and hydrocarbon density in the formation environment. The effects of these parameters were covered in an earlier paper (Davis 1987).

Oil density is a complex function and may require some evaluation of source rock geochemistry or knowledge of characteristics of oil produced from existing fields. Factors such as formation temperature, gas/oil ratios, composition and average density of the gas, and the gravity of the oil must be determined. In order to model migration paths accurately, it may be necessary to examine a range of oil and gas mixtures if there is considerable uncertainty as to the nature of the hydrocarbons. Engineering input may be helpful in resolving problems in determining the ultimate density of the gas/oil mixture.

Buoyancy drive

The term 'buoyancy drive' describes the tendency of hydrocarbons to rise to the top of the water column because of density differences. Buoyancy drive was analysed in some detail by Davis (1987), so the mathematical basis will not be re-examined here. In order to evaluate the drive, it is necessary to know (a) the densities of the fluids involved and (b) the changes in elevation of the top of the aquifer from point to point.

Maps of the elevation of the top of the aquifer are not necessarily the same as a structure map or as a map of elevation of the top of a stratigraphic unit. The top of the aquifer is defined here as the effective top to which hydrocarbons may rise before being prevented by threshold capillary pressures from rising further. In some cases of complex geology this horizon may not be easily determined. It is the horizon along which water and buoyancy drives will cause the oil to migrate. If this surface cannot be satisfactorily determined, then not only is the hydrodynamic model indeterminate, but discovery of many traps by any means is not feasible. Effective hydrodynamic analysis is more often limited by lack of geological knowledge than by lack of analytical methods.

Threshold capillary pressures

Threshold capillary pressures are the minimal pressures required to force an oil globule into and through the largest pore throats available in a mass of water wet rock. Geological input for this phase of hydrodynamic analysis includes information on formation temperature, hydrocarbon characteristics, and water salinity in order to determine the oil/water/rock interaction in the aquifer environment. It also requires some measure of the size and shape of the pore throats. Pore throat size and shape are often treated as a function of permeability (i.e. Thomas et al. 1967; Berg 1975). This avoids the necessity of otherwise doing a great deal

of petrographic work, but seems an oversimplification of such an important factor. Determination of pore throat data and their relationship to porosity, permeability, and threshold capillary pressures are an area in need of more research.

Little more can be said about determination of threshold capillary pressures. Present methods may not allow definition of the absolute effectiveness of stratigraphic traps. They do allow determination of the *relative* effectiveness of facies changes in trapping hydrocarbons.

The composite picture

Once all three mechanisms of oil migration and entrapment are defined, a composite picture can be compiled. Two types of maps can be constructed which summarize the data.

(a) A vector map which shows the direction and magnitude of the sum of the water and buoyancy drives and how these will interact with structure and stratigraphy.

(b) A contour map which shows the height of the hydrocarbon column which might be trapped behind stratigraphic barriers of fining sediment. Calculating the height of the column is a simple process of determining the ratio of threshold capillary pressure to resultant water plus buoyancy drive and converting the ratio to height of oil column for a specific oil or gas density.

In the process of generating the two maps listed above, the data become available to compile other useful contour maps including:

> (a) oil/water equipotential gradients (useful in doing volumetric calculations on structural traps);
> (b) threshold capillary pressures (useful for determining the area of balance between trapping facies and acceptable permeabilities for production);
> (c) recharge/discharge volumes (a quick check for correlating anomalous traps with anomalous model parameters).

A hypothetical case is offered as an example of a vector map analysis. Fig. 8 shows the elevation of the top of an aquifer. The aquifer dips northeasterly with a sharp, monoclinal increase in dip about halfway across the map. The aquifer is assumed to have a 24% porosity everywhere except in the northwest corner where porosity is 3%. A constant northward potentiometric surface gradient and constant salinity, and thermal gradient are also assumed. Figure 8 shows vectors of the resultant water plus buoyancy drive superposed on the structural contours.

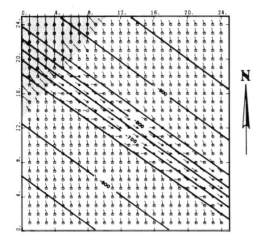

Fig. 8. Water plus buoyancy drive vectors overlain on a hypothetical model of structure and porosity. Vector lengths are proportional to the log of the drive. Structural contours are in feet above msl. Porosity is 24% in clear areas and 3% in the patterned area near the northwest corner of the map. The potentiometric surface gradient is due north.

It is often useful in analysis to label the vectors by colour or symbol to provide additional information on hydrodynamic factors. It is not possible to use colour, which is the easiest to analyse, in the format of this publication. Therefore, a vector code identified by the shape of the vector base is used here to indicate the following.

(1) When the water/oil equipotential surface dips greater than the top of the aquifer. There is no structural interference with the oil drive and, consequently, no structural type trap. (Denoted by a circle base.)

(2) When the water/oil equipotential surface dips less than the top of the aquifer, but buoyancy lifts the oil out of the potential trap. This could be called a 'near miss' which might be worth further study and refinement of the geological model. (Denoted by a diamond base.)

(3) When the dip of the top of the aquifer is great enough to cause a diversion of the hydrocarbons. This would be a necessary, but not sufficient, requirement to indicate a structural trap. (Denoted by a triangular base.)

Where the third condition exists on Figure 8 the vector lengths and directions are modified to reflect the resultant movement of hydrocarbons as they are forced into a new path by structural interference. Should the top of the aquifer develop sufficient closure by curving across the direction of hydrocarbon movement to form a barrier, a trap would result. In Figure 8 the increase in monoclinal dip is interfering with oil movement (Case 3) and forcing the oil toward the low porosity zone which, in combination with the monoclinal dip, may form a trap.

Figure 8 also shows another effect of the change in porosity in the northwest corner of the map. The direction of water drive in areas of permeability gradients must vary in accordance with the tangent law enunciated by Hubbert (1940, pp. 874–882). For this model permeabilities were varied directly with porosities in accord with a very simple logarithmic relationship:

$$K = 10^{(n-10)/6}$$

The potentiometric gradient was assumed constant and was not corrected for variations in permeability. Instead, on Figure 8 the correction in drive direction has been made to a first approximation using Hubbert's tangent law. This procedure was used here to show more dramatically the effects of permeability variations. Along the boundary between 24% and 3% porosities, the corrected vectors point into the area of 3% porosity. The example shows that changes in porosity (permeability) could potentially divert migrating oil globules into a stratigraphic trap. A simple embayment in the area of porosity gradients, given the resultant drive, would serve. The variation in drive directions means a complete barrier to northward flow would not be necessary.

Figure 8 also shows that changes in porosity can markedly increase or decrease the water drive which can, in turn, change the characteristics of a potential structural trap. The increase in ground water gradients in the northwest corner of Figure 8 caused by the reduced permeability eliminates the effect of the increase in monoclinal dip. It is entirely conceivable that a potential structural trap could be ineffective because a decrease in porosity allowed the hydrocarbon to escape. In oil fields with porosity gradients, the oil–water interface of entrapped petroleum can be expected to have varying gradients and directions of dip.

In spite of the problems in defining exact numerical values of pore throat radii, useable values of oil or gas column height can be obtained using hydrodynamic modelling methods. Using permeability/pore size relationships, it is simple to calculate potential heights of trapped oil columns which are probably accurate relative to each other. In other words, current methods allow analysts to determine which, among several possible traps, will hold the most oil.

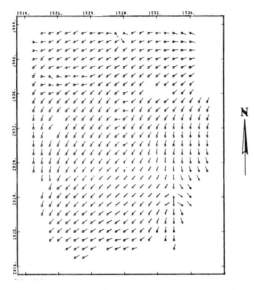

Fig. 9. Water plus buoyancy drive vector map of the first text example (Figs 2, 3, and 7). Vector lengths are proportional to the log of the drive. Regional structural dips are southwest.

The data shown in Figs 2, 3 and 7 are used to illustrate an analysis of oil column heights. Regional structural dips in that model are to the southwest, opposite to the water drives, making a structural trap unlikely. Figure 9 shows the vector directions of oil drive at nodes on a 1000 foot spacing. Figures 2 and 3 suggest the possibility of a stratigraphic trap. Figure 10 shows the height of oil column theoretically necessary to penetrate the porosity gradients given the magnitude and direction of the water plus buoyancy drive. At this point the analyst must decide whether the migration regime, as shown on Fig. 9, is favourable for migration of petroleum into any of the potential traps. Some of the barriers to migration can easily be bypassed by diversion of hydrocarbons around the barriers. Others offer sufficient 'closure' to form a trap. Patterned areas on Fig. 10 indicate where both the drive vector (Fig. 9) and the capillary resistance (Fig. 10) are favourable for oil accumulation. The analyst must also decide at which point in the porosity gradient area a

producible accumulation may occur. If a well is drilled in the extremely fine grained sediments an oil show may occur but permeabilities will be too small to allow production.

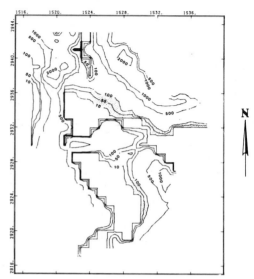

Fig. 10. Height (feet) of oil column necessary to penetrate sediments in direction of vector drives shown in Fig. 9. Areas where capillary resistance in concert with drive vectors (Fig. 9) is most likely to result in an oil accumulation are patterned.

Conclusions

The most common uses of hydrodynamic analysis at present are as an aid in evaluating the potential height of an oil column in a stratigraphic trap and in determining whether an oil accumulation is likely to be displaced from a structural high. However, the technology exists to make far more extensive uses of the science. The method can be used to evaluate all levels of geologic subtlety. As an example of one of the more unconsidered instances, it is not uncommon for the interaction of aquifer dip and oil/water equipotential surface to cause the *base* of a syncline to exhibit trap potential. In nearly all cases the pattern of hydrocarbon migration does not allow oil to enter the trap. However, it is likely that such a trap does exist somewhere in the world. The conditions required for the trap favour creation of a large instead of a small accumulation. Such a trap, unless found fortuitously, will only be found by analysing the dynamics of hydrocarbon migration patterns.

Areal or basinwide analyses can be used to

evaluate migration flow paths and their relationship to potential traps. Trap efficiency can also be evaluated with this approach. 'Trap efficiency' refers to the size of the area which can provide oil to fill a trap. If a trap lies on the 'downstream' side of a large trapping feature, then hydrodynamic analysis can be used to tell whether or not any oil may be able to reach the downgradient trap. A large trap athwart the direction of regional flow would retain the hydrocarbon and divert it to a spillpoint. In effect, the regional flow of hydrocarbon would be concentrated into a narrow corridor, leaving only water to fill most of the downstream structures.

Hydrodynamic analysis can also be a valuable tool in performing sensitivity checks on geological models. Such a check would highlight those geological factors to which hydrocarbon entrapment is most sensitive. The geologist could then direct most of his time and effort into refining the geological data of most importance.

Most of the theoretical bases for complete hydrodynamic analyses of petroleum basins and individual prospects have been formulated. A number of practical problems remain which are contingent upon progress in geological methods. If the geologist can adequately define the subsurface parameters, the methodology exists to perform hydrodynamic analysis. In cases where there is insufficient subsurface data to construct a hydrodynamic model, there is likewise insufficient data to predict whether or not a hydrocarbon trap exists. It cannot be stressed too strongly that successful hydrodynamic analysis is only as good as the geological model used. If the geology does not identify the spill or leakage points in a trap, for instance, hydrodynamic analysis may not resolve them. The situation is best described by the computer dictum, 'garbage in—garbage out'. As efforts to locate more subtle traps intensify, the geologist is going to have to construct much more sophisticated geological models and will find hydrodynamic analysis a valuable tool for testing those geological models.

References

BERG, R. R. 1975. Capillary pressures in stratigraphic traps. *Bulletin of the American Association of Petroleum Geologists*, **59**, 939–956.

BREDEHOEFT, J. D. & PAPADOPULOS, I. S. 1965. Rates of vertical groundwater movement estimated from the earth's thermal profile. *Water Resources Research*, **1**, 325–328.

DAVIS, R. W. 1987 Analysis of hydrodynamic factors in petroleum migration and entrapment. *Bulletin of the American Association of Petroleum Geologists*, **71**, 643–649.

FERRIS, J. G., KNOWLES, D. B., BROWN, R. H. & STALLMAN, R. W. 1962. *Theory of aquifer tests*. US Geological Survey Water Supply Paper 1536-E.

GIES, R. M. 1984. Case history for a major Alberta deep basin gas trap: The Cadomin Formation. *American Association of Petroleum Geologists, Memoir*, **38**, 115–140.

HITCHON, B. 1969a Fluid flow in the Western Canada Sedimentary Basin. I. Effect of topography. *Water Resources Research*, **5**, 186–195.

—— 1969b, Fluid flow in the Western Canada Sedimentary Basin. 2. Effect of Geology. *Water Resources Research*, **5**, 460–469.

HUBBERT, M. K. 1940. The theory of ground-water motion. *The Journal of Geology*, **48**, 785–944.

—— 1953. Entrapment of petroleum of petroleum under hydrodynamic conditions: *Bulletin of the American Association of Petroleum Geologists*, **37**, 1954–2026.

NEUZIL, C. E. 1983. Erosional unloading and fluid pressures in hydraulically "tight" rocks. *Journal of Geology*, **91**, 179–193.

—— 1986. Groundwater flow in low-permeability environments. *Water Resources Research* **22**, 1163–1195.

SCHOWALTER, T. T. 1979. Mechanics of secondary hydrocarbon migration and entrapment. *Bulletin of the American Association of Petroleum Geologists*, **63**, 723–760.

THOMAS, L. K., KATZ, D. L. & TEK, M. R. 1967. Threshold pressure phenomena in porous media. *Society of Petroleum Engineers, Preprint Paper No. SPE 1816*.

TOTH, J. 1963. A theoretical analysis of groundwater flow in small drainage basins. *Journal of Geophysical Research*, **68**, 4795–4812.

Phase-controlled molecular fractionations in migrating petroleum charges

STEPHEN LARTER & NIGEL MILLS[1]

Newcastle Research Group (NRG) in Fossil Fuels and Environmental Geochemistry, Drummond Building, The University, Newcastle upon Tyne, NE1 7RU

[1] *Present address: Saga Petroleum A/S, Maries Vei 20, Postboks 9, 1322 Høvik, Norway*

Abstract: An experimental approach has been used to make a preliminary examination of the possible effects of partitioning processes distributing molecular markers between vapour and liquid petroleum phases during petroleum migration in the system gas-saturated oil/condensate-saturated gas. These effects are discussed with reference to several traditional petroleum geochemical parameters. Some source and maturity related geochemical parameters are dispersed from the inherited source values when a petroleum system undergoes a transition from a single phase gas-condensate to a two phase system and it is probable that partitioning is one of the many processes that contribute to the scattered data sets observed in natural systems particularly those associated with gas/condensates.

In addition to affecting gross composition and gas and gasoline range composition, physical partitioning of petroleum components between vapour and liquid phases in deep reservoirs containing rich gas condensate vapour and a liquid petroleum phase in equilibrium can affect even high molecular weight components, selectively partitioning lower homologues into the vapour phase. This may disperse to various degrees any source inherited signature in source facies or maturity interpreted from parameters such as aromatic hydrocarbon distribution, gasoline range parameters, or sterane carbon number distribution.

Migration history, or the frequency and extent of multiple equilibrations between vapour and liquid phases in deep high pressure reservoirs followed by selective release of one phase to a shallower position to produce a two phase petroleum system is proposed as an important variable in the geochemical interpretation of once equilibrated petroleum systems. It is concluded that phase related effects must be considered in addition to accepted source and maturity considerations as controlling the molecular composition of two-phase petroleums derived from single-phase precursors. This is particularly so for rich gas condensate systems.

In addition to widely accepted source facies and maturation controls on the composition of petroleum expelled from source rocks (cf. Tissot & Welte 1984), both primary and secondary migration can have a dramatic role on the physical and chemical properties of a migrating petroleum charge (England & Mackenzie 1989; Larter 1990). In this article we do not discuss primary migration efficiency controls on petroleum composition but focus instead on assessment of the significance of partitioning of geochemical markers between subsurface gas and liquid petroleum phases.

From high pressured, high temperature source rocks the petroleum charge migrates under water potential and buoyant forces to cooler and lower pressured regimes in carrier systems or traps (England *et al.* 1987). The physical properties of gas saturated oil phases and condensate saturated gas phases are such that the solubilities of the minor phase in the major phase (i.e. gas in oil, or, condensate in gas) decrease under

temperature and primarily pressure controls (Zhuze *et al.* 1962; Zhuze & Ushakova, 1981; Price *et al.* 1983; England *et al.* 1987). Unfortunately, as Thompson (1988) and Zhuze and coworkers have pointed out, the distribution of individual components between liquid and vapour phases cannot be theoretically predicted for complex multi-component petroleum mixtures with any useful accuracy (for a molecular petroleum geochemist) at this time so that component fractionations between petroleum phases must be determined experimentally.

Although we cannot conclusively assess the gas to oil ratios (GOR) of the charges of petroleum expelled from source rocks, pyrolysis-gas chromatography data suggests that the GOR of initially generated petroleums in source rocks are typically below 1.0 kg/kg (gas/oil), equivalent to about 5000 scf/bbl for source rocks in general and are typically below 0.25 kg/kg or about 1200 scf/bbl for rich oil-prone source

From England, W. A. & Fleet, A. J. (eds), *Petroleum Migration*
Geological Society, Special Publication No. 59, pp. 137–147.

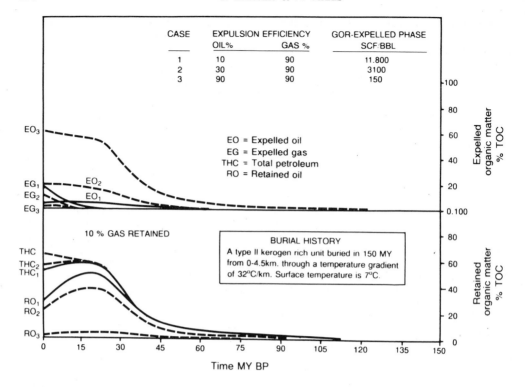

Fig. 1. A kinetic model of hydrocarbon generation and destruction. Based on Tissot & Espitalie (1975) the gas oil ratio of the expelled petroleum is a function of expulsion efficiency. Modified after Larter (1990) the figure shows yields of total petroleum (THC) expelled petroleum (EO, EG) and oil retained (RO) in the source rock.

rocks (England & Mackenzie 1989; Larter & Horsfield, in prep.). The actual GOR of the expelled phase will depend on the degree of oil cracking in the source rock which in turn is controlled by the source rock expulsion efficiency (Fig. 1) and may potentially vary between 12 000 and 150 scf/bbl. With efficient expulsion of oil at temperatures below 160°C from rich oil-prone source beds (cf. Cooles *et al.* 1986), even deeply buried oil-prone source rocks will produce a net expelled petroleum charge with only minor amounts of gas. With low expulsion efficiencies even Type II kerogen-rich source rocks would produce gas-condensates.

Equation of state calculations (Ungerer *et al.* 1983) and empirical studies of petroleum fluids (Standing 1947; Glasø 1980; England *et al.* 1987; England & Mackenzie 1989) suggest that under typical source rock pressure and temperature conditions of say, 150°C and 400–500 bar (or much greater) the expelled phases will most likely be single phase oil or single phase gas condensate fluids.

With decrease in pressure and temperature

along a migration path, or due to carrier bed or reservoir inversion, at some point, a dew point or a bubble point would be reached. This would result in the formation of a gas saturated oil fluid and a condensate saturated gas fluid. Partitioning of hydrocarbons and non-hydrocarbons between the two phases occurs depending on the vapour–liquid equilibrium constants for each component (Thompson 1987, 1988). As the minor phase may not reach critical pore saturation levels to enable mobility in a carrier bed and as saturated oil and gas densities diverge markedly in the shallow portions of basins (England *et al.* 1987), this fractionation, most marked in light and intermediate species (gas–diesel range), results in measurably changed compositions in the final trapped fluids. The most obvious changes in properties relate to densities and minor phase equilibrium saturations of the saturated liquid and gas phases. These changing properties, which will be the primary controls on the volumetrics and gas/oil ratios of coexisting petroleum phases (England & Mackenzie 1989), may also be accompanied by molecular level fractionations affecting geochemical ratios applied by petroleum geochemists to problems of source facies and maturity determination.

Knowledge of the degree of this molecular level fractionation is important, as otherwise it is impossible to interpret correctly geochemical data obtained from condensates and other light petroleum fractions where component concentrations may be controlled by phase equilibrium considerations rather than solely by inherited source characteristics. The range of fractionations observed may complicate the interpretation of in-field petroleum compositional variations in terms of traditional maturation/facies scenarios. In that sense it is surprising that the base level controls on the molecular composition of crude oils have not been further evaluated and we suggest that the rather long term belief in the accurate transmissibility of source characteristics by a migrating petroleum charge through to a reservoired petroleum may require further rethinking. We report a preliminary set of experiments aimed at developing a better understanding of these phase related effects.

Experimental details

An experimental approach has been used here to study the fractionation of petroleums due to changes in pressure and temperature resulting in one or two petroleum phase transitions. A light black oil from a North Sea clastic reservoir was compressed with excess solution gas to provide a rich gas condensate (GOR = 3 kg gas/kg oil). This was performed in a conventional reservoir engineering PVT setup (pressure, volume, temperature monitoring experimental set up) at 150°C and 1000 bar (mature source rock or deep reservoir conditions). The petroleums were then separated along a pressure/temperature migration route in the PVT apparatus, the two phase fluids (gas and oil or condensate) being sampled at 'equilibrium' at experimental pressure and temperature for a range of pressures and temperatures corresponding to subsurface conditions equivalent to reservoirs at 3 km and 2 km.

The original field separator oil and solution gas samples (Snorre Field, Norwegian Sector) were recombined to attain a single phase. Most of the oil had evaporated at approximately 700 bar but there was still a small amount of liquid left so the pressure was increased to 1000 bar. At this pressure virtually all of the sample was in the gas phase but there was still a very tiny, thickish deposit which would not evaporate (possible result of gas deasphaltation). For safety reasons the pressure was not increased beyond this point. Disregarding the presence of the deposit, this sample represented a 'single phase' at 150°C/1000 bar.

The single phase was sampled and flashed to standard conditions to give oil sample 1 and gas sample 1.

The first reduction of the single phase sample was made to 110°C/330 bar. The sample was allowed to reach 'equilibrium' (3 hrs) at this temperature/pressure combination and samples were taken of the resulting gas and oil phases. These were then flashed to standard conditions yielding gas sample 2 and oil sample 2 from the oil phase and gas sample 3 and the condensate ('oil' sample 3) from the gas phase.

The gas phase at this temperature/pressure combination was then drawn off at constant pressure (330 bar) leaving the oil phase at 330 bar.

The second reduction was then made on the oil phase to 70°C/166 bar. Two phases resulted and each was sampled and flashed to standard conditions. This gave gas sample 4 and oil sample 4 from the oil phase and gas sample 5 from the gas phase (no condensate).

A new recombination was then performed on the separator samples in a black oil cell and the 'original' sample was then reduced directly to 70°C/166 bar and allowed to reach 'equilibrium' (15 minutes with agitation). Two phases resulted and each was sampled and flashed to standard conditions. This gave gas sample 6 and oil sample 5 from the oil phase and gas sample 7 from the gas phase (no condensate could be sampled from this phase).

This meant we had two sets of samples representing 70°C/166 bar: one taken from temperature/pressure reduction directly from the 'original' sample (gases 6 and 7 and oil 5) and one taken via 110°C and 330 bar with consequent loss of some material (gases 4 + 5 and oil 4). The gases were all sampled in 1000 cc or 300 cc steel sampling bottles at atmospheric conditions. Table 1 and Fig. 2 list some relevant details from this part of the study.

These hypothetical reservoirs were experimentally tested to provide a variety of 'surface' gases and oils through separators. Sample identities and experimental summaries are shown in Table 1 and Fig. 2. Additional experiments were also performed to examine the effects of remigration (of the liquid phases from 3 km 'traps' (110°C/330 bar) to 2 km 'traps' (70°C/166 bar). Thus gas 5, oil 4 and gas 4

Table 1. *Sample information*

T/P Combination (°C/bar)	Resulting samples	GOR (Sm³/m³)	Density (kg/m³)	API Gravity
150/1000				
Gas phase:	Gas 1	3707.1		
	Oil 1		845.2	35.9
110/330				
Oil phase:	Gas 2	151.5		
	Oil 2		886.0	28.2
Gas phase:	Gas 3	7906		
	Oil 3		831.6	38.7
70/166 (via 110/330)				
Oil phase:	Gas 4	96.6		
	Oil 4		883.4	28.7
Gas phase:	Gas 5			
70/166 (direct)				
Oil phase:	Gas 6	1551.1		
	Oil 5		857.4	33.6
Gas phase:	Gas 7			

represent the surface fluids obtained when the oil leg of the 3 km 'reservoir' was 'spilled' to the 2 km situation, i.e. the liquid in the PVT cell from the '3 km' experiment was sampled at constant pressure then P and T were reduced to the new conditions.

The fluids (gas and oils/condensates) were then analysed for gross and isotopic composition, molecular parameters and in particular for the effects of fractionation on basic geochemical tracers such as alkylaromatic hydrocarbon and biomarker alkane distributions using standard procedures.

Results and discussion

The largest fractionations observed are at the gross compound class level in the C_{15+} fractions of the various oils and condensates (Fig. 3, Table 2). The largest fractionations are between the oil and condensate in the intermediate reservoir (oil 3 (condensate), oil 2) with the vaporized C_{15+} components in the gas phase being enriched in saturated hydrocarbons as reported previously (Thompson 1979, 1988; Price *et al.* 1983). Further the saturated hydrocarbon content varies as a function of 'migration history' in that the oil 4 (oil leg spill from 3 km reservoir) and oil 5 (oil leg produced from the original gas condensate) have significantly different saturated/aromatic

hydrocarbon ratios. It should be pointed out that this compositional variation with sample history has resulted from only a single stage fractionation. In a migrating petroleum charge in a carrier bed, continually reducing P–T conditions and the presence of local stratigraphic or structural micro traps would result in multiple trapping/spilling scenarios with greatly exaggerated fractionation effects.

Figure 4 shows the variation in some gasoline hydrocarbon parameters. As described by Thompson (1987, 1988), hydrocarbons in solution in a gas phase (or derived therefrom) are enriched in saturated hydrocarbons relative to aromatic hydrocarbons, paraffins being preferentially vaporized over naphthenes. This is clearly seen in the data of Fig. 4 which shows enhanced toluene/nC_7 ratios in oil 2 compared to oil 3 and so on. Again similar P–T reservoirs (oil 4, 5) contain different composition petroleums indicating that migration history is an important variable. These results raise the question of the validity of interpreting small variations in gasoline range hydrocarbon distributions as maturity variations in petroleum systems which have reached saturation in the minor phase during migration. Table 2 shows that coeval petroleums may have a range of heptane contents (a maturity related parameter adopted from the work of Thompson 1979) of from 24% to 35%, oil legs appearing consistently 'less mature' (lower heptane

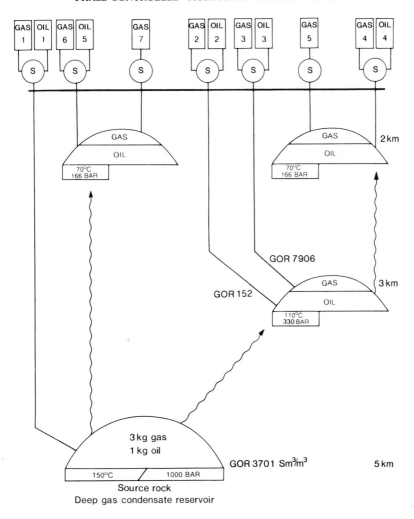

Fig. 2. Sample identification chart. The diagram represents schematic reservoirs, the various fluids being flashed to ambient conditions through separators (S).

content) than gas condensate caps. Repeat analyses of oils suggest that typically heptane values have standard deviations of 2%.

Table 2 shows isoprenoid hydrocarbon data. Pr/Ph ratios typically can have standard deviations as small as 0.02 but the scatter in the data shows no clear pattern and major fractionations between oils 2 and 3 are not evident.

Thompson (1988 and references therein) has shown by review and experimental data that the fractionation of C_{6+} saturated hydrocarbon components, between oil and gas phases is strongly influenced (promoted) by the presence of aromatic hydrocarbons which are in turn preferentially retained in the liquid phase. Our

data cannot be used to indicate large scale fractions for the alkylphenanthrenes distributed between condensate and oil fractions in the higher pressure simulations though fractionation may be occurring.

In line with vapour pressure considerations naphthalene and to a lesser extent, phenanthrene homologues are selectively partitioned between vapour and liquid phases. Figure 5 shows high resolution gas chromatograms of liquid chromatographically separated aromatic fractions of the oils. The most obvious variations are the selective partitioning of lower molecular weight components (alkyl-naphthalenes) compared to phenanthrenes into the condensate (oil 3) and the oil derived

Table 2. *C15+ compositional data*

	Sat. %	Aro. %	Asph. %	NSO %	Sat./Aro.	Non-HC %	Pr/Ph	Pr/nC$_{17}$	Ph/nC$_{18}$	Heptane Content*
Oil 1	83.4	9.6	0.8	6.3	8.6	7.0	1.70	0.68	0.45	35.6
Oil 2	79.9	11.0	4.1	5.0	7.2	9.1	1.75	0.58	0.43	24.2
Oil 3	89.4	5.4	0.5	5.0	16.7	5.0	1.72	0.69	0.46	33.6
Oil 4	73.4	14.8	2.7	9.2	4.9	11.8	1.59	0.54	0.37	24.9
Oil 5	79.7	11.2	2.6	6.5	7.1	9.1	1.84	0.56	0.36	26.7

* Modified after Thompson (1979)
Heptane content: n heptane as a % of total hydrocarbons eluting after nC$_6$ up to nC$_7$

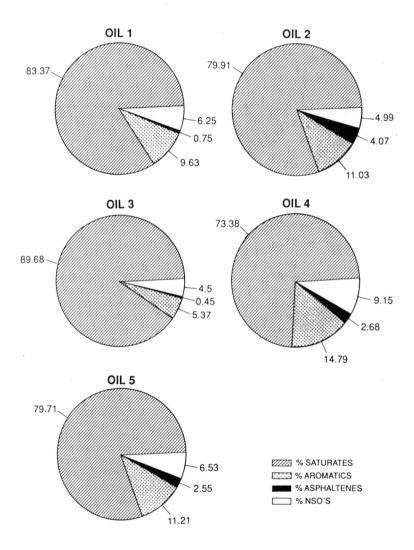

Fig. 3. Group type compositions for the oils identified in Fig. 2. The data were obtained by liquid chromatography and gravimetry.

directly from the deep condensate (oil 5). Similar trends were seen in the n-alkane envelopes of the whole oil chromatograms. Large differences are seen in the aromatic hydrocarbon fingerprints of coeval oils 2, 3 and oils 4 and 5 which though in similar condition reservoirs have suffered different migration histories. Figure 5 also suggests selective partitioning of phenanthrene compared to methyl-phenanthrenes into oils associated with gas phases.

Table 3. *Alkyl phenanthrene ratio data*

Oil No.	$\dfrac{2MP}{3MP}$	$\dfrac{3+2}{9+1}$	$\dfrac{9}{1}$	MPI 1*	R_o calc.*
1	1.02	0.61	1.07	0.59	0.73
2	1.11	0.58	0.91	0.61	0.75
3	0.87	0.72	1.11	0.64	0.77
4	1.14	0.52	1.03	0.55	0.71
5	0.87	0.67	1.04	0.66	0.78

* After Radke (1988)

Table 3 shows quantitative alkyl-phenanthrene data for the various oils and the condensate (oil 3); the methyl phenanthrene index (MPI 1) (Radke 1988 and references therein) shows minor variations between the oils data. Typically MPI 1 measurements have a standard deviation of 0.03. These variations are therefore small and are not considered significant for this data set. Whether multiple fractionations (condensate exsolution and oil leg spilling) would increase the dispersion of the MPI determined source 'maturities' needs to be tested. Horstad *et al.* (1990) indicate that aromatic hydrocarbon maturity parameter estimates from depth equivalent DST (drill stem test) and core extract samples can produce significant maturity estimate differences and it has been suggested that these differences might be related to fractionation effects during well testing though this remains unproven.

The selective fractionation of successive homologues between gas and liquid phases is

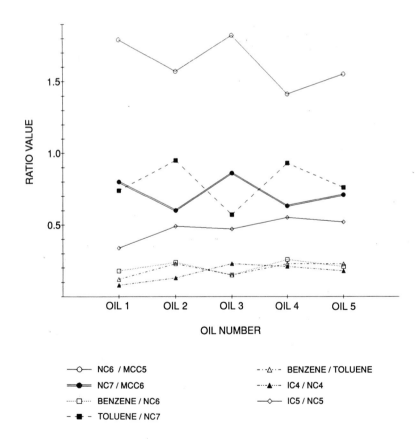

Fig. 4. Variation in gasoline range hydrocarbon ratios for fluids identified in Fig. 2.

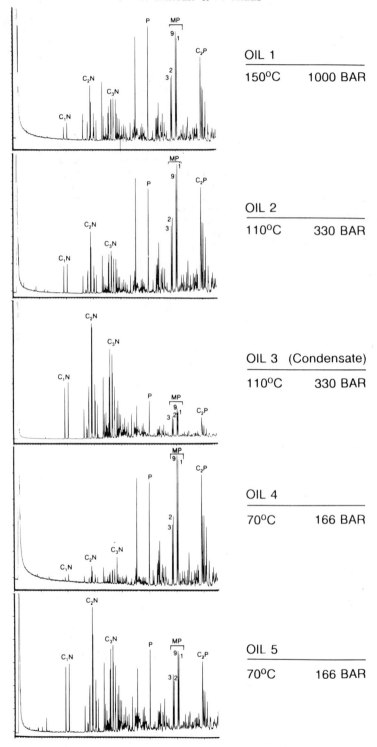

Fig. 5. Partial high resolution gas chromatograms of separated aromatic fractions of oils identified in Fig. 2. GC analyses were performed using a thick film DB5 WCOT silica column programmed at 2°C per minute. Compound identities are as follows: C_1N, methylnaphthalenes; C_2N, dimethyl and ethylnaphthalenes etc; P, phenanthrene; MP, methyl phenanthrenes; C_2P, dimethyl phenanthrenes.

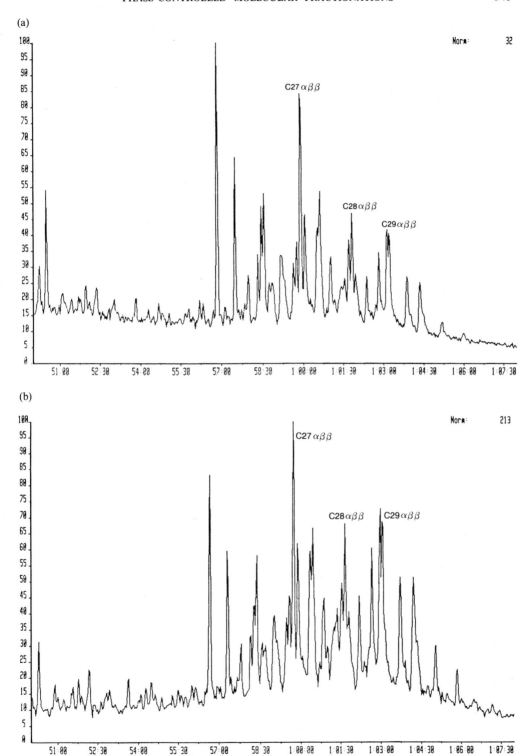

Fig. 6. M/z 217 mass chromatograms of oils 3 (**a**) and 2 (**b**). C_{27}–C_{29} 5α, $14\beta,17\beta$ steranes are identified. The analyses were performed using a Hewlett-Packard MSD GC-MS system with a temperature programmed OVI column and the quadrupole mass filter in SID mode.

also observed in the biomarker alkanes. Figure 6 shows mass fragmentograms (m/z 217) of oil 3 (condensate) (Fig. 6a) and oil 2 (Fig. 6b) which suggest selective partitioning of the C_{27} homologues of the steranes into the vapour phase compared to the C_{29} homologues. The two mass chromatograms show a general preference for lower molecular weight species in the condensate relative to the oil.

It appears, therefore, that physical partitioning of components between vapour and liquid phases can affect quite high molecular weight components selectively partitioning the lower homologues into the vapour phase. This may disperse source inherited signatures in organic facies or maturity parameters. Gross compound type parameters such as saturated hydrocarbon/aromatic hydrocarbon ratios are severely affected (Table 2) and n-paraffin concentrations in the gasoline range used as maturity indicators (e.g. Thompson's (1979) heptane index data) may also be dispersed with C_6+ petroleums in vapour phases being enriched in paraffins (Thompson 1988) and therefore appearing 'more mature'. In our experiments no clear trends relating to pristane/phytane ratio or methyl phenanthrene index were evident.

Our experiments, involving only a single step fractionation in the absence of a petroleum saturated water phase, show degrees of dispersion which would not be problematic at all in a general frontier basin analysis where only general source characteristics are to be inferred from petroleum composition. Larger scale multiple stage fractionations might be expected to be observed in migrating petroleum charges involved in continuous equilibration and spilling from small stratigraphic traps and this might affect even coarse interpretations. This will need to be tested experimentally.

Only a single set of experiments was carried out so our results can certainly be criticized as to the extent of their significance. Certainly we cannot test our data exhaustively by statistical means. Further some aspects of our study, particularly isotopic analysis of sampled gases, indicated that sampling procedures may have caused fractionation effects beyond those caused by simple phase partitioning. Additionally the equilibration times may not have been adequate. Our experience has been that these superficially simple experiments are difficult to perform. Much further work is obviously required in this field before we can be sure of the extent of fractionation by phase partitioning under natural conditions. Comparisons of the fractionations measured with typical standard deviations for the analyses reported above do

suggest that some of the fractionations are real.

Even the single step fractionations examined here cause data dispersion (from the source value-oil 1) in the 'maturity parameter' heptane content (Table 3) at a level that is beginning to be problematic for the interpretation of in-field maturity variations in petroleum columns though our data base is too small to establish this. It is considered likely that these phase equilibrium controlled effects are a significant further contributor to the general data dispersion observed in many subsurface geochemical data sets, particularly those relating to light components. We conclude that in interpreting geochemical data from gas–condensate systems great care and much further study must be undertaken before traditional source facies and maturity parameters can be confidently and successfully interpreted. Following the views expressed by Price et al. (1983), England & Price (1990) and Thompson (1988), plus our own observations of high molecular weight materials ($C_{30}+$) in rich natural (Viking Graben, N. Sea) and artificial condensates (oil 3, this study) we conclude that many condensates result purely from physical processes (i.e. solution of oil legs in migrating gas; cf. Glassman et al. 1989) rather than exclusively as a primary product expelled from high maturity source rock systems.

Conclusions

Physical partitioning of components between vapour and liquid phases primarily under pressure control can affect even quite high molecular weight components, selectively partitioning the lower homologues into the vapour phase. This can potentially disperse any source inherited signature in source facies or maturity parameters such as gross aromatic hydrocarbon distribution, or sterane carbon number distribution. Other parameters such as MPI appear relatively unaffected. Gross compound type parameters such as saturated hydrocarbon/aromatic hydrocarbon ratio are severely affected and paraffin concentrations in the gasoline range used as maturity indicators may also be dispersed by these phase related fractionations.

In our experiments, which involve only a single step fractionation, the degree of dispersion on parameters such as methylphenanthrene index are small but with multiple fractionations might still be significant in the case of in-field petroleum column analysis. Larger magnitude, multiple stage fractionations might be expected to be observed in migrating petroleum charges involved in continuous equilibration and spilling from small stratigraphic traps. These potentially large fractionations might affect even coarse petroleum geochemical inferences of source character made from reservoired petroleums that have reached minor phase saturation levels during migration. In particular the interpretation of gas ·condensate systems using traditional approaches would appear to be potentially fraught with problems. Migration history, relating to the possibility of phase equilibration between vapour and liquid phases in deep high pressure reservoirs then selective release of one phase, seems potentially to be an important variable in the interpretation of equilibrated petroleum systems. We plan to investigate further experimentally better equilibrated multiple stage fractionation effects and also to examine how the presence of a third phase (petroleum saturated water) affects the fractionations we have observed to date.

We thank Saga Petroleum A/S for support and permission to publish. The GC and GC-MS analyses were carried out by I. Harrison and P. Donohoe (NRG) and the aromatic hydrocarbon analyses were performed by D. Karlsen (U. Oslo). PVT analyses were carried out be GECO. Drafting was performed by L. Ravdal and J. Slettebø (Saga) and the manuscript was prepared by Y. Hall (NRG). We are grateful to W. A. England and B. W. Bramley for helpful discussions.

References

Cooles, G. P., Mackenzie, A. S. & Quigley, T. M. 1986. Calculation of petroleum masses generated and expelled from source rocks. *Journal of Organic Geochemistry*, **10**, 235–246.

England, W. A. & Mackenzie, A. S. 1989. Some aspects of the organic geochemistry of petroleum fluids. *Geologische Rundschau* **V78**, 1, 274–288.

—— & Price, I. 1990. Geochemistry and condensate reservoirs. *In: The development of condensate fields*. IBC Technical Services, London, 1988.

——, Mackenzie, A. S., Mann, D. M. & Quigley, T. M. 1987. The movement and entrapment of petroleum fluids in the subsurface. *Journal of the Geological Society, London*, **144**, 327–347.

Glasø, O. 1980. Generalised pressure-volume-temperature correlations. *Journal of Petroleum Technology*, **32**, 785–795.

Glassman, J. R., Clark, R. A., Larter, S. R., Briedis, N. A. & Lundegard, P. D. 1989. Diagenesis and hydrocarbon accumulations, Brent Sandstone (Jurassic), Bergen High Area, North Sea. *Bulletin of the American Association of Petroleum Geologists*, **V73**, 1341–1360.

Horstad, I., Larter, S. R., Dypvik, H., Aagaard, P., Bjørnvik, A. M., Johansen, P. E. & Eriksen, S. 1990. Degradation and maturity controls on oil field petroleum column heterogeneity in the Gullfaks field, Norwegian N. Sea. *Organic Geochemistry*, **16**, 497–510.

Larter, S. R. 1990. The molecular characterisation of kerogen—applications to primary and secondary migration studies and to maturation modelling. *In*: Truswell, E. (ed.) *Proceedings of the 7th World Palynological Congress*, August 1988, Brisbane. Reviews in Palaeobotany and Palynology, **65**, 379–391.

Price, L. C., Wenger, L. M., Ging, T. & Blount, C. W. 1983. Solubility of crude oil in methane as a function of pressure and temperature. *Journal of Organic Geochemistry*, **4**, 201–221.

Radke, M. 1988. Application of aromatic compounds as maturity indicators in source rocks and crude oils. *Marine and Petroleum Geology*, **5**, 224–236.

Standing, M. B. 1947. A pressure-volume-temperature correlation for mixtures of California oils and gases. *In*: Watts, E. V. (ed.) *Proceedings, Spring Meeting of the AAPG Pacific Coast District, Division of Production*, 275–287.

Thompson, K. F. M. 1979. Light hydrocarbons in subsurface sediments. *Geochimica et Cosmochimica Acta*, **43**, 657–672.

—— 1987. Fractionated aromatic petroleums and the generation of gas condensates. *Journal of Organic Geochemistry*, **11**, 573–590.

—— 1988. Gas-condensate migration and oil fractionation in deltaic systems. *Marine and Petroleum Geology*, **5**, 237–246.

Tissot, B. P. & Espitalie, J. 1975. L'Evolution thermique de la matiere organique des sediments; Application d'une simulation methematique. *Revue Institut Français du Pétrole*, **30**, 743–777.

—— & Welte, D. H. 1984. *Petroleum Formation and Occurrence*. (2nd Edition). Springer-Verlag, New York.

Ungerer, P., Behar, E. & Discamps, D. 1983. Tentative calculation of the volume expansion of organic matter during hydrocarbon genesis from geochemistry data. Implications for petroleum migration. *In*: Bjørøy, M. (ed.) *Advances in Organic Geochemistry 1981*. John Wiley & Sons, Chichester, 129–135.

Zhuze, T. P. & Ushakova, G. S. 1981. Dependency of the phase behaviour of petroleum-natural gas systems upon their composition at high pressures. *Zeitschrift für angewandte Geologie*, **27**, 37–49.

——, —— & Yushkevich, G. N. 1962. The influence of pressures and temperatures on the content and properties of condensate in the gas phase of gas-oil deposits. *Geochemistry*, **8**, 797–806.

Geochromatography in petroleum migration: a review

B. M. KROOSS,[1] L. BROTHERS,[2] M. H. ENGEL[2]

[1] Institute of Petroleum and Organic Geochemistry KFA-Jülich, PO Box 1913, D-5170 Jülich, F.R. Germany.

[2] School of Geology and Geophysics, The University of Oklahoma, 100 E. Boyd Street, Norman, OK 73019, USA

Abstract: Numerous works have been published during recent years discussing the occurrence and effects of chromatographic processes with respect to petroleum migration. 'Geochromatography' has been, and still remains, a controversial topic. Based on the principles and conventions of separation science a definition of geochromatography is proposed, followed by a discussion of the relevance and potential occurrence of chromatographic type fractionation effects during primary and secondary petroleum migration. The evaluation of the relevant literature suggests that chromatographic processes in natural systems involving liquid mobile phase can result only in compound class fractionation (saturates, aromatics, polars) whereas molecular fractionation may be expected to occur with gaseous mobile phases. Problems associated with the recognition, experimental verification and quantification of geochromatography are discussed in detail.

The explanation and interpretation of compositional differences and variations of petroleum and bitumen has been one of the major challenges since the establishment of organic geochemistry. The three main factors that are generally considered to control the composition of petroleum in the subsurface are:

(1) genetic differences, due to different types of source materials and different maturity stages in combination with mixing processes during migration;
(2) microbial or thermal alteration in reservoirs, biodegradation in reservoirs with temperatures up to 70–80°C or thermal cracking in high temperature reservoirs;
(3) fractionation effects related to migration, as a consequence of migration of an initially more or less homogeneous mixture of petroleum components.

Essentially, the problem of petroleum composition has not been resolved and will probably never be resolved completely, because in most cases the three factors cannot be differentiated unequivocally.

With respect to the migration-related compositional variations of petroleum, the idea of chromatographic-type fractionation processes is relatively old and can be traced back into the last century. Thus, D. T. Day (1897) in a treatise on the origin of Pennsylvania petroleum states that:

by experimental work it may easily be demonstrated that if we saturate a limestone such as the Trenton limestone with the oils characteristic of that rock and exert slight pressure upon it, so that it may flow upward through finely divided clay, it is easy to change it in its color to oils similar in appearance to the Pennsylvania oils, the oil which first filters through being lightest in color and the following growing darker.

In the same passage the author emphasizes

... the ease by which sulphur compounds and unsaturated compounds can be removed from petroleum by the use of aluminum chloride....

These statements which can be considered as the first records of the idea of geochromatography precede the historical experiments of M. Tswett with plant pigments which are nowadays referred to as the 'discovery' of chromatography.

In the following years separation processes of this kind in geological systems, which we will term here 'geochromatography', have been postulated and investigated by different authors. Various different terms have been used in this context, some of which are listed below.

(1) Filtering (Day 1897; Mileshina et al. 1959; Mileshina & Safonova (1963)
(2) Relocation alterations by adsorption (Hodgson & Baker 1959)
(3) Upward chromatography (Smith et al. 1959)
(4) Geological chromatography (Nagy 1960)
(5) Chromatography as a natural geological process (Ritchie 1966)

From England, W. A. & Fleet, A. J. (eds), *Petroleum Migration*
Geological Society, Special Publication No. 59, pp. 149–163.

(6) Geochromatography (Seifert & Moldowan 1981; Carlson & Chamberlain 1986)

The principal problem related to geochromatography remains the question if and under which conditions migration of fluids in the subsurface will result in fractionation effects similar to those obtained in laboratory chromatographic systems.

Research on compositional fractionation of petroleum during migration has proceeded mainly along two lines.

(1) Case history studies: Collecting data and information from natural systems in well-defined geological situations (e.g. Seifert & Moldowan 1981; Leythaeuser *et al.* 1983, 1984; Macko & Quick 1986)
(2) Experimental approaches: (e.g Roper *et al.* 1958; Safronova *et al.* 1972; Safronova 1981; Chakhmakhchev & Stepanova 1976; Chakhmakhchev *et al.* 1981, 1982, 1985; Carlson & Chamberlain 1986; Bonilla 1985; Bonilla & Engel 1986, 1988; Brothers 1989)

Relatively few approaches have considered the theoretical aspects of geochromatography. Besides the historical paper of Nagy (1960), who advocates the application of the chromatographic plate theory (Martin & Synge 1941) to subsurface fluid flow and furnishes a 'theoretical proof of geological chromatography', the quantitative stable isotope fractionation model for methane presented by Fuex (1980) and the approach of Lafargue & Barker (1988) to quantify the effects of water-washing on petroleum composition should be mentioned in this context.

Prerequisites for chromatographic effects are certainly met in a number of geological settings and most workers dealing with this phenomenon conclude that geochromatography potentially plays a role in the migration of subsurface fluids. On the other hand, despite indisputable progress, many essential problems remain to be solved, particularly regarding the quantification of geochromatographic processes and a reliable assessment of its importance as compared to other possible fractionation processes.

What is geochromatography?

One of the essential prerequisites for a better understanding and systematic treatment of fractionation effects related to subsurface migration is to have a clear definition of the term 'geochromatography'. In this context it seems reasonable to follow the established standards and termin-ology of analytical chromatography and separation science. According to the official IUPAC and ASTM recommendations *a chromatographic system consists of two immiscible phases one of which is stationary and the other mobile (liquid, gas, supercritical fluid) and streaming over or past the stationary phase* (Jönsson 1987). Components within a chromatographic system will be separated partly or completely if they have different affinities for the stationary and mobile phases. As shown schematically in Fig. 1, the chromatographic separation of chemical compounds occurring under these conditions is a continuous process which can be imagined as a large number of subsequent discrete equilibration and transport steps. This point of view is also reflected in the chromatographic plate theory of Martin & Synge (1941).

In contrast to laboratory systems which are usually composed of only two phases, geological systems are in many cases multiphase systems. Accordingly we propose the following definition of a *geochromatographic* system:

A geochromatographic system consists of two or more immiscible phases, one or more of which are stationary phases and at least one of which is a mobile phase.

This definition is comprehensive. It includes liquid–solid adsorption chromatography as well as, for instance, gas–solid, gas–liquid or even supercritical fluid chromatography, all of which have the same underlying concept.

It should be noted here that the above definition does not include fractionation processes resulting from mobility differences of different species within a single phase under the influence of external field gradients. Examples for these monophasic fractionation processes, which involve no partitioning between different phases are diffusion (chemical potential gradient), thermal diffusion (temperature gradient), or gravity segregation (gravitational field). These processes which may be of importance over shorter-range migration distances (up to the kilometre range) have been studied by several researchers (e.g. Schulte 1980; Hirschberg 1984; Montel & Gouel 1985; Costesèque 1987; England *et al.* 1987; Krooss 1987; Krooss & Leythaeuser 1988; Krooss *et al.* 1988).

Essentials of chromatography

From the physico-chemical point of view chromatography is a separation process resulting from differences in the partition

MOBILE PHASE   discrete process

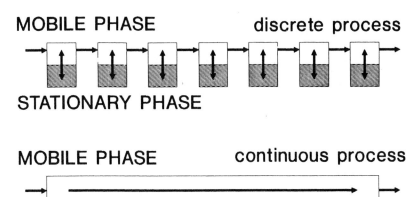

STATIONARY PHASE

MOBILE PHASE   continuous process

STATIONARY PHASE

after Karger et al. (1973)

Fig. 1. Schematic representation of a chromatographic process (after Karger *et al.* 1973).

coefficients of individual compounds between a stationary and a mobile phase. The partition coefficients are determined by the chemical potentials (μ_i) of the solutes (i) in the two phases:

$$\mu_i = \mu^0_i + RT \ln a_i \qquad (1)$$

where a_i is the solute activity in the phase under consideration and μ^0_i is the solute chemical potential under some standard state. The partition coefficient for a given solute between two phases (indicated by ' and ") is then

$$K_i = \exp\left(\frac{\mu_i^{0''} - \mu_i^{0'}}{RT}\right) \qquad (2)$$

Essentially, chromatographic separation of solutes will occur if different solute molecules interact with the stationary phase with different intensity. The efficiency of the fractionation depends mainly on the degree of differences in partition coefficients, the relative amounts of mobile and stationary phases (capacity), and the number of fractionation steps as expressed, for instance, by the number of theoretical plates.

Classification of chromatographic processes and techniques

Several different classification systems are used

in the literature to refer to the different chromatographic processes and techniques that have been developed over the past years. A brief overview of the terminology and official nomenclature is given by Jönsson (1987).

Table 1. *Classification of chromatographic processes according to mobile and stationary phases*

	Stationary phase	
Mobile phase	solid	liquid
liquid	LSC	LLC
gas	GSC	GLC
supercritical fluid	SFC	

The first system of classification is based on the stationary and mobile phases involved in a chromatographic process. The conceivable combinations of phases are given in Table 1. Undoubtedly, the most widely used types of chromatography in analytical chemistry are gas–liquid chromatography (GLC) and liquid–solid chromatography (LSC). Supercritical fluid chromatography (SFC) is at present undergoing rapid development. The term 'reverse phase chromatography' denotes a type of LSC where the mobile phase has a higher polarity than the stationary phase. Because of the common techniques of immobilization and cross-bonding of stationary phases in GLC the term 'liquid'stationary phase tends to have a more historical meaning. On the other hand, the usage of cross-linked polymers as stationary phases reveals possible parallels to geochromatographic processes where the 'geopolymer' kerogen may act as a stationary phase.

The second system of classification is based on the type of interaction of solutes and stationary phase causing retention and fractionation in a particular chromatographic process. The classes comprise:

(1) adsorption chromatography, based on interactions between solutes and a solid surface;
(2) partition chromatography, separation due to partitioning of solutes between a stationary and a mobile phase;
(3) ion exchange chromatography, involving ion exchange between an electrolyte solution and a functionalized stationary phase;
(4) size exclusion chromatography, using steric properties of solutes and stationary phase (gel in gel permeation chromatography, molecular sieve) for chromatographic separation;
(5) hydrophobic interaction chromatography, taking advantage of hydrophobic interaction of amphophilic solutes with specially designed stationary phases.

Finally, chromatographic processes can be subdivided according to the method of development into elution, frontal, and displacement chromatography.

In *elution chromatography* a small volume of sample is introduced into the flowing mobile phase and separated into component peaks that travel at different rates. In *frontal chromatography* a continuous flow of sample enters the system. Only the most quickly migrating content is partly separated from the other compounds in the mixture. The chromatographic effect manifests itself by shifts in the relative concentration of compounds in the region of the migrating front.

Elution and frontal chromatography are illustrated schematically in Fig. 2. The concentration profiles for an equimolar mixture of two components with different partition coefficients, K, were modelled numerically and plotted for these two types of chromatographic processes. The diagrams show the calculated concentration profiles at different times and also the time-resolved process in terms of simulated chromatograms.

Displacement chromatography is a variant of elution chromatography where solvents with an increased affinity towards the stationary phase are subsequently passed through the system. The solvent molecules compete with the previously sorbed compounds for the active surface sites and eventually displace them. Displacement chromatography is, for instance, widely used in organic geochemistry for the group separation of petroleum compounds on silica/alumina columns.

Specific aspects of geochromatography

Phases in geochromatography

The different possible phases involved in geochromatographic processes are shown in Table 2. Thus, minerals and kerogen constitute the solid stationary phases in geological systems whereas sorbed films of pore water and bitumen are liquid stationary phases.

Table 2. *Phases in geochromatographic systems*

Stationary phases	
solids	minerals, kerogen
liquids	sorbed films of pore water and bitumen
Mobile phases	
gases and supercritical fluids	natural gases (methane, CO_2, nitrogen, light hydrocarbon mixtures, water vapour)
liquids	pore water (brines), petroleum fluids

Mobile phases in geochromatographic processes can be natural gases, water and petroleum fluids (crude oil). In analogy to supercritical fluid chromatography (SFC) 'supercritical fluids' like, for instance, carbon dioxide might be listed here as an intermediate between gaseous and liquid mobile phases. However, this subdivision is somewhat arbitrary when considering that methane, for example, is also well above its critical temperature (190.2 K) and pressure (45.6 atm) under geological conditions. Most of the reported work on geochromatographic processes deals either with gas–solid or liquid–solid systems. Despite the occurrence of 'supercritical' CO_2 in the subsurface, no clear evidence has been reported for its role in geochromatographic processes.

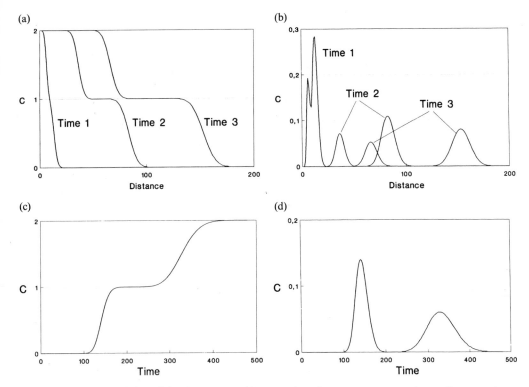

Fig. 2. Numerical simulation of the chromatographic separation of a two component mixture. Concentration profiles at different times for (**a**) frontal and (**b**) elution chromatography. Time-resolved chromatograms for (**c**) frontal and (**d**) elution chromatography.

The process of water-washing and its effect on the composition of petroleum, as investigated experimentally by Lafargue & Barker (1988) can be classified as a liquid–liquid chromatographic process. This type of separation is used in countercurrent extraction and would be expected to occur in natural systems if two immiscible liquid phases (petroleum and water) flow past each other.

Interactions in geochromatography

The types of solute–stationary phase interactions potentially occurring in geochromatographic processes are adsorption, partitioning, ionic exchange and size exclusion. Carlson & Chamberlain (1986) discuss the probabilities for these different chromatographic mechanisms to occur in petroleum migration and conclude that adsorptive interactions are most likely to be responsible for chromatographic effects involving unfunctionalized steroid hydrocarbons. Adsorption is also considered by most other researchers to be the predominant mechanism.

Partitioning of non-polar compounds into kerogen and sorbed hydrocarbon phases resulting in reversed-phase geochromatography is refuted by Carlson & Chamberlain (1986) due to solvent-strength considerations. On the other hand, little information is available on the partitioning of petroleum components between coexisting phases in the subsurface and the treatment of associated fractionation effects remains largely hypothetical. Partitioning may also play a role in molecular migration processes, which are not chromatographic according to our above definition. Thomas (1989) investigated the role of partitioning in determining the relative fluxes of light hydrocarbons migrating by diffusion.

Ion exchange processes are expected mainly in inorganic geochemistry, but they may be of importance also in the migration of carboxylic acids which interact strongly with the mineral surface as shown for instance by Barth et al. (1988).

Size exclusion effects as a possible mechanism in geochromatographic fractionation can be envisaged with respect to the properties of claystones to act as semipermeable membranes for petroleum filtering through them. Although substantiated to a certain extent by experimental work (Mileshina & Safonova 1963) the significance of this effect during petroleum migration remains to be established. Opportunities for size-exclusion fractionation appear to be best during primary migration in fine-grained clastic source rocks.

The occurrence and importance of hydrophobic interactions in petroleum migration remains to be investigated.

Types of development in geochromatography

In terms of a classification by mechanisms most workers consider frontal chromatography as the most likely process in geological systems. This mechanism requires a more or less continuously generated stream of petroleum proceeding along a migration pathway. The petroleum compounds in the frontal region of this migration current are exposed to unimpregnated mineral surfaces which selectively remove those species from the mobile phase that have the highest affinity with resepct to sorption. This theory involves certain problems regarding the total sorptive capacity along the migration pathway. Fractionation effects in frontal chromatography are restricted to the migrating front and once the sorptive capacity of the stationary phase is exhausted little further fractionation will occur. In addition, mixing of initial charges and subsequent charges of petroleum in reservoir accumulations can be expected to annihilate previous compositional fractionations.

Elution chromatography in geological systems can be expected if a mobile carrier phase passes through a region containing components that can be carried away in solution. This process can be imagined to occur if gas migrates through formations containing indigenous hydrocarbon or petroleum accumulations. The hypothetical mechanisms associated with a two-phase flow in such systems are discussed by Thompson (1987). An assessment of such processes in geological time and space requires detailed knowledge of a specific system and a reconstruction of the geological history including petroleum migration.

Displacement chromatography in laboratory systems is associated with rapid and significant changes in mobile phase properties. With respect to petroleum migration, maturity related changes in the composition of the migrating petroleum are very likely to occur in geological time. Thus, the proportion of light hydrocarbons

increases with increasing maturity of the source material. Due to their lower polarity, chances for the less polar, late-stage petroleum to displace earlier generated and migrated polar compounds are low. One other effect, however, which can be associated with compositional changes in the migrating petroleum fluids is the precipitation of asphaltenes along the migration pathway.

Potential occurrence of geochromatographic processes in petroleum migration

In the following passages we will attempt to give a brief overview of the potential occurrence of geochromatographic processes in petroleum migration based on observations and experimental evidence. We consider the different combinations of stationary and mobile phases and differentiate between primary and secondary migration. Observed concentration trends in natural systems can be interpreted in terms of different fractionation mechanisms and no ultimate proof can be given. From the geological environment and the reconstructed geological history it may be possible, however, to conclude which chromatographic process may have prevailed.

Liquid–solid geochromatography

The investigation of this mechanism can be traced back to the works of Day (1897) and Gilpin & Cram (1908). Experiments with petroleum and water-free mineral phases reportedly yielded significant fractionation in terms of compound classes (saturate, aromatic, NSO) (e.g. Safronova et al. 1972; Mileshina et al. 1959; Mileshina & Safonova 1963). Molecular fractionation of sterane isomers was observed and quantified by Carlson & Chamberlain (1986) on a water-free montmorillonite stationary phase. Similar effects were observed by Fan & Philp (1987) and Jiang et al. (1988) on water-free alumina columns for various biomarker compounds. The results of Safronova et al. (1972) and Carlson & Chamberlain (1986) show that the presence of only small quantities of water practically eliminates any molecular fractionation effects of saturated hydrocarbons. The effect of water in liquid–solid geochromatographic processes is demonstrated by Brothers (1989) and Brothers et al. (1990). Passage of an *n*-decane-based synthetic oil through a bed of dry montmorillonite/quartz mixture resulted in a complete and irreversible sorption of quinoline and significant retention of the aromatic compounds. Molecular fractionation occurred only within the aromatic compound class, whereas the relative concentrations of saturates remained essentially unaffected. The same experiment performed with a moderately wet stationary phase (4.7 wt% water) yielded a drastically reduced molecular fractionation of the aromatic compounds. The quinoline was still retained completely on the column but a small percentage (6%) could be extracted from the column with methylene chloride. Quinoline and other basic nitrogen compounds are abundant in oil shale retorts and pyrolyzates but show a more limited occurrence in oils (Brothers *et al.* 1990).

Chakhmakhchev *et al.* (1981) detected no changes in *n*-alkane/isoprenoid ratios after petroleum percolation through sand/clay and sand/limestone mixtures and concluded that the 'sorption processes accompanying migration do not appear to alter the original genetically produced hydrocarbon ratios in the oils'. Safronova *et al.* (1972) state that chromatographic separation is substantially reduced in wet cores but is not eliminated completely. They conclude, however, that this reduction can be compensated by large migration distances in natural systems. This argument remains to be substantiated. As an experimental verification appears essentially impossible, theoretical considerations and numerical calculations based on chromatographic concepts could reveal its validity provided the migration pathways, palaeofluxes and petroleum quantities involved can be reconstructed with sufficient precision.

At present it seems to be established that liquid–solid geochromatography can produce compound class (saturate, aromatic, NSO) fractionation accompanied by the corresponding isotope fractionation ('piggy-back isotope fractionation') due to isotopic differences of the compound classes. Experimental evidence for this has been presented by Bonilla (1985) and Bonilla & Engel (1986, 1988), who also report slight isotopic shifts within the compound classes depending on migration distance.

Besides the experimental evidence of compound class fractionation effects, which is often not completely unambiguous, the liquid–solid geochromatographic mechanism is suggested by field observations related to the primary and secondary migration of petroleum. Thus, source rock extracts usually have higher percentages of aromatic and polar compounds than the corresponding reservoir oils (cf. Tissot & Pelet 1971). Furthermore, an increase in paraffinicity with migration distance, as reported by Silverman (1965) for the Qiriqire oilfield in Venezuela and also pointed out by Ritchie (1966, 1973) can be interpreted as a consequence of geochromatographic fractionation (liquid–solid, gas–solid, or gas–liquid) provided alternative fractionation effects (e.g. gravity segregation) can be dismissed.

Seifert & Moldowan (1978, 1981) investigated in great detail the biological marker composition of crude oils and established a number of 'maturation specific' and 'migration specific' compound ratios. Based on field observations and elution sequences on water-free chromatographic columns (alumina with *n*-hexane solvent) they report a 'greater mobility of 5β-steranes over 5α-steranes' (Seifert & Moldowan 1978). According to their 'biomarker migration index' (BMII) concept (Seifert & Moldowan 1981), the preferential retention of the 5α(H), 14α(H),17α(H)-steranes as compared to the 5α(H),14β(H),17β(H) isomers results in a relative enrichment of the ββ compounds in the oil with increasing migration distance and thus indicates the occurrence of geochromatography. This concept is modified but essentially supported by the experimental work of Carlson & Chamberlain (1986) on water-free montmorillonite systems. Hoffmann *et al.* (1984) in a study of biological markers in oils from the Mahakam Delta do not observe this expected enrichment of ββ-steranes in migrated oils. They report, however, an enrichment in monoaromatic hydrocarbons relative to triaromatics which they tentatively ascribe to migration effects.

In conclusion, liquid–solid fractionation effects are the most favoured processes with respect to geochromatography and experimental as well as observational data indicate their occurrence during petroleum migration. However, due to the limited understanding of petroleum migration, it is usually not possible to eliminate alternative geochromatographic or non-geochromatographic fractionation mechanisms.

Liquid–liquid geochromatography

In contrast to liquid–solid geochromatography the liquid–liquid chromatographic mechanism has received only limited attention. The two phases of interest during petroleum migration are petroleum and pore water, each of which may, theoretically, act as a stationary or mobile phase, depending on the wettability and relative permeability regime. Following our definition, the water-washing experiments of Lafargue & Barker (1988) and the process termed 'aqueous fractionation' by Thompson (1989) would be classified as liquid–liquid geochromatography. This fractionation process is mainly considered for secondàry migration as it requires an active hydrodynamic regime to bring about significant fractionation. The experimental results of Lafargue & Barker (1988) reveal that, due to their enhanced aqueous solubility, the lower molecular weight (C_{15-}) saturate components of crude oils, are affected by water-washing. In the (C_{15+}) fraction water washing affects only the aromatic and sulphur-bearing compounds, especially dibenzothiophene. Removal of the more soluble light hydrocarbon compounds produces a decrease in API gravity of the remaining oil. Thompson (1989) addresses the possible implications of 'aqueous fractionation' on chemical maturity parameters.

Liquid–liquid geochromatography with respect to primary migration is apparently judged to be of minor importance, due to the low permeability of source rocks and the relatively small amounts of water involved during the principal phase of petroleum generation.

Gas–solid geochromatography

This class of potential geochromatographic processes involves a mobile gas phase which can be envisaged as composed principally of methane, or carbon dioxide. Minerals and kerogen represent the solid stationary phases during petroleum migration. Experimental approaches to this geochromatographic mechanism were made mainly with respect to the investigation of isotopic fractionations during natural gas migration (e.g. Colombo *et al.* 1965, 1966; May *et al.* 1968). The importance of these isotopic fractionation effects in natural systems is still disputed (cf. Fuex 1980) because the water content of sedimentary rocks renders a gas–liquid fractionation effect more probable. Contradictory results make the judgement somewhat difficult. Thus, Galimov (1967) states that 'heavy methane ($^{13}CH_4$) migrates at higher velocity than light methane ($^{12}CH_4$) in water-wet sand', a finding that is also supported by May *et al.* (1968) for dry media (molecular sieve and sandstone), whereas Colombo *et al.* (1966) report an isotopic fractionation of 3 per mil with the light isotope fraction ($^{12}CH_4$) migrating faster.

Gas–solid geochromatography during primary migration may be envisaged where a significant amount of gas is generated which can act as a carrier phase for higher molecular weight compounds. In most cases, however, in addition to the solid stationary phases (mineral matrix, kerogen) liquid phases will also be present and in equilibrium with a mobile gas phase and the process will be gas–liquid geochromato-

graphy. Another important aspect is the total amount of gas generated as compared to the amount of higher molecular-weight hydrocarbon compounds and this relates to the question whether the gas phase is a mobile phase over extended periods of time or whether molecular diffusion within a largely stagnant gas phase produces the observed fractionation processes.

Gas–liquid geochromatography

As pointed out above, due to the presence of liquid phases in practically all geological environments of relevance for petroleum geochemistry, any process involving gas-phase fractionation effects will correspond essentially to gas–liquid geochromatography. The relevant gas phases are the same as in gas–solid geochromatography. Sorbed films of water or bitumen can be envisaged as stationary phases. Experimental evidence of the occurrence of gas–liquid geochromatography during secondary gas migration was presented by Roper *et al.* (1958). Mixtures of methane and propane were fractionated on passage through reservoir cores and packed columns of crushed reservoir rock impregnated with Ventura crude oil.

In a wider sense the 'separation-migration' mechanism proposed by Silverman (1965) can be interpreted as a gas–liquid geochromatographic mechanism because it involves the escape of an oil-associated cap gas into which the more volatile compounds of the associated petroleum liquid partitioned. Subsequent retrograde condensation at shallower depth and lower pressures are supposed to result in the formation of reservoirs containing preferentially lighter petroleum components. Strictly speaking, the phase from which retrograde condensation occurs in Silverman's (1965) separation–migration model is a supercritical one. Thus, this proposed mechanism represents a SFC type geochromatographic process.

Leythaeuser *et al.* (1987) invoke a gas phase transport of hydrocarbons from source rocks containing type III kerogen to explain significant molecular fractionation effects in the C_{15}–C_{30} range. Evidently, the gas-prone type III kerogen provides good prerequisites for a gas-phase transport of petroleum compounds. As pointed out above, a clear definition between gas–solid and gas–liquid fractionation processes is not possible in geological systems. Thus, the kerogen as well as an adsorbed liquid bitumen phase or both may have acted as stationary phases in this case.

Equilibria between gas and liquid phases in

petroleum accumulations are routinely investigated in the field of petroleum engineering because they are of economic importance for the development of production strategies. The relevance of these gas–liquid equilibria for compositional changes of petroleums and condensates during migration has been examined by various authors (e.g. Zhuze *et al.* 1961, 1977; Thompson 1987; England & Mackenzie 1989; Larter & Mills, this volume). One of the main problems remains the assessment of the 'separation step' (Thompson (1987), i.e. the movement of the saturated gas phase away from the liquid phase in natural systems in geological time, which is a prerequisite for geochromatographic processes.

The results of multiple equilibration experiments as reported, for example, by Thompson (1987) indicated that, in contrast to liquid–solid and liquid–liquid chromatographic processes, the geochromatographic processes involving a mobile gas phase have a clear potential of producing molecular fractionations of hydrocarbons.

Quantitative aspects of geochromatographic processes

While the qualitative changes of petroleum during migration have been recognized relatively early and are essentially agreed upon by many researchers, the establishment of the mechanisms and the quantification of the observed fractionation processes still poses severe problems.

Quantitative approaches to geochromatography have mainly dealt with an assessment of the relative extent of fractionation of different compound classes or even individual compounds. Only very little progress has been made with respect to an absolute quantification.

Relative extent of fractionation processes

Seifert & Moldowan (1981) in their study entitled 'Paleoreconstruction by biological markers' define the 'biomarker migration index' (BMII) as a measure of geochromatographic influence on C_{29} sterane isomer composition in oils. In their conclusion the authors state that

> In spite of the encouraging results on a relative scale, it is not possible to translate the BMII data into absolute migration distances because, at present, the effect of the inorganic matrix is unknown.

The work of Carlson & Chamberlain (1986) constitutes a further important step in the assessment of the selectivity of geochromatographic effects. Their adsorption free energies (AFE), although measured in a very idealized system furnish a means to predict the directions of compositional fractionation of biomarker compounds (compounds with a very high source-specificity and usefulness for correlation purposes) during migration in geochromatographic systems. Although water reduces the sorption capacity and, thus, the fractionation capability of mineral phases drastically, the AFE remain essentially unaffected by increasing water content (Carlson & Chamberlain 1986) and the elution sequence under constant chromatographic conditions remains the same. Similarly, the works of Fan & Philp (1987) and Jiang *et al.* (1988) have established elution sequences of other biomarker compounds relevant in correlation and migration studies.

All of these works, in a sense, correspond to a refinement of the qualitative experience, formulated already by Day (1897), that systematic changes in petroleum composition occur during migration. The results are particularly important for a differentiation between possible migration- and maturity-induced variations in petroleum composition. Even if an absolute quantification of the geochromatographic effects is not possible one can estimate in which direction biomarker compositions would be shifted by migration effects. A brief summary of the results of these studies on relative fractionation parameters is given in Table 3.

Absolute quantification of geochromatographic processes

A quantitative treatment of geochromatography in terms of migration distance and evaluation of absolute losses during migration seems to be out of reach at present. Such a quantification requires the resolution of geochromatographic processes in time and space and can be envisaged only in the context of a comprehensive basin model including primary and secondary migration. One essential problem is the extrapolation of laboratory data to natural systems and geological time. A general statement frequently made in this context is that migration distance in natural systems must make up for the low fractionation capacity of geochromatographic systems, but a quantitative or even semiquantitative test for this postulate in terms of required migration distances has not been presented so far. On the other hand analytical data, for instance

Table 3. *Summary of the relative compositional fractionations during petroleum migration reported in the literature*

Mobility
Decreases with increasing molecular weight
decreases with increasing interaction with stationary phase

Retention of compound groups
Asphaltenes > resins (NSO) > > naphthenes > paraffins
(e.g. Ritchie 1973; Silverman 1965; Mileshina 1963)

Biomarkers ·
C_{29} steranes: (Seifert & Moldowan 1981)
5α, 14β, 17β 20R migrates faster than 5α, 14α, 17α 20R and 5α, 14α, 17α 20S

Steroid biomarkers: (AFE values of 23 compounds on montmorillonite by Carlson & Chamberlain 1986)
C_{27} mono-, and C_{26} di-, and triaromatic steroids: log (retention volume) increases with aromatic carbon number.

14β, 17β steranes move more rapidly through montmorillonite clay than isomeric 5α, 14α, 17α (20R) steranes

20S steroids have consistently smaller retention volumes (smaller AFE values) than corresponding 20R isomers

---

by Leythaeuser *et al.* (1987), Schwarzkopf (1987) and Rückheim (1989), show significant compositional differences of petroleums and rock extracts even in locally restricted systems.

The following considerations outline possible approaches to an absolute quantification of geochromatographic processes and the assessment of their importance in petroleum migration.

First, it is essential to recognize the similarities and differences between analytical and preparative laboratory chromatography and geochromatography. As indicated earlier, one of the specific features of geochromatography is the high degree of nonideality of the chromatographic system involved, which prohibits a detailed theoretical treatment. One other aspect is the essential difference in flow regimes. Elution chromatography in laboratory setups is carried out with a more or less constant flow rate of an excess of mobile phase. The quantity of solutes in the system is usually much lower than the total sorption capacity of the chromatographic system. While mobile phase flow in laboratory systems is usually stationary and may be considered as plug flow to a first approximation, petroleum migration almost certainly involves different flow regimes.

Flow regimes in petroleum migration. Movement of petroleum fluids in the subsurface is generally considered to be buoyancy-driven during secondary migration. During primary migration in fine-grained source rocks, buyancy forces usually will be too small to cause separate phase movement and an additional driving force is required (e.g. compaction or pressure build-up

due to volumetric changes of rock components). Particularly during the initial stages of primary migration the generated petroleum is finely dispersed, and mineral contacts and sorption effects will accordingly be most significant. Later stages of primary migration and expulsion require a saturation of the conducting pore spaces involving a concentration or gathering (focusing) process. If increasing quantities of migrating petroleum pass along the same migration avenue (interpreted here as a chromatographic column), fractionation effects will decrease or disappear, similar to frontal chromatography, after passage of the concentration front. It remains to be established to what extent fractionation effects expected (and observed?) during the initial phase of primary migration will be expressed in the composition of the expelled petroleum or if the onset of the principal phase of expulsion will also lead to a compositional homogenization of the petroleum.

England *et al.* (1987) conclude that in the early stages of accumulation petroleum flows through probably only a small fraction of the gross pore volume of the carrier beds. This is supported by laboratory experiments by Dembicki & Anderson (1989) which indicate that buoyancy-driven flow of petroleum during secondary migration occurs along few selected conduits. Such a flow regime would reduce the losses of petroleum during secondary migration. At the same time the migrating petroleum would interact only with a relatively small fraction of the total mineral surface area in a sedimentary basin. Under these circumstances fractionation effects during this stage of migration can be expected to be minimal.

Concepts for a quantitative treatment. A quantitative treatment of chromatographic processes during large-scale migration in natural systems should be based on the theories of multicomponent chromatography. A recent review by Rudzinski (1987) depicts the inherent difficulties still associated with these approaches. Theoretical treatments of the most commonly used laboratory chromatographic techniques are given by Karger *et al.* (1973).

Because of the lack of experimental data, the complexity of natural geochromatographic systems and other uncertainties a relatively crude model may serve as a first approach to describe and quantify the principal features of a geochromatographic separation or fractionation. The main purpose of such a model would be to:

(1) put the fundamental concepts of geochromatography into an organized form;
(2) indicate the relative importance of different parameters and the sensitivity of the model with respect to variations of these parameters;
(3) serve as a starting point for more refined models;
(4) prompt and direct future experimental work.

The quantitative models by Fuex (1980), Lafargue & Barker (1988) and the approach of Ballentyne *et al.* (1989) are very promising attempts to apply the concepts of separation science to the problems of petroleum migration. Even if an exact quantification cannot be obtained, these calculations are useful tools for establishing the limitations of fractionation processes in geological systems.

To demonstrate the main factors affecting chromatographic fractionation we have set up a simple numerical model for sorptive fractionation of a multicomponent mixture. The basic concept of this model is a one-step equilibration process of a multicomponent mixture with a stationary phase. In terms of petroleum geochemistry this might correspond either to a small volume of rock into which a certain amount of petroleum (multicomponent mixture) is introduced or to a whole sedimentary basin receiving a certain input of petroleum. The amount of petroleum in the system exceeds the amount of available sorption sites which implies simply that not all the generated petroleum is lost on the way, or, in other words, the system is oversaturated. A 'sorptivity factor' gives the ratio of available sorption sites versus total amount of sorbate molecules, which, due to the above assumption, is less than 1. Each molecular species of the multicomponent mixture is attributed a

certain affinity for sorption on the stationary phase. A convenient measure for this sorption affinity is the adsorption free energy (AFE) as determined by Carlson & Chamberlain (1986) for different steroid biomarker compounds. In liquid–liquid geochromatogrphy (water-washing) this affinity corresponds to the aqueous solubility of a given petroleum compound. The probability for a particular component to be sorbed is calculated from its affinity and its relative concentration (mole fraction) in the mixture. The equilibration process is numerically broken down into a large number of steps to avoid instabilities of the algorithms.

In Fig. 3 the results of two sample calculations with this simple model are shown. Both calculations were performed for a mixture of 15 components with different sorption affinities. While in the first run an equimolar starting mixture was assumed with sorption affinities in the same order of magnitude, the second run was performed with one component (component 1) in excess by three orders of magnitude and an affinity about two orders of magnitude higher than the other components (Table 4). Comparison of the initial and final relative composition of the mixtures shows a significant fractionation effect in the first calculation (Fig. 3a) whereas the fractionation effect is almost totally quenched in the second run due to the excess of the high affinity component. Sample calculations like the one shown here can be used to estimate the order of magnitude of fractionation effects that can be expected in geochromatographic systems. It is evident that one critical point in these calculations is the total sorption capacity for petroleum components within a geological system or along a suspected migration pathway. This relates to the yet unanswered question of the amount of petroleum lost during migration and the mode of relocation of petroleum (individual, dispersed stringers or few major migration avenues; early or late focusing of migration currents) during different stages of primary and secondary migration.

Conclusion

The existence of compositional differences between source rock bitumen and reservoir petroleum is now generally recognized and has been confirmed repeatedly. The reasons and mechanisms causing these observed effects, however, are still largely unresolved. Essentially only a limited number of fractionation mechanisms exist, all of which have been recognized and

(a)   (b)

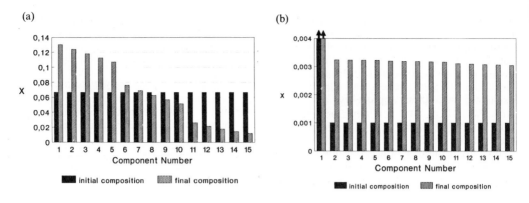

Fig. 3. Numerical simulation of one-step fractionation of a multicomponent mixture; **(a)** equimolar mixture of components with similar affinities; **(b)** excess of one high-affinity component (No. 1, out of range) and its influence on the fractionation of lower-affinity components. (*X*: molar fraction or relative concentration).

Table 4. *Input parameters for multicomponent fractionation model*

Run 1		Run 2	
Relative abundance	Sorption affinity	Relative abundance	Sorption affinity
1	1.1	1000	100
1	1.2	1	1.2
1	1.3	1	1.3
1	1.4	1	1.4
1	1.5	1	1.5
1	2.2	1	2.2
1	2.4	1	2.4
1	2.6	1	2.6
1	2.8	1	2.8
1	3.0	1	3.0
1	4.4	1	4.4
1	4.8	1	4.8
1	5.2	1	5.2
1	5.6	1	5.6
1	6.0	1	6.0

Sorptivity factor: 0.7
Total amount of sorbate: 10000

treated in the field of separation science. With respect to the mathematical treatment it is useful to classify the separation processes into two groups.

(1) Multiphase separation processes (chromatographic processsses) requiring at least one mobile fluid phase and involving a combination of equilibration and bulk phase transport.
(2) Single phase separation processes, (diffusion, thermal diffusion, gravity segretation) which may occur also in different phases but do not require bulk movement of a phase.

Experimental evidence (e.g. Carlson & Chamberlain 1986; Brothers 1989) indicates that

in (water-wet) natural systems geochromatographic processes involving liquid mobile hydrocarbon phases will result only in changes in compound class composition. Molecular fractionation of individual hydrocarbon compounds, as observed in water-free, idealized liquid–solid laboratory systems is usually quenched in the presence of water and an extrapolation of these results to natural systems is difficult. Geochromatographic processes involving gas/solid or gas/liquid partitioning, however, appear to have some potential of creating molecular fractionation processes.

In this paper we have attempted to demonstrate the usefulness of a systematization of the likely fractionation processes in petroleum migration with respect to a comprehensive mathematical tratment. With the increased tendency to include migration processes into the numerical simulation of sedimentary basins it can be hoped that the boundary conditions to model geochromatographic processes will become available. But, as recent modeling approaches have shown, even with relatively crude assumptions it is possible to obtain estimates regarding the efficiency and importance of chromatographic fractionations in geologic systems. One important input parameter for these models is the type and intensity of interaction of different petroleum components with the relevant stationary phases. Substantial experimental effort is still required to investigate and quantify these interactions, particularly with respect to the polar petroleum components.

We wish to thank K.F. Thompson and A.L. Mann for their careful reviews and helpful comments. The editorial comments and suggestions by W.A. England are also gratefully acknowledged. B. Krooss thanks his colleagues at ICH-5 for exposing themselves to an earlier version of this presentation and giving helpful hints.

References

BALLENTYNE, C. J., DEAK, J., DOVENY, P., HORVARTH, F., O'NIONS, R. K. & OXBURGH, E. R. 1989. Systematics of hydrocarbon associated rare gases in the Parmonian Basin. *Terra Abstracts*, **1**, 52–53.

BARTH, T., BORGUND, A. E., HOPLAND, A. L. & GRAUE, A. 1988. Volatile organic acids produced during kerogen maturation—Amounts, composition and role in migration of oil. *Organic Geochemistry*, **13**, 461–465.

BONILLA, J. V. 1985. *Laboratory simulation of crude oil migration: possible implications for oil exploration.* Thesis, University of Oklahoma.

—— & ENGEL, M. H. 1986. Chemical and isotopic redistribution of hydrocarbon during migration: Laboratory simulation experiments. *Organic Geochemistry*, **10**, 181–190.

—— & —— 1988. Chemical alteration of crude oils during simulated migration through quartz and clay minerals. *In*: MATTAVELLI, L. & NOVELLI, L. (eds) *Advances in Organic Geochemistry 1987. Organic Geochemistry*, **13**, 503–512.

BROTHERS, L. 1989. *The effects of fluid flow through porous media on the distribution of organic compounds in a synthetic crude oil.* MS Thesis, University of Oklahoma.

——, ENGEL, M. H. & KROOSS, B. M. 1990. The effects of fluid flow through porous media on the distribution of organic compounds in a synthetic crude oil. *Organic Geochemistry*, (in press).

CARLSON, R. M. K. & CHAMBERLAIN, D. E. 1986. Steroid biomarker-clay mineral adsorption free energies: Implications to petroleum migration indices. *In*: LEYTHAEUSER, D. & RULLKÖTTER, J. (eds) *Advances in Organic Geochemistry 1985*. Pergamon, London, 163–180.

CHAKHMAKHCHEV, V. A. & STEPANOVA, G. S. 1976. Influence of adsorption-chromatographic effects on the composition of oils during secondary processes of migration. *Geologiya Nefti i Gaza*, **7**, 37–42. (in Russian).

——, PUNANOVA, S. A. & ZHARKOV, N. I. 1981. Fil'tratsiya nefti i preobrasovaniye yeye sostava v poristykh sredakh (po resul'tatam eksperimental'-nykh issledovaniy). *Geologiya Nefti i Gaza*, **11**, 23–28. (in Russian). Transl.: Percolation of oil and changes in its composition in porous media (based on experimental studies). *International Geology Reviews*, **25**, (1983), **10**, 1223–1228.

——, ——, TVERDOVA, R. A., ZHARKOV, N. I. & RYABOVA, G. M. 1982. Laboratory studies of the composition changes in petroleum filtering through porous media. *Geokhimiya*, **7**, 1035–1042.

——, BURKOVA, V. N., ZHARKOV, N. I., PUNANOVA, S. A., SEREBRENNIKOVA, O. V. & TITOV, V. I. 1985. Composition changes in vanadyl porphyrins in oil filtering through porous media. *Geokhimiya*, **3**, 381–386.

COLOMBO, U., GAZZARRINI, F., SIRONI, G., GONFIANTINI, R. & TONGIORGI, E. 1965. Carbon isotope composition of individual hydrocarbons from Italian natural gases. *Nature*, **205**, 4978, 1303–1304.

——, ——, GONFIANTINI, R., SIRONI, G. & TONGIORGI, E. 1966. Measurements of C13/C12 isotope ratios on Italian natural gases and their geochemical interpretation. *In*: *Advances in Organic Geochemistry 1964*. Pergamon Press, Oxford, 279–292.

COSTESÈQUE, P., HRIDABBA, M. & SAHORES, J. 1987. Possibility of differentiation of the hydrocarbons by thermogravitational diffusion in a crude oil stored in porous medium. *Comptes Rendus de'l Académie des Sciences Paris*, **304**, Ser. II, 1069–1074.

DAY, D. T. 1897. A Suggestion as to the origin of Pennsylvania petroleum. *Proceedings of the American Philosophical Society*, **36**, 112–115.

DEMBICKI, H. & ANDERSON, M. J. 1989. Secondary migration of oil: Experiments supporting efficient movement of separate, buoyant oil phase along limited conduits. *American Association of Petroleum Geologists Bulletin*, **73**, 1018–1021.

ENGLAND, W. A. & MACKENZIE, A. S. 1989. Some aspects of the organic geochemistry of petroleum fluids. *Geologische Rundschau*, **78**, 291–303.

——, ——, MANN, D. M. & QUIGLEY, T. M. 1987. The movement and entrapment of petroleum fluids in the subsurface. *Journal of the Geological Society*,

London, **144**, 327–343.

FAN ZHAO-AN & PHILP, R. P. 1987 Laboratory biomarker fractionations and implications for migration studies. *Organic Geochemistry,* **11**, 169–175.

FUEX, A. N. 1980. Experimental evidence against an appreciable isotopic fractionation of methane during migration. *In:* DOUGLAS, A. G. & MAXWELL, J. R. (eds) *Advances in Organic Geochemistry 1979.* Pergamon, London, 725–732.

GALIMOV, E. M. 1967. Carbon isotopic variation in methane flowing through wet rocks. *Geochkimiya,* **12**, 1504–1505.

GILPIN, J. E. & CRAM, M. P. 1908. The fractionation of crude petroleum by capillary diffusion. *American Chemical Journal,* **40**, 495–537.

HIRSCHBERG, A. 1984. The role of asphaltenes in compositional grading of a reservoir's fluid column. *Society of Petroleum Engineers of AIME, SPE 13171, 59th Annual Technical Conference, Houston, Texas, September 16–19, 1984,* 1–10.

HODGSON, G. W. & BAKER, B. L. 1959. Geochemical aspects of petroleum migration in Pembina, Redwater, Joffre, and Lloydminster oil fields of Alberta and Saskatchewan, Canada. *American Association of Petroleum Geologists Bulletin,* **43**, 311–328.

HOFFMANN, C. F., MACKENZIE, A. S., LEWIS, C. A., MAXWELL, J. R., OUDIN, J. L., DURAND, B. & VANDENBROUCKE, M. 1984. A biological marker study of coals, shales and oils from the Mahakam Delta, Kalimanta, Indonesia. *Chemical Geology,* **42**, 1–23.

JIANG ZHUSHENG, PHILP, R. P. & LEWIS, C. A. 1988. Fractionation of biological markers in crude oils during migration and the effects on correlation and maturation parameters. *In:* MATTAVELLI, L. & NOVELLI L. (eds) *Advances in Organic Geochemistry 1987. Organic Chemistry,* **13**, 561–571.

JÖNSSON, J. A. 1987. Common concepts of chromatography. *In:* JÖNSSON, J. A. (ed.), *Chromatographic Theory and Basic Principles.* Chromatographic Science Series, **38**, Marcel Dekker, New York, 1–25.

KARGER, B. L., SNYDER, L. R. & HORVATH, C. 1973. *An Introduction to Separation Science.* Wiley & Sons, NY.

KROOSS, B. M. 1987. Experimental investigation of the molecular migration of C_1–C_6 hydrocarbons: Kinetics of hydrocarbon release from source rocks. *Organic Geochemistry,* **13**, 513–523.

—— & LEYTHAEUSER, D. 1988. Experimental measurements of the diffusion parameters of light hydrocarbons in water-saturated sedimentary rocks—II. Results and geochemical significance. *Organic Geochemistry,* **12**, 91–108.

——, —— & SCHAEFER, R. G. 1988. Light hydrocarbon diffusion in a caprock. *Chemical Geology,* **71**, 65–76.

LAFARGUE, E. & BARKER, C. 1988. Effect of water washing on crude oil compositions. *American Association of Petroleum Geologists Bulletin,* **72**, 263–276.

LARTER, S. & MILLS, N. 1991. Phase-controlled molecular fractionations in migrating petroleum charges. *This volume.*

LEYTHAEUSER, D., BJOROY, M., MACKENZIE, A. S., SCHAEFER, R. G. & ALTEBÄUMER, F. J. 1983. Recognition of migration and its effects within two coreholds in shale/sandstone sequences from Svalbard, Norway. *In:* BJOROY, M. *et al.* (eds) *Advances in Organic Geochemistry 1981.* John Wiley, Chichester, 136–146.

——, RADKE, M. & SCHAEFER, R. G. 1984. Efficiency of petroleum expulsion from shale source rocks. *Nature,* **311**, 745–748.

——, SCHAEFER, R. G. & RADKE, M. 1987. On the primary migration of petroleum. *In: Proceedings of the 12th World Petroleum Congress, Houston 1987,* **2**, John Wiley & Sons, London, 227–236.

MACKO, S. A. & QUICK, R. S. 1986. A geochemical study of oil migration at source rock reservoir contacts: stable isotopes *In:* LEYTHAEUSER, D. & RULLKÜTTER, J. (eds) *Advances in Organic Geochemistry 1985 Organic Geochemistry 10,* 199–205.

MARTIN, A. J. P. & SYNGE, R. L. M. 1941. A new form of chromatogram employing two liquid phases. I—A theory of chromatography. II—Application of the micro-determination of the higher monoamino-acids in proteins. *Biochemical Journal,* **35**, 1358–1368.

MAY, F., FREUND, W., MÜLLER, E. P. & DOSTAL, K. P. 1968. Model experiments on isotope fractionation of natural gas components during migration. *Zeitschrift für Angewandte Geologie,* **14**, 376–381.

MILESHINA, A. G. & SAFONOVA, G. I. 1963. Variation in the chemical composition of oils under the influence of the absorbing properties of clayey rocks. *Geologiya Nefti i Gaza,* **8**. (in Russian). Translation: McLean. 1963 Petroleum Geology, **7**, 410–414.

——, —— & KANAYEVA, N. A. 1959. The effect of mineralogic composition of rocks on petroleum filtering through them. *Petroleum Geology,* **3**, 124–129.

MONTEL, F. & GOUEL, P. L. 1985. *Prediction of compositional grading in a reservoir fluid column.* Society of Petroleum Engineers. Report No. SPE 14410.

NAGY, B 1960. Review of the chromatographic "plate" theory with reference to fluid flow in rocks and sediments. *Geochimca et Cosmochimica Acta,* **19**, 289–296.

RITCHIE, A. S. 1966. Chromatography as a natural process in geology. *Advances in Chromatography,* **3**, 119–134.

—— 1973. Fractionation of hydrocarbons and carbon isotopes during migration of oil. *In: First International Geochemical Congress,* **4**, Book 2, Nauka, Moscow, 134–141. (in Russian)

ROPER, W. A., DOSCHER, T. & KOBAYASHI, R. 1958. Evidence of chromatographic effect during flow of gases through oilfield cores. *Journal of Petroleum Technology,* March 1958, 61–63.

RUDZINSKI, W. 1987. Retention in liquid chromatography. *In:* JÖNSSON, J. A. (ed.) *Chromatographic Theory and Basic Principles.* Chromatographic

Science Series, **38**, Marcel Dekker, New York, 245–312.

RÜCKHEIM, J. 1989. *Migrations- und Akkumulations-geschichte der Erdöle des nördlichen Oberrheingra-bens und deren Beziehung zur Diagenese des klas-tischen Speichergesteine.* Thesis, RWTH Aachen, FRG.

SAFRONOVA, T. P. 1981. Der Einfluß von Migrations-prozessen auf die Herausbildung der Erdölzusam-mensetzung. *Zeitschrift für Angewandte Geologie,* **27**, 41–44. (in German).

——, ZHUZE, T. P. & SUSHILIN, A. V. 1972. The effect of chromatographic separation during the mi-gration of oil hydrocarbon mixtures in the gas phase through rocks. *Doklady Akademiya Nauk SSSR,* **207**, 193–196 (Russian).

SCHULTE, A. M. 1980. Compositional variations within a hydrocarbon column due to gravity. *American Institute of Mining, Metallurgical, and Petroleum Engineers, Inc., 55th Annual Fall Technical Con-ference, Dallas, Texas, September 21–24, 1980,* 1–10.

SCHWARZKOPF, T. 1987. *Herkunft und Migration des Erdöls in ausgewählten Dogger beta Lagerstätten des Gifhorner Troges: Wechselwirkungen zwischen Kohlenwasserstoffgenese und Sandsteindiagenese.* Thesis, RWTH Aachen, FRG.

SEIFERT, W. K. & MOLDOWAN, J. M. 1978. Appli-cations of steranes, terpanes and monoaromatics to the maturation, migration and source of crude oils. *Geochimica et Cosmochimica Acta,* **42**, 77–95.

—— & —— 1981. Paleoreconstruction by biological markers. Geochimica et Cosmochimica Acta 45, 783–794.

SILVERMAN S. R. (1965) Migration and segregation of ʼ oil and gas. *In: Fluids in Subsurface Environ-ments—A Symposium.* American Association of Petroleum Geologists Memoir, **4**, 52–65.

SMITH, H. M., DUNNING, H. N., RALL, H. T. & BALL, J. S. 1959. Keys to the mystery of crude oil. *Proceedings of the American Petroleum Institute,* **39**, 3, 433–465.

THOMAS, M. M. 1989. Comments on calculation of diffusion coefficients from hydrocarbon concen-tration profiles in rocks. *Bulletin of the American Association of Petroleum Geologists,* **73**, 787–791.

THOMPSON, K. F. M. 1987. Fractionated aromatic petroleums and the generation of gas condensates. *Organic Geochemistry,* **11**, 573–590.

—— 1989. Comparison of characteristics attributed to migration in petroleums reservoired in clastic and carbonate sequences in the U.S. Gulf Coast re-gion. *14th International Meeting on Organic Geo-chemistry, Paris, Sept. 18–22, 1989,* Abstract No. 20.

TISSOT, B. & PELET, R. 1971. Nouvelles données sur les mécanismes de genèse et de migration du pétrole, simulation mathématique et application á la pro-spection. *Proceedings of the 8th World Petroleum Congress, Moscow,* **2**, 35–46.

ZHUZE, T. P. & BUROVA, E. G. 1977. Influence of different processes of primary migration of hydro-carbons on the composition of petroleums in reservoirs. *In*: CAMPOS, R. & GONI, J. (eds) *Ad-vances in Organic Geochemistry 1975. Proceedings of the 7th International Meeting, Madrid, 16.–19.09.1975.* ENADIMSA. 493–499.

—— & USHKEVICH, G. H. 1961. Characteristics of the gas and liquid phases of petroleum condensate system at great depth. *Neftjanoye Khozjaistvo,* **7**, 30–34.

PART III:
CASE STUDIES

Hydrocarbon generation, migration, alteration, entrapment and mixing in the Central and Northern North Sea

P. C. BARNARD & M. A. BASTOW

The Robertson Group plc, Llandudno, Gwynedd, LL30 1SA, UK

Abstract: There is little doubt that sediments of Upper Jurassic to lowermost Lower Cretaceous age, particularly those of the Kimmeridge Clay Formation and its equivalents, are the source of the vast bulk of the oil in the Central and Northern North Sea. This is true both of oil in conventional Mesozoic plays and in Tertiary clastic reservoirs. The timing of oil generation and migration ranges from the Late Cretaceous through to the present.

The oil productivity of the Kimmeridge Clay Formation in the Central Graben, Moray Firth and South Viking, Graben areas of the North Sea is estimated to be in excess of 250 billion barrels, of which about 25% is accounted for in accumulations discovered to date. There is strong areal differentiation between oil trapped in Mesozoic reservoirs as opposed to Upper Cretaceous Chalk and Tertiary reservoirs.

This paper is particularly concerned with hydrocarbons in Tertiary reservoirs, which have a wide range of compositions and appear to have complex accumulation and degradation histories. Migration to Chalk and younger Tertiary reservoirs occurs often through thick shale sequences. The evidence for the timing and mechanism of this process is considered and a preferred model for the vertical migration and subsequent lateral migration, mixing and degradation is proposed.

The North Sea oil province after 20 years of exploration and production remains a very active oil exploration area. During this time the major tectonic and structural factors controlling sedimentation of potential petroleum reservoir and source rocks have been investigated and are now fairly well understood. As a result of this intensive effort new discoveries continue to be made and new play concepts developed. This paper attempts to synthesise data on the distribution and hydrocarbon productivity of the major oil source rock unit, the Kimmeridge Clay Formation and its equivalents, over an area of the Central and Northern North Sea between 56°N and 60°N. Comparison with known in-place volumes of oil and gas shows that, in the sub-basins throughout this area there is always a substantial excess of generated over discovered hydrocarbons. There is considerable variation in the physical and chemical properties of those hydrocarbons, particularly when they are found in the younger Tertiary age clastic reservoirs. Exploration plays based on these Tertiary sandstones need to consider a model which accounts for that variation and predicts the likely physical and chemical properties of hydrocarbons in any prospect identified.

The major North Sea oil source rock

The Upper Jurassic–Lower Cretaceous age Kimmeridge Clay Formation and its equivalents are

the major source of oil in the Central and Northern North Sea (Fig. 1), as has been described and documented in numerous publications and studies over the past 10–15 years (Fuller 1975; Oudin 1976; Brooks & Thusu 1977; Barnard & Cooper 1981; Barnard *et al.* 1981; Demaison *et al.* 1983; Cornford *et al.* 1983; Goff 1983; Pearson & Watkins 1983; Reitsema 1983; Cooper and Barnard 1984; Field 1985; Huc *et al.* 1985; Thomas *et al.* 1985; Baird 1986; Cornford 1990). This statement, however, needs considerable clarification if we are to define where in the North Sea the Kimmeridge Clay Formation is effective as the major oil source rock. No regional scale maps of the Kimmeridge Clay Formation and its equivalents showing its distribution, thickness and sedimentological and organic geochemical facies variations have been published. The isopach map shown in Fig. 2 is based on over 600 wells in the area (Robertson Group 1985, 1987, 1988, 1989, 1990) and seismic data where on rare occasions intra-Jurassic reflectors approximating to the base of the Kimmeridge Clay Formation are recognized. This map shows that the major graben bounding faults are not the only features controlling the thickness of deposited or preserved Kimmeridge Clay and that there are significant intra-graben highs separating the depocentres. The mapped thickness variations, the curvilinear nature of many of the bounding faults at base Cretaceous level and in-house structural analysis strongly suggest that there is a small but significant wrench component in the

From England, W. A. & Fleet, A. J. (eds), *Petroleum Migration*
Geological Society, Special Publication No. 59, pp. 167–190.

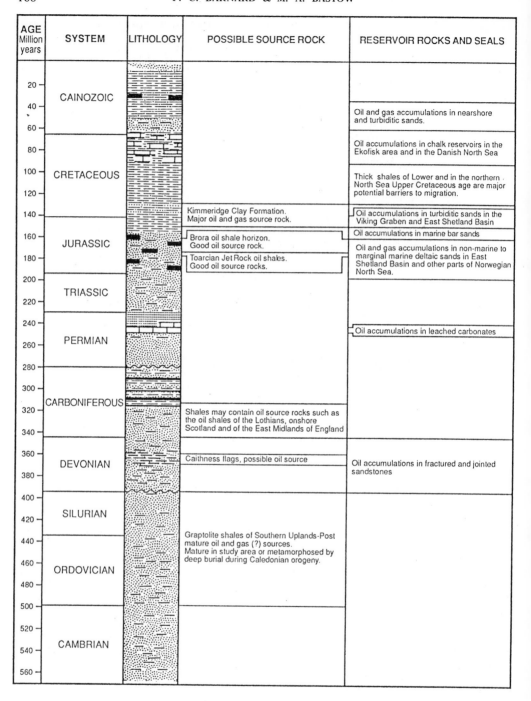

AGE Million years	SYSTEM	LITHOLOGY	POSSIBLE SOURCE ROCK	RESERVOIR ROCKS AND SEALS
20–60	CAINOZOIC			Oil and gas accumulations in nearshore and turbiditic sands.
80	CRETACEOUS			Oil accumulations in chalk reservoirs in the Ekofisk area and in the Danish North Sea
100–120	CRETACEOUS			Thick shales of Lower and in the northern North Sea Upper Cretaceous age are major potential barriers to migration.
140			Kimmeridge Clay Formation. Major oil and gas source rock.	Oil accumulations in turbiditic sands in the Viking Graben and East Shetland Basin
160	JURASSIC		Brora oil shale horizon. Good oil source rock.	Oil accumulations in marine bar sands
180	JURASSIC		Toarcian Jet Rock oil shales. Good oil source rocks.	Oil and gas accumulations in non-marine to marginal marine deltaic sands in East Shetland Basin and other parts of Norwegian North Sea.
200–220	TRIASSIC			
240–280	PERMIAN			Oil accumulations in leached carbonates
300–340	CARBONIFEROUS		Shales may contain oil source rocks such as the oil shales of the Lothians, onshore Scotland and of the East Midlands of England	
360–380	DEVONIAN		Caithness flags, possible oil source	Oil accumulations in fractured and jointed sandstones
400–420	SILURIAN			
440–480	ORDOVICIAN		Graptolite shales of Southern Uplands-Post mature oil and gas (?) sources. Mature in study area or metamorphosed by deep burial during Caledonian orogeny.	
500–560	CAMBRIAN			

Fig. 1. Stratigraphic summary of potential source rocks, reservoirs and seals in the Central and Northern North Sea.

formation of the graben systems which has important implications for models of the amount of extension in the system, the associated heat flux and the later renewed movement and subsidence over these deep seated faults.

Geochemical data for around 220 wells are

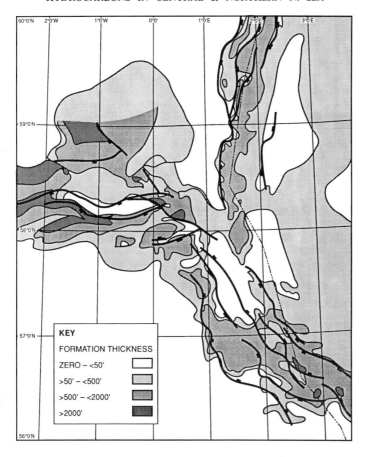

Fig. 2. Kimmeridge Clay Formation isopach.

available to us as a result of major regional geological studies of the area described in this paper (Robertson Group, 1985, 1987, 1988, 1989, 1990). This database has enabled us to map the organic facies associated with the sediments shown in the isopach map (Fig. 2) together with the level of maturation achieved by the sediments (Fig. 3). It can be seen that the top of the Kimmeridge Clay Formation at present shows marked variations in maturity, from immature through oil mature and to gas mature in the deepest parts of the basin. This map does not indicate the maturity at the base of the Kimmeridge Clay Formation since, although in principle knowing the thickness variation in the Kimmeridge Clay Formation it should be trivial to extrapolate to maturity at the base of the Formation, analytical data show that maturity profiles are frequently not linear in the source rocks in the deeper parts of the graben system, for reasons which are not immediately obvious. The extrapolation difficulties are such that we have not been able satisfactorily to model the

maturation history of the base of the Kimmeridge Clay Formation and thus in this paper all reference is to the maturation of the top of the source rock interval. The implications for hydrocarbon generation in the source rock are discussed quantitatively. Thus, in some areas where Fig. 3 suggests it is immature, the Kimmeridge Clay Formation may be oil mature at its base and in other areas where Fig. 3 suggests it is oil mature, it may be gas mature at its base. The map does not give any indication of the time when the Kimmeridge Clay Formation entered the oil window or the source area from which oil or gas was available to migrate.

Kimmeridge Clay Formation oil productivity

It is possible to combine geochemical data on source bed richness (total organic carbon and Rock-Eval pyrolysis) and maturity, thickness and kerogen facies to arrive at an estimation of the oil productivity of a source rock. These oil

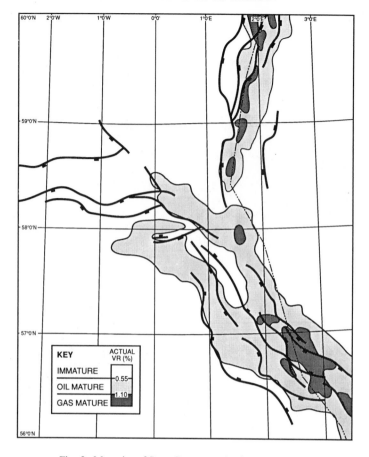

Fig. 3. Maturity of Base Cretaceous horizon at present.

productivities are often expressed volumetrically as barrels of oil per acre foot (bbl/acre ft), or where thickness is known, as barrels of oil per acre (bbl/acre) of source rock. The productivities are intended to identify the volume of oil rather than gas generated from the source rocks during the maturation process. There is, however, some uncertainty as to how much cracking of oil to gas or direct generation of gas from the kerogen occurs.

This procedure has been routinely applied by the geochemistry division of Robertson Group in five regional North Sea source rock studies between 1985 and 1990 and the results are shown as a map of oil productivity in the area under discussion (Fig. 4). This shows a very distinct differentiation across the study area between areas of poor or negligible hydrocarbon productivity and highly prolific basinal areas or 'hydrocarbon kitchens'. These relatively localized kitchen areas are the main hydrocarbon generative areas and are the source of the bulk of the oil

found in the North Sea basin. For the purposes of the discussion in this paper these sub-basins have been named as the Ekofisk–Montrose kitchen, a large area of deep, thick source rock with very high hydrocarbon productivity, the Buchan–Wytch Ground Graben kitchen, the major source area for much of the Outer Moray Firth oil fields, the Fisher Bank Basin, a relatively localised but still prolific kitchen area, and the South Viking Graben, which is a long, ribbon-like feature of highly productive hydrocarbon source rocks. The total volume of oil generated in each of these areas has been estimated and is shown in Table 1.

Timing of oil and gas generation

An important factor in understanding the likely hydrocarbon type and charge in a prospective structure is the ability to predict when the source rock entered the oil window and when the result-

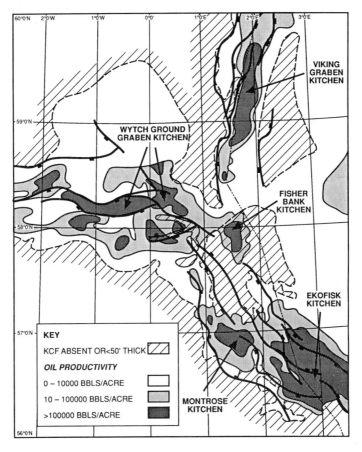

Fig. 4. Oil productivity of Kimmeridge Clay Formation to date.

Table 1. *Hydrocarbon reserves and source rock productivity*

Kitchen area	Reserves	Mesozoic		Tertiary		Total		Estimated productivity of oil
		Oil	Gas	Oil	Gas	Oil	Gas	
Ekofisk/Montrose	Recoverable	1.3	1.7	7.0*	18.4*			
	In-place	3.8	2.1	20.9*	22.1*	24.7	24.2	171
Buchan/ Wytch Ground Graben	Recoverable	3.0	0.6	–	–			
	In-place	9.1	0.7	–	–	9.1	0.7	42
Fisher Bank	Recoverable	0.1	0.6	0.5	1.6			
	In-place	0.4	0.7	1.4	1.9	1.8	2.6	13
South Viking Graben	Recoverable	6.7	26.0	1.6	15.8			
	In-place	20.0	31.1	4.9	18.9	24.9	50.0	120

Units: oil, billion barrels; Gas, trillion cubic feet

* Tertiary includes Upper Cretaceous Chalk reservoirs

Conversion factor for recoverable reserves to in-place reserves: oil, 3.33; gas, 1.25

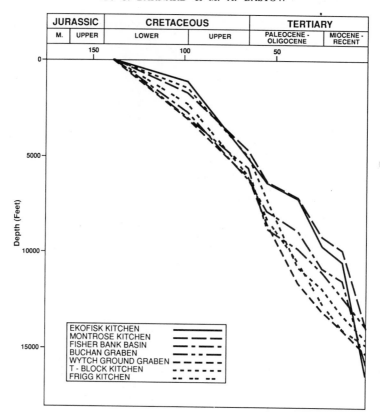

Fig. 5. Burial history curves for selected points in the kitchen areas.

ant generated hydrocarbons were available to migrate. Such modelling is a difficult exercise requiring an understanding of the thermal history of the source rock and the relationship of oil generation and expulsion to temperature and time. A number of authors have carried out such calibrations on a regional scale (Goff 1983; Baird 1986; Dahl *et al.* 1987; Doligez *et al.* 1987) particularly in the Viking Graben area.

Such studies are initiated by an analysis of the subsidence profile of wells in the depocentres and kitchen areas. Profiles for 7 synthetic well sections based on seismic data interpretation in basinal situations are shown in Fig. 5. The time–temperature integral approach (Lopatin 1971; Waples 1980; Royden *et al.* 1980; Goff 1983; Wood 1988) provides a reasonable approximation and has been the usual method in such studies. Our calculations are modified to take account of sediment compaction and time-variable sediment–seawater interface temperatures. The geothermal gradients seen in the study area (typically around 33°C/km) are generally below average for sedimentary basins worldwide. For

modelling purposes the associated heatflow is the important parameter and this we have considered has been approximately constant through time, a not unreasonable assumption for modelling the maturation of Upper Jurassic source rocks in the North Sea (Leadholm *et al.* 1985) since amounts of extension are now recognized to be quite small (Thorne & Watts 1989) and maturation is achieved relatively late in the history of the basin.

Simple observation of the subsidence profiles (Fig. 5) shows that the time of entry of the top of the source rock interval into the critical depth/temperature range corresponding to significant generation of oil varies from about 65 to 30 Ma before present (South Viking Graben kitchen to Ekofisk kitchen). This, however, does not allow for the fact that the base of the source rock, will have entered the oil generation window even earlier, depending on the thickness of the Kimmeridge Clay Formation.

Our modelling of the development of maturity in the Kimmeridge Clay Formation has shown that over much of the Central and Northern

Table 2. *Timing of initiation and peak hydrocarbon generation in the basinal areas of the Central and Northern North Sea*

Hydrocarbon zone Basinal area	Initiation of oil generation		Peak oil generation		Initiation of gas generation	
	Base	Top	Base	Top	Base	Top
Ekofisk	Lower Palaeocene	Lower Eocene	Lower Oligocene	Oligocene– Miocene	Upper Miocene	*
West Montrose	Lower Eocene	Middle Eocene	Oligocene– Miocene	Lower Miocene	*	*
Buchan Graben	Maastrichtian	Upper Palaeocene	Lower Eocene	Middle Miocene	Lower Miocene	*
Wytch Ground Graben	Campanian	Lower Palaeocene	Lower Eocene	Middle Eocene	Lower Oligocene	Upper Oligocene
Fisher Bank	Maastrichtian	Lower Palaeocene	Middle Eocene	Upper Eocene	Lower Miocene	Pliocene
T-Block	Upper Palaeocene	Lower Eocene	Middle Eocene	Upper Eocene	Middle Miocene	Lower Miocene
Frigg	Maastrichtian	Upper Palaeocene	Middle Eocene	Middle Eocene	Upper Oligocene	Lower Miocene

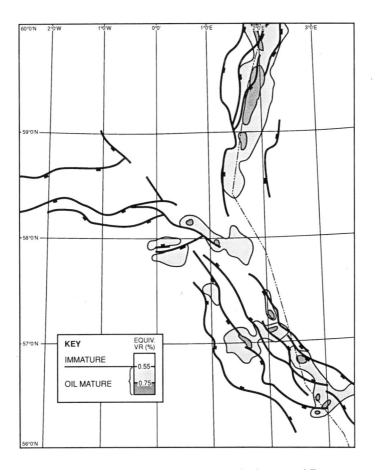

Fig. 6. Modelled maturity of Base Cretaceous horizon at end Eocene.

North Sea oil generation was initiated in the latest Cretaceous to Early Tertiary (Table 2) and that the source rocks had reached a sufficient level of maturity for major oil generation and consequent expulsion to occur by the end of the Eocene, as has also been demonstrated by reservoir diagenetic studies (Liewig *et al.* 1987, Burley *et al.* 1989). A map showing modelled vitrinite reflectivity equivalent maturity for the top of the source rock unit at this time is shown in Fig. 6, but again it needs to be realised that the base of the unit will be significantly more mature and the areal extent of mature source rock greater. Large volumes of oil would have been available both to fill Jurassic structures and collect at culminations awaiting the opportunity to migrate vertically by the end of the Eocene. Our modelling also shows that, coincident with the generally enhanced sedimentation rates in the mid–late Tertiary, significant gas generation in the Kimmeridge Clay Formation occurred when higher formation temperatures were attained, beginning in the Late Oligocene to Miocene (Eggen 1984).

The migration problem

The migration of hydrocarbons from source rock to accumulation has been described as a two part process; an initial primary migration within the fine grained argillaceous source bed sequence and a secondary migration within the relatively coarse grained arenaceous conduit or carrier bed. The mechanism by which these migration processes take place has been much discussed but in many aspects considerable doubt still remains. The following discussion attempts to describe the important mechanisms effective in the North Sea Basins. There is unlikely to be conclusive proof of the exact mechanisms operative regionally because of the difficulty of simulating in the laboratory processes taking place naturally over geological time.

Before discussing the details of migration mechanisms, however, it is necessary to describe the nature and scale of the migration problem. Hydrocarbon accumulations in the North Sea can be divided into those in which there is a close proximity between the likely source and the

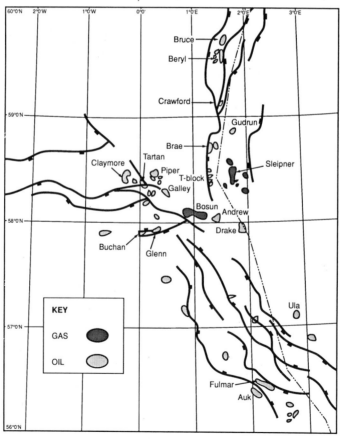

Fig. 7. Hydrocarbon accumulations in Mesozoic (Pre-Chalk) reservoirs.

reservoir (e.g. Ula, Piper or Fulmar sands acting as reservoirs for oil generated in the enclosing or adjacent Kimmeridge Clay Formation, Turner *et al.* 1984; Brae Formation sands as proximal turbidite deposits receiving hydrocarbons from the distal equivalent Kimmeridge Clay Formation, Leythauser *et al.* 1988*a,b*; Mackenzie *et al.* 1987, 1988; Middle Jurassic sands faulted against Upper Jurassic source rocks) and those in which the source and reservoir are widely separated in depth and time (e.g. Ekofisk/Tor Chalks, Munns 1985; Forties Formation sandstones, Cornford 1990; Frigg Formation sandstones).

The distribution of hydrocarbon accumulations in the North Sea is shown in Figs 7 and 8, in which they have been differentiated by whether they occur in Mesozoic (pre-Chalk reservoirs) or in Chalk and Tertiary reservoirs. The volumes of oil trapped in these different systems have been calculated using published and scouted reserve estimates and are shown in Table 1, where they are distinguished on the

same geographic basis as was used in defining the major kitchen areas. This exercise was completed by converting the recoverable reserves quoted in the literature to an in-place volume estimate using approximate recovery factors of 30% for oil and 80% for gas.

This is instructive in that it shows that, if one assumes that exploration has so far been equally effective at finding hydrocarbons in reservoirs of all ages, then there is marked regional variation in the extent of vertical migration into Chalk and younger reservoirs as opposed to relative lateral migration into Jurassic traps. This is particularly obvious in the Central Graben where the vast bulk of oil and gas is trapped in the Chalk and Tertiary reservoirs. Given that the Viking Graben is apparently a very prolific kitchen area, the relative lack of hydrocarbons in the Tertiary can be viewed as a result of either a much more effective trapping system within the Jurassic or a much less effective trapping system in the Tertiary.

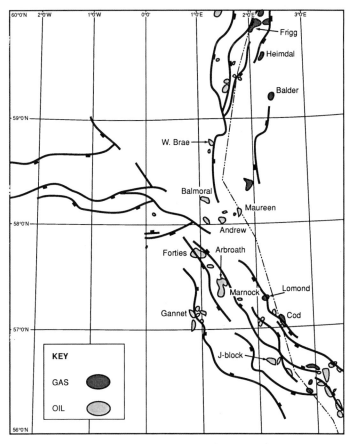

Fig. 8. Hydrocarbon accumulations in Tertiary and Chalk Reservoirs.

Primary migration

Primary migration is the process of mobilization and migration of oil within the fine grained source rock to reach the point where it can then move relatively freely in the more porous and permeable carrier bed (Chapman 1972). Fine-grained sediments when deposited are water wet, that is all the pore spaces are filled with water. During burial the sediment particles are compacted together and water is expelled. The porosity decreases from initially very high values of about 60% in a mud to values of 5–10% in a mudstone at depth of say 9000 ft under normal lithostatic pressure, while permeabilities which are initially low become extremely low.

Hydrocarbons typically start to be liberated in source rocks at around 7500 ft to 9000 ft in the North Sea basin and initially saturate the kerogen from which they are derived. Many hypotheses have been proposed to explain how the liquid hydrocarbons, which are by nature hydrophobic and contain at the early stage of generation relatively very large molecules, can pass through very small pore throats against very high capillary pressures at a significant rate in geological time. The balance of evidence is strongly against either simple diffusion through pore throats or water phase solution. The two most likely mechanisms are migration by diffusion along continuous oil wet kerogen laminae or sheets (such as are observed to exist in rich source rocks) (Ungerer et al. 1984), and high-pressure microfracturing of the fabric of the source rock (Palciauskas & Domenico 1980; Ozkaya 1988a,b), which produces a chaotic micro-structure in the sediment the effects of which have been seen under the microscope in polished rock sections. The discussion in this paper is not attempting to present or discuss the evidence for these processes but we do believe that both are observable in the Kimmeridge Clay Formation in the North Sea basin.

The simplest migration phenomena which can be fairly easily understood are provided by fields such as the Brae complex where proximal turbidites pass laterally into distal fine grained source beds. Migration is presumably driven by the fluid potential gradient existing at the time of oil generation laterally towards the more permeable and coarser grained facies. However, in the South Viking Graben area, there is a strong asymmetry in the basin morphology which has persisted through time and the whole section in the Jurassic rift is sealed by thick Cretaceous shales. Oil generated in the Kimmeridge Clay east of the major north–south synclinal axis could not migrate vertically across the bedding

in the effectively sealed and highly pressured system. In this situation we suggest that migration will preferentially occur laterally, though in a generally updip direction, in a continuous oil wet organic kerogen network assisted by pressure. Porosities in these source beds calculated from petrophysical analysis are at least 10% and maybe up to 15%, even when at more than 9000 ft depth. Broad structural culmination in shales at base Cretaceous level therefore could become reservoirs for large volumes of oil and act as gathering points for hydrocarbons awaiting a trigger to undergo vertically long distance migration to more porous beds. Such culminations can be mapped throughout the North Sea basin and when drilled may appear as shows at the source rock level.

Secondary migration

Secondary migration is generally defined as the process by which hydrocarbons migrate through the relatively porous and permeable conduit or carrier beds, which at culminations may act as reservoirs.

The possible mechanisms for secondary migration have been thoroughly reviewed in the literature (Momper 1978; Schowalter 1979; Jones 1981; Durand 1988) and it is generally concluded that the secondary migration process begins only after sufficient oil has filled the pore spaces in the conduit at its surface interface with the fine grained source rock. Secondary migration of hydrocarbons within the carrier bed is controlled primarily by the buoyancy of the hydrocarbon phase, by the hydrodynamic flow, driven by gradual compaction and porosity loss in the sediments with increasing burial, and is resisted by capillary pressures (England et al. 1987). A sufficient amount of oil is needed for buoyancy to overcome the impeding capillary pressure at the pore throats in the conduit rock. Once this point is reached, secondary migration takes place relatively rapidly and in a discrete and localised pathway (Dembicki & Anderson 1989), controlled principally by buoyancy and to a lesser extent by permeability. Fluid conditions within the relatively shallow Tertiary sandstones at the time of oil migration in the North Sea would have been essentially hydrostatic (Chiarelli & Richy 1984), and therefore the direction of migration would be controlled mainly by the buoyancy of the hydrocarbons and the permeability of the rocks through which the hydrocarbons passed. In the deeper Mesozoic section, the net force acting on the hydrocarbons would have incorporated a greater contribution from the fluid potential gradient, which was influenced by the distribution of relatively impermeable seals and the distribution of overpressures.

In situations where there is a clearly defined interface between source bed and conduit/carrier bed the mechanism defined above is simple to comprehend. However, in much of the North Sea such a simple situation does not exist. As previously described, the Kimmeridge Clay Formation is often separated from the underlying reservoir (e.g. the Brent Sand Formation in the Brent/Statfjord Fields) by a thick Heather Formation shale unit. In much of the Central and Viking Grabens, the oil is located in Upper Cretaceous or Tertiary reservoirs, separated from the Upper Jurassic source beds by the great thicknesses of shale. The situation has been described by various authors but few have tried to explain the problem, and the migration of Jurassic-sourced hydrocarbons into the Paleocene and Eocene remains poorly understood (Parsley 1986), although it is considered that the basin structure and fault pattern are important controls. Goff (1983) analysed the pressure conditions in the Frigg area and concluded that, while microfracturing could assist in migrating gas through the lower part of the Cretaceous shales, it was not a valid mechanism in the shallower section (less than 3500 m), so it was postulated that aqueous solution occurred in the shallower part followed by exsolution of gas in the Tertiary sandstones. This mechanism seems unlikely enough for gas but of course there are many instances where major amounts of oil have managed to migrate to the Tertiary through thick shale sections as in the Frigg Field. It is therefore necessary to consider an alternative, more plausible mechanism. We suggest that tectonically produced faults and fractures and differential compaction over basement highs provide a more realistic mechanism for this major migration phenomenon. This approach was described by Leonard (1984) who showed that where there was seismic evidence of fault connection between identified kitchen areas and chalk structures, the chance of exploration success was 40% compared to less than 10% where there was no fault conduit. Bjorlykke et al. (1988) suggest that oil migration into shallower reservoirs was not associated with large quantities of pore water from the deeper parts of the basin and that migration took place along narrow pathways where high oil saturations and high relative oil permeabilities could be established.

Evidence of vertically migrating hydrocarbons has frequently been seen in the North Sea most often as velocity sags on seismic records over massively leaking structures and fields (e.g. Block 25/2, Buhrig 1989). We interpret these as instances of active present day migration or leakage, while what we are seeking are previously active vertical conduits. Our approach to the problem has been to identify where hydrocarbons have been generated and then accumulated within the Jurassic system, prior to migration into the Tertiary.

Where migration conduits within the Mesozoic terminated against faults, it is assumed that some leakage of hydrocarbons into the fault plane may have occurred, but that the bulk of the hydrocarbons would have been diverted to follow the structural dip along the plane of the fault towards structural culminations within the Jurassic section. At these culminations, more vigorous upwards migration would be promoted by the continuous availability of hydrocarbons and the pressure differential due to the oil column, thus creating hydrocarbon chimneys. Applications of these principles using detailed regional seismic maps has shown a number of major structural culminations spread through the South Viking Graben area. Many of these culminations are the result of deep seated tectonism and overlie Jurassic faults which have been reactivated by Early Tertiary movements or by differential compaction over basement high blocks. Maturation modelling studies indicate that source rocks will have generated sufficient oil to create hydrocarbon accumulations at Jurassic level beginning in the latest Cretaceous but mainly in the latest Palaeocene to Eocene, possibly to be released as a result of Early Tertiary fault movements. Although many of the faults appear to die out within the Cretaceous section, it is likely that zones of fractured rock extend into the Tertiary and that these fault-related fracture systems facilitated vertical migration. Fracturing of the Cretaceous occurs over the graben edge fault at the point of maximum flexuring (Cayley 1987). In the Central Graben diapiric salt has provided an additional effective mechanism for faulting and opening up of vertical paths for hydrocarbons sourced from Upper Jurassic shales to Chalk or younger reservoirs (Cayley 1987). In many cases the porous Chalk structures are themselves overlain by thick shales so that no further vertical or lateral migration has taken place. In the South Viking Graben and Moray Firth areas as well as in the more northerly part of the Central Graben there have been many more migration options open to hydrocarbons once they reach the relatively open Tertiary system.

The dimensions of these vertical migration

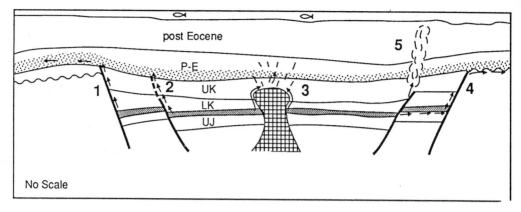

Fig. 9. Cartoon showing some possible secondary migration mechanisms from the Upper Jurassic to the Palaeocene and Eocene reservoirs.

1, Migration via graben bounding faults which cut the Palaeocene. Migration occurs from hydrocarbon chimneys close to the fault where hydrocarbons have accumulated before migrating up fault conduits.

2, Migration via fractures occurring above a deeper fault block which when reactivated, causes fracturing of the overlying strata.

3, Migration via fractures produced in response to halokinesis.

4, Migration via faults and then along unconformity surfaces possibly within the Palaeocene and Eocene.

5, 'Gas chimneys'.

N.B. Once in the carrier horizon/possible reservoir petroleum can migrate over short or long distances until a suitable culmination is encountered or it is lost by seepage at the surface.

paths or hydrocarbon chimneys overlying Jurassic culminations may be very limited and their functioning intermittent as pressure builds-up or faults move. Both Goff (1983) and Buhrig (1989) have described the pressure constraints necessary to migrate hydrocarbons vertically and these discussions both illustrate that very high overpressures generated and maintained by gas generation are necessary to initiate the process in a fully sealing shale section. However, the need is for a mechanism to migrate black oils vertically during the Early Tertiary and it is unlikely that massively high gas pressures existed at that time. We therefore are more inclined to the role of faults and fractured zones as the mechanism which initiates the chimney effect. A schematic representation of the possible mechanisms of secondary migration from Jurassic source beds to Tertiary conduits and reservoirs is shown in Fig. 9.

Oil composition

The composition of oils in Tertiary reservoirs is widely varying from low gravity oils in the 13–16° API range to light condensates in the 50° + API range. This range of gravities is greater than is seen generally in Jurassic reservoirs. The purpose of this discussion is to confirm the origin of the oil, to describe the variations in properties and to provide a mechanism which can account for the observed variations.

Figure 10 shows the variations in API gravity with depth for oils in the South Viking Graben area. It can be seen that there is a trend to lower API gravities with decreasing depth which is generally restricted to Tertiary reservoirs and to reservoir depths less than 6500 ft. Chromatography analysis of oils tested from Mesozoic and Tertiary reservoirs and of extracted oil shows in Tertiary reservoirs shows that Tertiary reservoired hydrocarbons are slightly richer in polar hydrocarbons and this is particularly pronounced in extracted oil shows (Fig. 11).

Gas chromatograms of a selection of oils from Tertiary reservoirs are shown in Fig. 12. Four of the five chromatograms represent analyses of the whole oil so that hydrocarbons from butane (when present) up to n-C_{32+} can be seen, while the last is of the alkane fraction of an oil show. The first chromatogram is similar to a typical North Sea light oil with a complete range of n-alkanes and an API gravity of 35.7°. The oil from UK3/25a-2, although having a gravity of 39.6° API, would appear to have had a complex alteration and accumulation history. The oils from the Balder Field (25.4° API) and UK9/19-4 (17.5° API) are also complex mixtures whose compositions we attribute to a combination of bacterial degradation, water flushing and probably gas stripping processes, all of which have

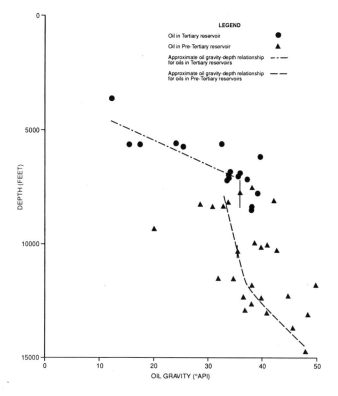

Fig. 10. Oil gravity (°API) versus reservoir depth: South Viking Graben area.

played a part in altering the more usual North Sea oil composition (Cornford *et al.* 1983; Brooks *et al.* 1984). The extracted oil stain from UK9/27-1 did not flow on test and has clearly been extensively degraded.

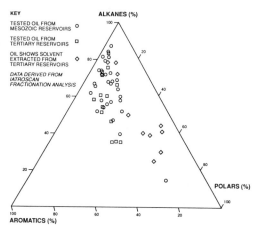

Fig. 11. Composition of oils and oil stains from Mesozoic and Tertiary reservoirs: South Viking Graben area.

Detailed oil–oil correlation studies, particularly using source specific biomarkers such as the relative abundance of C_{27}, C_{28} and C_{29} steranes (Fig. 13) and the presence of other markers such as C_{28} bisnor-28,30-hopane confirm that, despite the variation in physical and chemical properties seen in the Tertiary oils, they are derived from the same Upper Jurassic Kimmeridge Clay Formation source rocks as the bulk of the Jurassic reservoired oils (Mackenzie & Maxwell 1983; Cooper & Barnard 1984; Northam 1985; Schou *et al.* 1985; Cornford 1990).

The process of bacterial degradation or alteration of oils has profound affects on the physical and chemical properties of oils (Evans *et al.* 1971; Bailey *et al.* 1973a, b; Cornford *et al.* 1983; Connan 1984; Palmer 1984; Lafargue & Barker 1988) and these can be monitored in a number of different ways. From the practical and commercial point of view, degradation results in increased density (lower API gravity) and concomitant increase in viscosity. These two factors generally serve to reduce the producibility of oil from reservoirs, unless there are compensating factors such as increased reservoir permeability. It is an extremely fortunate fact that some of the

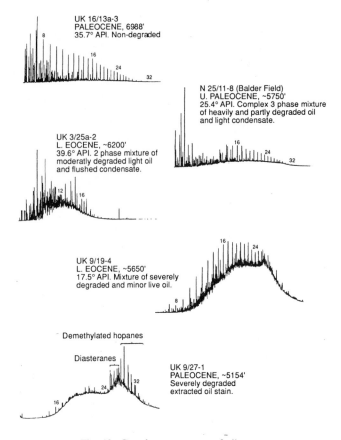

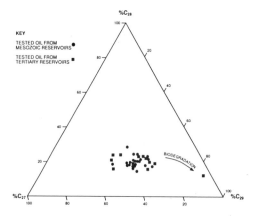

Fig. 12. Gas chromatograms of oils.

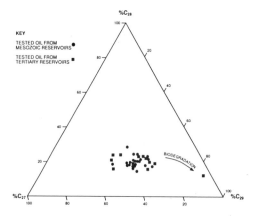

Fig. 13. Composition of C_{27}, C_{28} and C_{29} 5α(H)-14α(H)17α(H)20R steranes in oils: South Viking Graben area.

sandstone reservoirs in the Tertiary of the North Sea are remarkably porous and permeable. An illustration of the effect of reduction in produci-

bility is shown by plotting viscosity against temperature (Fig. 14), where the dramatic increase in viscosity with decreasing gravity can be seen. However, it should be remembered that all these oils flowed on test at reservoir temperatures at significant rates (The Bressay Field oil in UK3/27-1Z at present is unique in having flowed at over 3000 barrels oil per day under pump test although of only 12.5° API + with a viscosity at reservoir conditions of around 1000 centi-Stokes).

The effects of degradation on the geochemical characteristics of oils, including molecular biomarker compositions and carbon isotope ratios, has been noted by various authors (Reed 1977; Seifert & Moldowan 1979; Stahl 1980; Rullkotter & Wendisch 1982; Didyk *et al.* 1983; James & Burns 1984; Sofer 1984; Wehner *et al.* 1986) among whom Volkman *et al.* (1983) were able to identify characteristic changes in certain components of oils which were amenable to classification in the biomarker spectrum. Their classification has been expanded using a database of

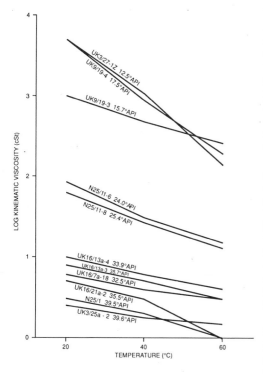

Fig. 14. Viscosity–temperature curves for representative oils from Tertiary reservoirs.

North Sea oils (Table 3) such that we are now able to elaborate the scheme to include a numerical scale which can be useful in mapping such changes and an API gravity or range of gravities to each level of degradation. There are complications of course to any such scheme which include the possibility of mixing of oils which have experienced different degrees of degradation and of adding to a degraded oil a light non-degraded oil. Both of these complications are seen to occur in the North Sea.

Of particular interest to us as exploration geochemists is the mechanism by which degraded oils reach their accumulation points and the dynamics of the degradation process. The complex variability of oil composition seen in degraded oils has been perplexing for a long time because the mechanisms to create such oils seemed so convoluted and in geological terms often rather unlikely. The constraints and traditional views of the degradation process need to be reviewed if we are to advance our understanding.

The process of degradation of an oil is believed to be carried out by bacteria which require oxygen for the metabolic process. The systematic changes seen in oil compositions are certainly consistent with a selective and progressive bacterial degradation process which requires the presence of large amounts of oxygen which is usually assumed to be found dissolved

Table 3. *Geochemical characteristics of low sulphur oils during progressive degradation in clastic reservoirs (after Volkman* et al. *1983)*

Level	Oil gravity (°API)	Chemical composition	Extent of biodegradation
1	>30	Typical paraffinic oil; abundance of n-alkanes	Not degraded
2	30–28	Low molecular weight normal alkanes removed	Minor
3	28–27	>90% normal alkanes removed	Moderate
4	27–26	Alkylcyclohexanes removed; isoprenoid alkanes reduced	Moderate
5	26–25	Isoprenoid alkanes (including pristane and phytane) removed	Moderate
6	25–24	C_{14}–C_{16} bicyclic alkanes removed	Moderate
7	24–22	>50% 5α(H)14α(H)17α(H)2OR steranes removed	Extensive
8	22–20	Steranes altered; demethylated hopanes abundant	Very extensive
9	20–15	Demethylated hopanes predominant, some alteration of diasteranes; no steranes	Severe
10	15–10	Demethylated hopanes removed	Extreme
11	10– 5	Alkane content very low	Extreme

in recharging meteoric waters passing through the oil bearing system. Since the volumes of degraded oil in the North Sea are very substantial, and since the degradation process can be shown on mass balance considerations to have removed more than 50% of the oil in some of the producible accumulations, then the volume of oxygen required and the scale of the recharging system can be seen to be very large and regional in extent.

Volkman et al (1983) also showed that there were significant depth constraints on the defined levels or degrees of degradation ranging from about 3000 ft depth of burial to around 1000 ft in the most degraded oils. In our view these depth limits are quite reasonable particularly bearing in mind that we have already shown oils were generated and available for vertical migration by middle to late Eocene time, when many of the Palaeocene and Eocene sands which now contain the oil were at burial depths of 1000 ft to 3000 ft. It is probably also highly relevant that unconformities associated with the updoming and continental separation taking place to the west and north of the North Sea as well as eustatic sea level changes provide the necessary basin margin uplifts for establishment of an hydrostatic head to drive circulating meteoric waters into the basin and thereby aiding the degradation process.

More difficult to understand is the observation that in many of the Tertiary oil accumulations, the very large volumes of oil (500 million to several billion barrels in place) are of constant composition throughout the reservoir both laterally and vertically. This uniformity in composition is not consistent with the conventional assumption that the degradation takes place at the oil-water interface or during filling of the reservoir (Cornford et al. 1986). Even if the process does occur during filling, variations in degree of degradation would be expected and since, as England et al. (1987) and England (1990) have shown, convective and diffusive mixing of oils in reservoirs is not generally significant, there is a major practical problem in degrading the oil at an oil-water interface either after accumulation or during reservoir filling. It is also to be noted that during the late Eocene–Oligocene, when oil was apparently migrating and at sufficiently shallow depths for the degradation process to occur, at least some of the stratigraphic traps in lower Tertiary sands were either not viable structures or the amplitude and crest of the structures have subsequently migrated due to basin tilt and drape-compaction of the underlying and enclosing shales over the basin margins (Robertson Group 1989).

We also recognize that if our calculations about the timing of vertical migration and degradation are correct, then the viscosities of the degraded oils at the shallow depth and low conduit temperatures encountered (1000–3000 ft, 20–30°C) would be such that they would be hardly mobile (Fig. 14). It is only the later burial which has raised reservoir temperatures and in many cases the admixture of a later light oil/condensate component that has mobilized the oil and carried it to the currently drilled structures.

Distribution of Early Tertiary sand bodies

Late Cretaceous tectonic events signalled a major change in sedimentation style in the North Sea. Major reworking of latest Cretaceous sediments was accompanied by massive transport of coarse clastics across the East Shetland Platform and over the shelf edge into the Viking and Central Grabens. The source area in the west overlying the present day Scottish mainland was, during the Early Tertiary, experiencing domal uplift associated with plate separation.

On the shelf the thickness and facies of the Lower Tertiary sediments are poorly known because they are not drilled. From structural and seismic considerations they are recognized as deltaic sediments with over the shelf edge down slope turbidites extending across the basin. The first pulse of sand, the Maureen Formation, is thick and extensive in the Viking Graben. Through the Palaeocene and Early Eocene, a combination of eustatic sea level changes and tectonic events led to a whole series of sands entering the basin as turbidity flows, but apart from the slightly younger Forties Formation, which extends more particularly southeastwards into the Central Graben, these are generally of progressively lesser extent and thickness. The important feature of the Maureen Formation is that it is the first sand in the Lower Tertiary and is deposited as a sheet sand dipping into the basin. It may, like all subsequent Lower Tertiary sands, pinch out in places on the shelf edge or further back upon the platform or have been truncated by later regressive phases. Figures 15 to 17 are net sand isopach maps for the major sand units in the Lower and Upper Palaeocene and Eocene. There are a number of other sand units known which are locally important as potential reservoirs in the area but are not relevant to this discussion.

An important feature of the Maureen Formation sands is that they are laterally extensive, overlie many of the suggested hydrocarbon

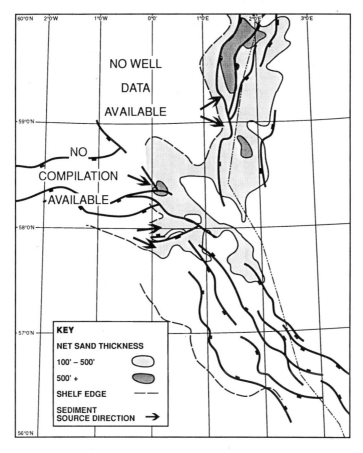

Fig. 15. Maureen Formation sand isopach.

chimney locations and, at the time of vertical oil migration at the end of the Eocene, would have led updip to the East Shetlands Shelf. Logically, therefore, the bulk of the migrating oil would have commenced lateral migration upon reaching the highly porous and permeable Maureen Formation sands. Furthermore, because of their sheet-like geometry and extensive development updip, migration would have been uninhibited towards the outcrop. The only significant exception to this uninterrupted progress appears to be in the Maureen Field where salt tectonics created a structural culmination and in this instance oil was trapped in the Maureen Formation.

Migration routes through the Maureen Formation into the Forties Formation, and higher into the Balder, Frigg and other sandstone formations, apart from in the Central Graben where the Maureen Formation constitutes the first major sand present, were probably relatively restricted. The possibility therefore is that the bulk of the oil reaching the Tertiary was directed out of the Tertiary basin within the Maureen

Formation, and a relatively small proportion, for some reasons of circumstance, migrated into stratigraphically higher levels and was trapped in the overlying more stratigraphically constrained sand bodies.

Oil degradation within the Tertiary system

The model we have proposed suggests that oil migration in the Tertiary sands was localised to discrete conduits, whose course was controlled by the structural dip and permeability within the conduit, and with flow rate dependent on the availability of oil to enter the conduit. Even if the flow was pulsed by the opening and closing of fault controlled chimneys, the $30° +$ API oil would have initially migrated sufficiently quickly in the conduit so that the flowing oil occupied a small volume of the conduit, possibly only a few metres deep and a few tens of metres wide. The sand body, through which the oil was migrating, however, would have been filled with water,

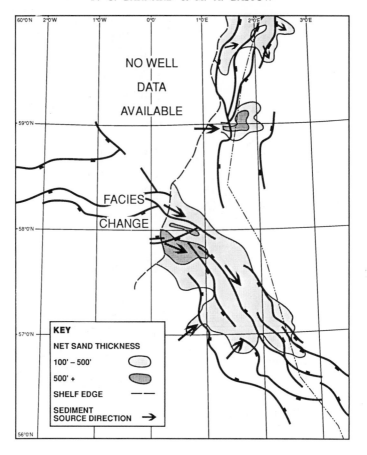

Fig. 16. Forties Formation sand isopach.

initially saline and containing no dissolved hydrocarbons. Oil migrating through the water wet system would therefore have partitioned at the oil/water interface some of the lighter and more water soluble hydrocarbons into the water phase, as described by Barker (1989). This would have resulted in changes in the composition of the oil passing through the conduit and would be described, if the oil were analysed geochemically, as water flushing.

The situation is complicated, however, because with eustatic sea level changes and basin margin doming, there would be a periodic counter-current circulation of fresh meteoric waters within the open sand system. As the migrating oil entered this oxygen-enriched, freshwater system, bacterial degradation would have been intensified, causing marked changes in the viscosity of the oil at the oil/water interface. Unless the oil flow rate then decreased, this

would have caused the migrating oil to spread out and engulf the initially degraded oil, resulting in mixing and remobilisation. In this way the volume of the migration conduit occupied by the oil would have increased and the composition of the oil rapidly changed. Ultimately, the oil within the updip end of the migration pathway would have become so viscous as to stop and plug the porosity in the sands, effectively cutting off meteoric water recharge. Any later migrating oil would therefore no longer have been exposed to fresh waters and degradation, and may even be trapped behind the tar plug. This process is illustrated schematically in Fig. 18. Such situations have been recognized in other petroleum provinces around the world and in some cases can lead to highly commercial accumulations. An example of such an accumulation is provided in the Lagunillas Field in Venezuela (see fig. 26 in Bockmuelen *et al.* 1983).

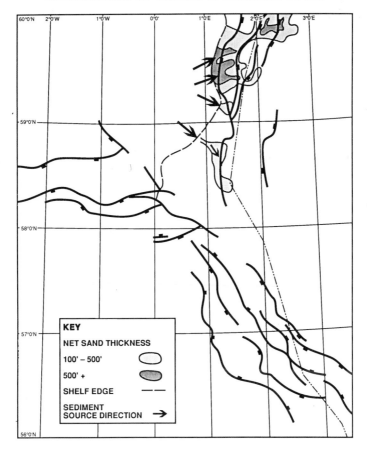

Fig. 17. Frigg Formation sand isopach.

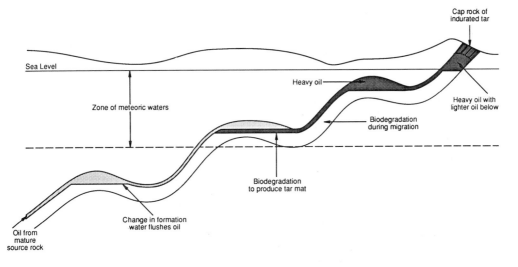

Fig. 18. Changes in oil composition induced during migration.

The interesting question which arises from this discussion is whether such analogues exist in the North Sea basin. It is a well recognized feature of many petroleum provinces that the ultimate updip traps contain large volumes of degraded oil (e.g. Madagascar; Venezuela; Llanos Basin, Colombia; Western Canada) (Demaison 1977). In the North Sea basin the longest updip migration proved to date is seen in the super giant Troll gas and oil field in the Norwegian Sector in which migration distances of at least 40 km have been postulated (Thomas *et al.* 1985) where as well as many tens of trillions of feet of gas in place there is proven reserves of over 5 billion barrels of variably but often heavily degraded oil (Brekke *et al.* 1981).

Implication of migration and degradation model

The model described in this paper is essentially that of a dynamic process in which the processes of alteration and degradation take place during the migration process within the conduit, not in the reservoir or accumulation (Chapman 1982). In the ultimate case, the product is tar sand, clogging all of the porosity, and possibly having the potential to trap lighter oil below. The model suggests that perhaps the bulk of oil generated and migrated into the Tertiary system has been lost by migration and degradation within the Maureen Formation. This may account for some of the difference between the estimated gross oil productivity of the Jurassic source rock system and that found to date (Table 1).

The model suggests that a proportion of the oil reaches the Upper Palaeocene and Eocene sands where it is degraded to varying degrees and at the time of migration/degradation, may have moved very sluggishly and not been effectively trapped in conventional structures. Most of the oils found to date have remigrated to their present location either by increasing reservoir temperature or by mixing with later generated and recently migrated light oils and condensates.

Using the scheme described earlier and the modified Volkman *et al.* (1983) diagram we have been able to classify oils and oil shows in the Viking Graben area as shown in Fig. 19. The cut-off for producibility in the area from the date obtained appears to be about 13° API below which viscosities at reservoir temperatures are too low.

An interesting feature of many of the Tertiary hydrocarbon accumulations at less than 6500 ft depth is that as well as containing degraded oil, they often also have a significant light hydrocar-

bon component. The light hydrocarbon may be dry gas, or gas and condensate. The relative proportions of gas, condensate and degraded, heavy oil are highly variable, so that an accumulation may consist largely of degraded oil with a relatively small and discrete gas cap (e.g. Bressay Field, Forth Field) degraded oil with condensate and gas (e.g. Heimdal Field) or mainly dry gas with producible condensate and a rim of heavy oil (e.g. Frigg Field, Heritier *et al.* 1979, 1981). In general, the gasier accumulations are concentrated to the east of the main synclinal axis of the Tertiary basin. The compositional variations seen suggest that there have been two phases of hydrocarbon generation in the basin, the first being the main phase of oil generation, and which was frequently exposed to water washing and bacterial degradation during migration within the Tertiary, and the second a later phase of dominantly light oil/condensate and gas generation, and which has in general been exposed only to minor water flushing.

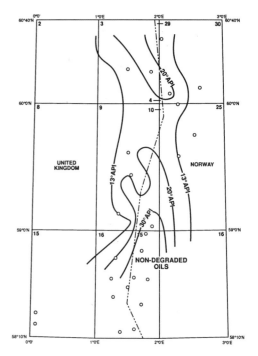

Fig. 19. Sketch map illustrating degree of degradation of first phase oil in Tertiary reservoirs: South Viking Graben area.

We wish to record the contributions made by our colleagues in the geochemistry division of The Robertson Group plc, which have assisted in the development of the ideas presented in this paper. In particular, we thank A. Collins, C. Darlington, A. Livsey and S.

Thompson. We also thank the directors of the Robertson Group plc for permission to publish this paper. We also wish to acknowledge the input of B. S. Cooper of B. S. Cooper and Associates, who has over the years regularly contributed novel geological interpretations of geochemical observations. We also thank A. Fleet, C. Cornford and one anonymous reviewer for their helpful comments and criticisms.

References

BAIRD, R. A. 1986. Maturation and source rock evaluation of Kimmeridge Clay, Norwegian North Sea. *Bulletin of the American Association of Petroleum Geologists,* **70,** 1–11.

BAILEY, N. J. L., KROUSE, H. R., EVANS, C. R. & ROGERS, M. A. 1973*a.* Alteration of crude oil by waters and bacteria: evidence from geochemical and isotopic studies. *Bulletin of the American Association of Petroleum Geologists,* **57,** 1276–1290.

BAILEY, N. J. L., JOBSON, A. M. & ROGERS, M. A. 1973*b.* Bacterial degradation of crude oil: comparison of field and experimental data. *Chemical Geology,* **11,** 203–221.

BARKER, C. 1989. Petroleum losses resulting from solution during migration through carrier beds. (Abstract) *Petroleum migration meeting, Geological Society, London, September 1989.*

BARNARD, P. C. & COOPER, B. S. 1981. Oils and source rocks of the North Sea area. *In* ILLING, L. V. & HOBSON, G. D. (eds) *Petroleum geology of the continental shelf of north-west Europe.* Heyden and Son, 19–33.

——, COLLINS, A. G. & COOPER, B. S. 1981. Identification and distribution of kerogen facies in a source rock horizon—examples from the North Sea Basin. *In*: BROOKS, J. (ed.) *Organic maturation and fossil fuel exploration.* Academic Press, London, 271–282.

BJORLYKKE, K., MO, A. & PALM, E. 1988. Modelling of thermal convection in sedimentary basins and its relevance to diagenetic reactions. *Marine and Petroleum Geology,* **5,** 338–351.

BOCKMUELEN, H., BARKER, C. and DICKEY, P. A. 1983. Geology and geochemistry of crude oils, Bolivar Coastal Fields, Venezuela. *Bulletin of the American Association of Petroleum Geologists,* **67,** 242–270.

BREKKE, T., PEGRUM, R. M. & WATTS, P. B. 1981. First exploration results in Block 31/2, offshore Norway. *In: Norwegian symposium on exploration* Norwegian Petroleum Society.

BROOKS, J. & THUSU, B. 1977. Oil-source rock identification and characterisation of the Jurassic sediments in the northern North Sea. *Chemical Geology,* **20,** 283–294.

——, CORNFORD, C., GIBBS, A. D. & NICHOLSON, J. 1984. Geologic controls on occurrence and composition of Tertiary heavy oils, Northern North Sea (Abstract). *Bulletin of the American Association of Petroleum Geologists,* **68,** 793.

BUHRIG, C. 1989. Geopressured Jurassic reservoirs in the Viking Graben: modelling and geologic significance. *Marine and Petroleum Geology,* **6,** 31–48.

BURLEY, S. D., MULLIS, J. & MATTER, A. 1989. Timing diagenesis in the Tartan reservoir (UK North Sea): constraints from combined cathodoluminescence microscopy and fluid inclusion studies. *Marine and Petroleum Geology,* **6,** 98–120.

CAYLEY, G. T. 1987. Hydrocarbon migration in the central North Sea. *In*: BROOKS, J. & GLENNIE, K. (eds) *Petroleum geology of northwest Europe.* Graham and Trotman, 549–555.

CHAPMAN, R. E. 1972. Primary migration of petroleum from clay source rocks. *Bulletin of the American Association of Petroleum Geologists,* **56,** 2185–2191.

——, 1982. Effects of oil and gas accumulation on water movement. *Bulletin of the American Association of Petroleum Geologists,* **66,** 368–378.

CHIARRELI, A. & RICHY, J. F. 1984. Hydrodynamic flow in sedimentary basins. *In*: DURAND, B. (ed.) *Thermal phenomena in sedimentary basins* Editions Technip, 175–187.

CONNAN, J. 1984. Biodegradation of crude oils in reservoirs. *In*: BROOKS, J. & WELTE, D. (eds), *Advances in petroleum geochemistry, Volume 1* Academic Press, 299–335.

COOPER, B. S. & BARNARD, P. C. 1984. Source rocks and oils of the Central and Northern North Sea. *In*: DEMAISON, G. & MURRIS, R. J. (eds) *Petroleum Geochemistry and Basin evaluation.* Memoir of the American Association of Petroleum Geologists, **35,** 303–315.

CORNFORD, C. 1990. Source rocks and hydrocarbons of the North Sea. *In*: GLENNIE, K. W. (ed.) *Introduction to the petroleum geology of the North Sea* (third edition). Blackwell Scientific Publications, 294–361.

——, MORROW, J. A., TURRINGTON, A., MILES, J. A. & BROOKS, J. 1983. Some geological controls on oil composition in the UK North Sea. *In*: BROOKS, J. (ed.) *Petroleum Geochemistry and Exploration of Europe.* Geological Society, London, Special Publication, **12,** 175–194.

——, NEEDHAM, C. E. J. & DE WALQUE, L. 1986. Geochemical habitat of North Sea oils and gases. *In*: SPENCER, A. M. *et al.* (eds) *Habitat of hydrocarbons on the Norwegian continental shelf.* Norwegian Petroleum Society, Graham and Trotman, 39–54.

DAHL, B., NYSAETHER, E., SPEERS, G. C. & YUKLER, A. 1987. Oseberg area-integrated basin modelling. *In*: BROOKS, J. & GLENNIE, K. (eds) *Petroleum geology of northwest Europe.* Graham and Trotman, 1029–1038.

DEMAISON, G. J. 1977. Tar sands and supergiant oil fields. *Bulletin of the American Association of Petroleum Geologists,* **61,** 1950–1961.

——, HOLCK, A. J. J., JONES, R. W. & MOORE, G. T. 1983. Predictive source bed stratigraphy: a guide to regional petroleum occurrence, North Sea basin and eastern North American continental margin. *In*: *Proceedings of the 11th World Petroleum Congress*, London 1983.

DEMBICKI, H. & ANDERSON, M. J. 1989. Secondary migration of oil: Experiments supporting efficient movement of separate buoyant oil phase along limited conduits. *Bulletin of the American Association of Petroleum Geologists*, **73**, 1018–1021.

DIDYK, B. M., SIMONEIT, B. R. T. & EGLINTON, G. 1983. Bitumen from Coalport tar tunnel. *Organic Geochemistry*, **5**, 99–109.

DOLIGEZ, B., UNGERER, P., CHENET, P. Y., BURRUS, J., BESSIS, F. & BESSEREAU, G. 1987. Numerical modelling of sedimentation, heat transfer, hydrocarbon formation and fluid migration in the Viking Graben, North Sea. *In*: BROOKS, J. & GLENNIE, K. (eds) *Petroleum geology of northwest Europe*. Graham & Trotman, 1039–1048.

DURAND, B. 1988. Understanding of HC migration in sedimentary basins (present state of knowledge). *Organic Geochemistry*, **13**, 445–459.

ENGLAND, W. 1990. The organic geochemistry of petroleum reservoirs. *Organic Geochemistry*, **16**, 415–475.

——, MACKENZIE, A. S., MANN, D. M. & QUIGLEY, T. M. 1987. The movement and entrapment of petroleum fluids in the subsurface. *Journal of the Geological Society, London*, **144**, 327–347.

EGGEN, S. 1984. Modelling of subsidence hydrocarbon generation and heat transport in the Norwegian North Sea. *In*: DURAND, B. (ed.) *Thermal phenomena in sedimentary basins*. Editions Technip, Paris, 271–283.

EVANS, C. R., ROGERS, M. A. & BAILEY, N. J. L. 1971. Evolution and alteration of petroleum in Western Canada. *Chemical Geology*, **8**, 147–170.

FIELD, J. D. 1985. Organic geochemistry in exploration of the northern North Sea. *In*: THOMAS, B. M. *et al.* (eds) *Petroleum geochemistry in exploration of the Norwegian Shelf*. Graham and Trotman, 39–57.

FULLER, J. G. C. M. 1975. Jurassic source rock potential and hydrocarbon correlation. *In*: *Proceedings of the symposium on Jurassic–northern North Sea*. Norwegian Petroleum Society.

GOFF, J. C. 1983. Hydrocarbon generation and migration from Jurassic source rocks in the E Shetland Basin and Viking Graben of the northern North Sea. *Journal of the Geological Society, London*, **140**, 445–474.

HERITIER, F. E., LOSSEL, P. & WATHNE, E. 1979. Frigg Field—a large submarine-fan trap in Lower Eocene rocks of North Sea Viking Graben. *Bulletin of the American Association of Petroleum Geologists*, **63**, 1999–2020.

HERITIER, F. E., LOSSEL, P. & WATHNE, E. 1981. The Frigg Gas Field. *In*: ILLING, L. V. & HOBSON, G. D. (eds) *Petroleum geology of the continental shelf of Northwest Europe*. Proceedings of the 2nd Conference of the Institute of Petroleum, Heyden & Son Ltd., London, 380–391.

HUC, A. Y., IRWIN, H. & SHOELL, M. 1985. Organic matter quality changes in an Upper Jurassic shale sequence from the Viking Graben. *In*: THOMAS, B. M. *et al.* (eds) *Petroleum geochemistry in exploration of the Norwegian Shelf*. Graham & Trotman, 179–184.

JAMES, A. T. & BURNS, B. J. 1984. Microbial alteration of subsurface natural gas accumulations. *Bulletin of the American Association of Petroleum Geologists*, **68**, 957–960.

JONES, R. W. 1981. Some mass balance constraints on migration mechanisms. *Bulletin of the American Association of Petroleum Geologists*, **65**, 103–122.

LAFARGUE, E. & BARKER, C. 1988. Effect of water washing on crude oil compositions. *Bulletin of the American Association of Petroleum Geologists*, **72**, 263–276.

LEADHOLM, R. H., HO, T. Y. & SAHAI, S. K. 1985. Heat flow, geothermal gradients and maturation modelling on the Norwegian continental shelf using computer methods. *In*: THOMAS, B. M. *et al.* (eds) *Petroleum geochemistry in exploration of the Norwegian shelf*. Graham and Trotman, 131–143.

LEONARD, R. 1984. Generation and migration of hydrocarbons on Southern Norwegian Shelf (Abstract). *Bulletin of the American Association of Petroleum Geologists*, **68**, 796.

LEYTHAEUSER, D., RADKE, M. & WILLSCH, H. 1988a. Geochemical effects of primary migration in Kimmeridge source rocks from Brae Field area, North Sea: molecular composition of alkylated nathalenes, phenanthrenes, benzo- and dibenzothiophenes. *Geochimica et Cosmochimica Acta*, **52**, 2879–2891.

——, SCHAEFER, R. G. & RADKE, M. 1988b. Geochemical effects of primary migration of petroleum in Kimmeridge source rocks from Brae Field area, North Sea: gross composition of C_{15+} soluble organic matter and molecular composition of C_{15+} saturated hydrocarbons. *Geochimica et Cosmochimica Acta*, **52**, 701–713.

LIEWIG, N., CLAUER, N. & SOMMER, F. 1987. Rb-Sr and K-Ar dating of diagenesis in Jurassic sandstone reservoir, North Sea. *Bulletin of the American Association of Petroleum Geologists*, **71**, 1467–1474.

LOPATIN, N. V., 1971. Temperature and geologic time as factors in coalification. *Akademii Nauk USSR Ser. Geol.*, **3**, 95–106 (in Russian).

MACKENZIE, A. S. & MAXWELL, J. R. 1983. Biological marker and isotope studies of North Sea crude oils and sediments. *Proceedings of the 11th World Petroleum Congress*, **2**, 45–56.

——, LEYTHAUSER, D., MULLER, P., QUIGLEY, T. M. & RADKE, M. 1988. The movement of hydrocarbons in shales. *Nature*, **331**, 63–65.

——, ——, ——, RADKE, M. & SCHAEFER, R. G. 1987. The expulsion of petroleum from Kimmeridge Clay source rocks in the area of the Brae Oilfield, UK continental shelf. *In*: BROOKS, J. & GLENNIE, K. W. (eds) *Petroleum geology of northwest Europe*. Graham and Trotman, London, 865–878.

MOMPER, J. A. 1978. Oil migration limitations suggested by geological and geochemical considerations. *American Association of Petroleum Geologists course note series*, **8**, B1–B60.

MUNNS, J. W. 1985. The Valhall Field: a geological review. *Marine and Petroleum Geology*, **2**, 23–43.

NORTHAM, M. A. 1985. Correlation of northern North Sea oils: the different facies of their Jurassic source. *In*; THOMAS, B. M. *et al.* (eds) *Petroleum geochemistry in exploration of the Norwegian Shelf*. Graham and Trotman, 93–100.

OUDIN, J. L. 1976. Etude geochimique du basin de la Mere du Nord. *Bulletin Centre de Recherches de Pau-SNPA*, **10**, 339–358.

OZKAYA, I. 1988a. A simple analysis of primary oil migration through oil-propagated fractures. *Marine and Petroleum Geology*, **5**, 170–174.

—— 1988b. A simple analysis of oil-induced fracturing in sedimentary rocks. *Marine and Petroleum Geology*, **5**, 293–297.

PALCIAUSKAS, V. V. & DOMENICO, P. A. 1980. Microfracture development in compacting sediments: relation to hydrocarbon-maturation kinetics. *Bulletin of the American Association of Petroleum Geologists*, **64**, 927–937.

PALMER, S. 1984. Effect of water washing on C_{15+} hydrocarbon fraction of crude oils from Northwest Palawan, Philippines. *Bulletin of the American Association of Petroleum Geologists*, **68**, 137–149.

PARSLEY, A. J. 1986. North Sea hydrocarbon plays. *In*: GLENNIE, K. W. *et al.* (eds) *Introduction to the pertroleum geology of the North Sea*. Blackwell Scientific Publications, 237–263.

PEARSON, M. J. & WATKINS, D. 1983. Organofacies and early maturation effects in Upper Jurassic sediments from the Inner Moray Firth Basin, North Sea. *In*: BROOKS, J. (ed.) *Petroleum Geochemistry and Exploration of Europe*. Geological Society, London, Special Publication, **12**, 147–160.

REED, W. E. 1977. Molecular compositions of weathered petroleum and comparison with its possible source. *Geochimica et Cosmochimica Acta*, **41**, 237–247.

REITSEMA, R. H. 1983. Geochemistry of north and south Brae areas, North Sea. *In*: BROOKS, J. (ed.) *Petroleum Geochemistry and Exploration of Europe*. Geological Society, London, Special Publication, **12**, 203–212.

ROBERTSON GROUP plc. 1985. *The Moray Firth area, North Sea: stratigraphy, petroleum geochemistry and petroleum geology*. Unpublished multiclient report.

—— 1987. *Central Graben, North Sea: stratigraphy, structure and petroleum geology of the Paleocene and Eocene*. Unpublished mutliclient report.

—— 1988. *Central Graben, North Sea: stratigraphy, structure and petroleum geochemistry of the Lower Cretaceous, Jurassic and older strata*. Unpublished multiclient report.

—— 1989. *South Viking Graben, North Sea: stratigraphy, structure and petroleum geochemistry of the Paleocene and Eocene*. Unpublished multiclient report.

—— 1990. *South Viking Graben, North Sea: stratigra-*

phy, structure and petroleum geochemistry of the Lower Cretaceous, Jurassic and older strata. Unpublished multiclient report.

ROYDEN, L., SCLATER, J. R. & VAN HERZEN, R. P. 1980. Continental margin subsidence and heat flow: important parameters in formation of petroleum hydrocarbon. *Bulletin of the American Association of Petroleum Geologists*, **64**, 173–187.

RULLKOTTER, J. & WENDISCH, D. 1982. Microbial alteration of 17α(H)-hopanes in Madagascar asphalts: removal of C-10 methyl group and ring opening. *Geochimica et Cosmochimica Acta*, **46**, 1545–1553.

SCHOWALTER, J. T. 1979. Mechanics of secondary hydrocarbon migrations and entrapment. *Bulletin of the American Association of Petroleum Geologists*, **63**, 723–760.

SCHOU, L., EGGEN, S. & SCHOELL, M. 1985. Oil-oil and oil-source rock correlation, northern North Sea. *In*: THOMAS, B. M. *et al.* (eds) *Petroleum geochemistry in exploration of the Norwegian Shelf*. Graham and Trotman, 101–120.

SEIFERT, W. K. & MOLDOWAN, J. M. 1979. The effects of biodegradation on steranes and terpanes in crude oils. *Geochimica et Cosmochimica Acta*, **43**, 111–126.

SOFER, Z. 1984. Stable carbon isotope compositions of crude oils: application to source depositional environments and petroleum alteration. *Bulletin of the American Association of Petroleum Geologists*, **68**, 31–49.

STAHL, W. J. 1980. Compositional changes and $^{13}C/^{12}C$ fractionations during the degradation of hydrocarbons by bacteria. *Geochimica et Cosmochimica Acta*, **44**, 1903–1907.

THOMAS, B. M., MØLLER-PEDERSEN, P., WHITAKER, M. F. & SHAW, N. D. 1985. Organic facies and hydrocarbon distributions in the Norwegian North Sea. *In*: THOMAS, B. M. *et al.* (eds) *Petroleum geochemistry in exploration of the Norwegian Shelf*. Graham and Trotman, 3–26.

THORNE, J. A. & WATTS, A. B. 1989. Quantitative analysis of North Sea subsidence. *Bulletin of the American Association of Petroleum Geologists*, **73**, 88–116.

TURNER, C. C., RICHARDS, P. C., SWALLOW, J. L. & GRIMSHAW, S. P. 1984. Upper Jurassic Stratigraphy and sedimentary facies in the central Outer Moray Firth Basin. *Marine and Petroleum Geology*, **1**, 105–117.

UNGERER, P., BESSIS, F., CHENET, P. Y., DURAND, B., NOGARET, E., CHIARELLI, A., OUDIN, J. L. & PERRIN, J. F. 1984. Geological and geochemical models in oil exploration: principles and practical examples. *In*: DEMAISON, G. & MURRIS, R. J. (eds) *Petroleum geochemistry and basin evaluation*. Memoir of the American Association of Petroleum Geologists, **35**, 53–77.

VOLKMAN, J. K., ALEXANDER, R., KAGI, R. I. & RULLKOTTER, J. 1983. Demethylated hopanes in crude oils and their applications to petroleum geochemistry. *Geochimica et Cosmochimica Acta*, **47**, 1033–1040.

WAPLES, D. W. 1980. Time and temperature in petro-

leum formation: application of Lopatins method
to petroleum exploration. *Bulletin of the American
Association of Petroleum Geologists,* **64**, 916–926.
WEHNER, H., TESCHNER, M. & BOSECKER, K. 1986.
Chemical reactions and stability of biomarkers
and stable isotope ratios during in vitro biodegradation of petroleum. *Organic Geochemistry,* **10**,

463–471.
WOOD, D. A. 1988. Relationships between thermal
maturity indices calculated using Arrhenius equation and Lopatin method: implications for petroleum exploration. *Bulletin of the American Association of Petroleum Geologists,* **72**, 115–134.

Contrasting characteristics attributed to migration observed in petroleums reservoired in clastic and carbonate sequences in the Gulf of Mexico region

K. F. M. THOMPSON

Geochemical and Environmental Research Group, Texas A&M University, College Station, TX 77840, USA

Present address: BP Exploration Inc., PO Box 4587, Houston, TX 77210, USA

Abstract: Systematic differences among two subsets of petroleums selected geographically and by reservoir type were attributed to lithologically-characteristic migration phenomena. Most oils in offshore Louisiana Tertiary clastic reservoirs are deficient in light ends, evidently because of evaporative fractionation. Residual oils are enriched in cycloalkanes and aromatic hydrocarbons. Spuriously low evident maturities based upon concepts of paraffinicity are observed. Numerous oils are both fractionated and biodegraded. Mesozoic carbonate-sealed reservoirs at the margins of the basin generally contain pristine oils with intact light ends. However, many oils appear to be depleted in preferentially water soluble compounds: benzene, toluene, light cycloalkanes and phenanthrene, evidently due to aqueous fractionation during intraformational migration.

Two suites of oils from the Gulf of Mexico region (Fig. 1) are compared. The first represents those occurring in sandstone reservoirs of Tertiary and Quaternary age, offshore Texas and Louisiana; the second, accumulations in carbonate reservoirs, or in reservoirs which are interbedded with or sealed by carbonates, occurring in eastern Texas, northern Louisiana, southern Mississippi and Alabama. The carbonate-sequence oils are dramatically different from the clastic-reservoired, as evidenced by three characteristics considered here: distributions of molecular weight, the nature of the gasoline composition, and aspects of the distribution of phenanthrene and the methylphenanthrenes. Differences of the two latter types bear upon inferences concerning the thermal histories of the oils, based on one hand upon the earlier work of the writer concerning gasoline paraffinicity (Thompson 1979, 1983), and on the other upon that of Radke *et al.* (1982*a, b*) on phenanthrenes. The comparisons made are not exhaustive. Further systematic differences exist.

Stratigraphy and source rocks

Tertiary

The offshore US Gulf Coast Tertiary–Quaternary section comprises sediment thicknesses ranging from 3 km to 17 km (east of the Mississippi delta and close to the Mexican border, respectively), generally exceeding 10 km. The sequence is progradational. Delivery of clastic sediment began during the Eocene in south Texas and depocentres migrated northwards during the Oligocene. After the middle Tertiary, sediment supply from the vicinity of the present Mississippi embayment was dominant. The sequences are regressive and progressively younger towards the basin (Cook & Bally 1975). The history of sedimentation at any point is generally one of continual reduction of water depth. Outer shelf and slope pro-delta shales occur at the greatest depths in the stratigraphic sequence, succeeded upwards by a sand-shale facies (delta fringe and plain) and by a massive, fluvial sand sequence (Curtis 1988).

Reservoir sands are invariably lensoid and compartmentalized (Morton *et al.* 1983), precluding the existence of regional carrier beds. Reservoir seals comprise interbedded mudstones and shales, as well as sealing fault surfaces in many cases. Sand/shale ratios range from 0.1 to 0.4 in approximately 70% of the productive sequence (Glaser & Jurasin 1971). A significant proportion of gas reservoirs in the onshore region are abnormally pressured, while the great majority of oil accumulations are normally pressured (Thompson 1976).

LaPlante (1974) determined the elemental H/C and O/C ratios of 96 kerogens from Oligocene, Miocene and Pliocene shales and mudstones representing all environments of deposition from neritic to bathyal. All were found to be of Type III, with no significant petroleum

From England, W. A. & Fleet, A. J. (eds), *Petroleum Migration*
Geological Society, Special Publication No. 59, pp. 191–205.

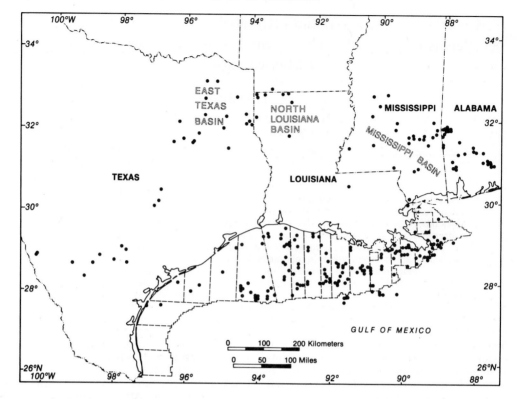

Fig. 1. Sampled locations representing Mesozoic oils from interior rift basins, and offshore Tertiary–Quaternary-reservoired oils.

generating capacity. Nevertheless, numerous authors have claimed the existence of Tertiary source rocks in the Gulf region (Silverman 1965; Dow 1984; Walters & Casa 1985). In the off-shore area, the claims have not been substantiated by analytical data. However, Sassen (1990) demonstrated the occurrence of Type II kerogens in the Sparta Group (Eocene) of southern Louisiana.

Mesozoic

The Mesozoic strata of southern Texas, and of the rift basins which bound the Gulf Coast region on the northwest (the East Texas Basin), the north (the North Louisiana Basin), and the northeast (the Mississippi Salt Basin) principally comprise carbonates with interbedded evaporites, shales and several sandstone units. In each of the sub-basins the oldest Mesozoic units are the Eagle Mills (Triassic) and the Louann Salt (Callovian). The three failed rifts are younger analogues of the Triassic rifts of the northeast-

ern seaboard of the United States. All represent the progressive development of the passive margin bounding the opening Atlantic Ocean. The Gulf of Mexico was virtually fully opened by the end of the Jurassic. According to Buffler (1984), the Callovian salt was divided by rifting in the region of the present basin center, and a brief period of formation of oceanic crust followed in the later Jurassic. Cooling and subsidence in the marginal regions resulted in the development of carbonate platforms surrounding a deep basin.

The failed rifts were thus principally shallow-water depositional sites. The Lower Cretaceous carbonates are clastic in nature and traceable over distances of hundreds of miles. Upper Cretaceous sedimentation throughout the Gulf region comprised chalks and marls, though a major siliciclastic sequence which includes the Woodbine, Tuscaloosa and Eagle Ford Formations was deposited in Louisiana and east Texas during the Cenomanian–Turonian interval. The total thickness of Mesozoic sediments approaches 6 km in the East Texas Basin, and exceeds 7 km in the Mississippi Salt Basin. The Middle Cretaceous unconformity is traceable across the entire Gulf of Mexico and surrounding shelves.

A Lower Cretaceous or Jurassic age was suggested for oils of the Flex Trend (distal offshore Louisiana) on the basis of carbon and sulphur isotopic ratios which are closely similar to values in oils occurring in the Campeche Shelf area (Thompson *et al.* 1990). The latter occur in reservoirs of Upper Jurassic to Palaeocene age (Peterson 1983). Thompson (1983) detailed H/C ratios representing a cored section of the Eagle Ford Formation in Vernon Parish, Louisiana, proving the occurrence of a 30 ft microlaminated shale interval bearing Type II kerogen. Sassen *et al.* (1987) documented source rocks in the Oxfordian Smackover Formation. Thompson (1990*a*) provided carbon and sulphur isotopic data representing Lower Cretaceous Type II kerogens in marls at DSDP Sites 535 and 540 in the eastern Gulf of Mexico. In terms of the ranges of both analytical parameters, it was shown that the kerogens match the distribution of values in oils in the offshore Gulf of Mexico region. Thus, the only sources which have been proven, or inferred by petroleum correlations, occur at the base of, or below, the Tertiary sequence.

Petroleum systems

Mesozoic petroleums

The rift basins exhibit at least three isotopically distinct petroleum systems which can be dated on the basis of reservoir ages. Thus, a family of oils occurring principally in Smackover (Oxfordian) reservoirs possesses a modal $\delta^{13}C(C_{15+}$ saturated hydrocarbon fraction) of $-24‰$ (Fig. 2). A second family with a mode at $-27.0‰$ occurs in Lower Cretaceous reservoirs, and a third family having a modal value of $-28.0‰$ occurs solely in Upper Cretaceous reservoirs. This clear tripartite division, based simultaneously on reservoir age and isotope ratios, demonstrates the subordinate role of vertical migration, and emphasizes the significance of intraformational migration on the flanks of the basins, as well as the effectiveness of regional carrier beds. Although oils of a single family are produced over substantial depth intervals in numerous salt domes (e.g. Erdman & Morris 1974), such are exceptions, regionally. Many salt domes in the rift regions are non-productive, attesting to their role as creators of conduits for vertical migration. The majority of petroleum accumulations are stratigraphically or anticlinally trapped.

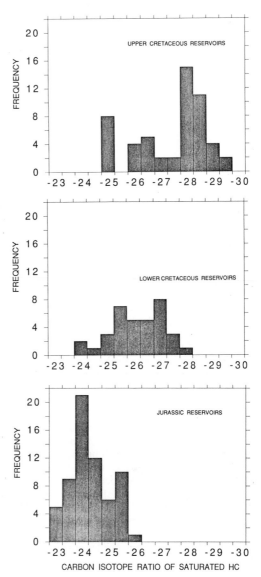

Fig. 2. Frequency distribution of carbon isotope ratio values representing Gulf Coast Mesozoic oils in the interior basins.

Intraformational migration. Intraformational migration is considered to be the hallmark of oils in Mesozoic reservoirs in the Gulf of Mexico region. It is suggested that in this regard they do not differ from the majority of oils disposed on homoclinal basin flanks, well removed from basin-centre sites where deep burial subjects those remaining to gasification and dissipation by tertiary migration.

There is substantial production from the Jurassic Smackover Formation on the northeast

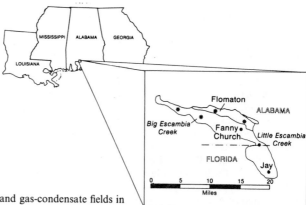

Fig. 3. Locations of selected oil and gas-condensate fields in Alabama.

flanks of the Mississippi Salt basin in southern Arkansas, central Mississippi and southwestern Alabama. An interesting example of intraformational migration is represented by the complex of fields shown in Fig. 3. Here, in southern Alabama, the source is the basinal Smackover Formation (Sassen *et al.* 1987). Migration has taken place within the Smackover Fm. which also comprises the reservoir. Evidently oil and then gas-condensate migrated along the same pathways during a protracted period of generational history. Figure 4 shows the relationship between two molecular ratios which serve as maturity parameters in *unfractionated* petroleums such as these. The exponential increase in n-nonane/n-nonadecane indexes the accumulation of light ends as thermal cracking proceeds. Linear increase in the ratio xylene/n-octane measures the accumulation of thermally stable, light aromatic hydrocarbons, a further index of maturity (Sawatsky *et al.* 1977; Thompson 1987). The contrast of maturities between oils (Fanny Church and Little Escambia) and gas-condensates (Flomaton and Big Escambia Creek) is demonstrative of continuing migration. The oils are of high maturity, but the gases represent thermal gas-condensates of extreme maturity. This interpretation suggests that the products of two distinct phases of generation migrated along closely associated pathways to the reservoir complex. A high degree of certainty attaches to this interpretation in the light of details relating degree of thermal transformation (conversion of oil to gas) provided for these and other Alabama petroleums by Claypool & Mancini (1989). They demonstrated that gas-condensates of lower degree of conversion than Flomaton (83%, reservoir temperature 132°C) occur at higher reservoir temperatures (e.g. Chunchulla, 48% conversion, reservoir temperature 160°C). This

strongly suggests that conversion of the Flomaton fluid took place deeper in the basin than present reservoir site. Protracted filling histories of individual fields are becoming widely recognized. An example from the Federal Republic of Germany was recently described by Leythaeuser & Rückheim (1989).

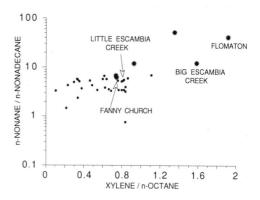

Fig. 4. Maturity parameters in Smackover oils and gas-condensates in Alabama.

Tertiary-reservoired petroleums

The extent to which the Tertiary and Quaternary reservoirs contain post-Mesozoic oils is unclear. Initial indication of the Lower Cretaceous age of a substantial fraction of them, the distal offshore Louisiana oils, was given by Thompson *et al.* (1990). Identities between complex fingerprints representing branched C_{20+} alkane patterns (Thompson & Kennicutt 1991) indicate, furthermore, that at least a significant proportion of the near-shore and coastal region oils are also of Cretaceous origin.

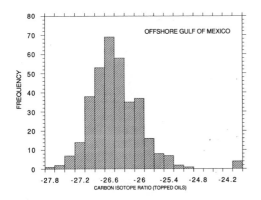

Fig. 5. Frequency distribution of carbon isotope ratio values representing offshore Gulf of Mexico oils in Tertiary–Quaternary reservoirs.

Figure 5 represents the distribution of carbon isotope ratios in the C_{15+} fractions of 333 offshore Gulf of Mexico oils. Although the breadth of the distribution indicates the presence of more than one family (within which a range of only approximately 1.5‰ would be expected), the majority of cases possess values closely similar to those of the Lower Cretaceous oils of the interior basins.

Migration in the Tertiary–Quaternary section. Temperatures at the base of the Tertiary are sufficiently high to have resulted in the gasification of any liquids once generated there, if retained at depth. Preservation of Palaeogene or older oils has therefore involved progressive, possibly stepwise, upward migration. Abundant fault pathways have facilitated this. Such faults are associated with gravitational sediment movements (growth faults) and the intrusion of stocks and diapirs of Jurassic salt. Evidence of vertical migration is provided by the repeated occurrence of near-identical carbon isotope ratios representing multiple, stacked oil accumulations in many structures of both salt dome and rollover anticlinal type (Thompson *et al.* 1990). The geographically random occurrence of isotopically heavy methane, generated by thermal cracking at depth, among gas fields bearing biogenic or low-maturity methane, also provides substantial evidence of vertical migration, as does the exponential relationship in the Gulf Coast between reservoir age and $\delta^{13}C$ (methane) (Thompson 1990*b*).

The concept of pervasive and continuous upward migration of gas which gives rise to the exponential relationship described above, combined with evidence of the loss of light ends from deep Tertiary-reservoired oils on a massive scale, gave rise to the concept of *evaporative fraction-*

ation (Thompson 1987, 1988). This term is applied to compositional alteration attendant upon phase change, particularly the transition from liquid to liquid-plus-vapour brought about by the addition of gas. A concept of gas stripping is suggested: the eventual removal of the light ends of accumulated oils by migrating gas.

Partial vaporization of oil by purely physical means results in the enrichment of the residual liquid in aromatic hydrocarbons, and in cycloalkanes relative to normal or branched alkanes of equal carbon number. Previous studies have described these effects in numerous oils of worldwide occurrence (Thompson 1988), and in Gulf Coast oils (Thompson & Kennicutt 1990). These papers examined the phenomena by employing compositional ratios expressing aromaticity (toluene/n-heptane and xylenes/n-octane), and paraffinicity (n-heptane/methylcyclohexane). The present study examines the mutual relationships of two further indices of paraffinicity which were termed the Heptane and Isoheptane Values (Thompson 1983). Such were shown to be responsive to catagenesis in sedimentary rocks, that is, they are indices of maturity when unaltered (Thompson 1979).

Maturity

Maturity indices of two types are reviewed, the first based upon gasoline range hydrocarbons, the second upon aromatic hydrocarbons of intermediate molecular weight. Evaluations of these parameters as they are expressed in Gulf Coast petroleums are presented, as well as estimates of equivalent vitrinite reflectance values based on literature data. Experimental results are discussed which show that all of the parameters are subject to secondary modification by non-thermal processes, and therefore must be interpreted with caution.

Gasoline range indices

Thompson (1979) showed that progressive increase in the ratio of normal and branched alkanes to cycloalkanes takes place in both Type II and Type III kerogens during catagenesis. Two ratios were defined to index these changes:

(1) Heptane Value $(H) = 100$(n-heptane)%/ (b.p. range 80.7–100.9°C), where the denominator includes the range cyclohexane to methycyclohexane, less 1,*cis*-2-dimethylcyclopentane;

(2) Isoheptane Value (I) = (2-methylhexane + 3-methylhexane)/(three isomers of dimethyl-cyclopentane: 1, *cis*-3, 1, *trans*-3 and 1, *trans*-2).

It has been suggested (Lijmbach 1989) that the relative proportions of paraffins and napth-thenes in the gasoline range are significantly dependant upon the nature of the source rock. Lijmbach claimed that 'terrestrial' sources generate naphthenic gasolines while those representing 'marine' sources are more paraffinic. The hypothesis requires verification in source rock extract data.

The trend illustrated in Fig. 6 is termed a 'generation curve', represented by the equation:

$$H = 24.27\log_{10}I + 19.08.$$

The curve is employed as a standard with which to compare petroleum analytical data, on the basis of the inference that paired values of H and I which fail to lie on the curve are indicative of alteration. In this sense the generation curve is considered to be akin to that termed 'concordia' which assesses the gain or loss of ^{208}Pb in lead–uranium dating.

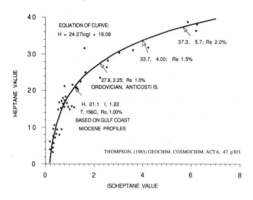

Fig. 6. Generation curve representing the evolution of Heptane and Isopheptane Values in Type II (aliphatic) kerogens.

The interpretation of the indices H and I in terms of equivalent vitrinite reflectance remains uncertain for lack of appropriate data. The highest values of both H and I in Fig. 6 represent measurements made in the well ARCO Anticosti No. 1 (Thompson 1983). The penetrated section comprised Ordovician carbonates containing marine planktonic kerogens. An extensive suite of reflectance measurements representing the section was generated by Bertrand & Héroux (1987). The Anticosti well yielded reflectance curves for graptolite, chitinozoan, scolecodont

and pyrobitumen particles. Jacob (1985) determined elsewhere the relationship between the reflectance of vitrinite and pyrobitumen, when both are present simultaneously. Employing this relationship, Bertrand & Héroux developed a hypothetical vitrinite reflectance curve for Anticosti No. 1 which proved to be almost coincident with the scolecodont curve. Values $(R_s\%)$ from the latter were therefore employed to annotate Fig. 6, associating paraffinicity measurements with reflectance values by depth. These data, however imperfect, are the only existing guide to the interpretation of elevated, conformable, par-affinicity values in petroleums.

The parameters H and I are covariant with other measures of the relative concentrations of paraffins and naphthenes. This is illustrated by the data of Table 1, representing 350 analyses of offshore Gulf Coast petroleums in Tertiary and Quaternary reservoirs. The latter are the subject of this and other reports (Thompson *et al.* 1990; Thompson & Kennicutt 1990a, 1991; Thompson 1990*b*).

Table 1. *Pearson correlation coefficients relating indices of paraffinicity in Tertiary–Quaternary reservoired oils in the Gulf Coast offshore region.*

	C^*	I	F^*	H
C	1.00	0.75	0.98	0.79
I	–	1.00	0.78	0.66
F	–	–	1.00	0.81
H	–	–	–	1.00

* C = (n-hexane + n-heptane)/(cyclohexane + meth-ylcyclohexane).
* F = n-heptane/methylcyclohexane (Heptane Ratio).

Methylphenanthrene indices

Indices of maturity based on concentrations of phenanthrene and methylphenanthrene isomers were described by Radke *et al.* (1982*a*, *b*). Of the numerous ratios defined, only one is considered here: MPI1, which is of value in defining maturity levels in petroleums (Boreham *et al.* 1988; Leythaeuser *et al.* 1988; Radke 1988). A related index, MPI3, was introduced by Garrigues (1985). Definitions are as follows:

$$(1) \ \mathrm{MPI1} = \frac{1.5(3M + 2M)}{(P + 9M + 1M)}$$

$$(2) \ \mathrm{MPI3} = \frac{(3M + 2M)}{(9M + 1M)}$$

where 'P' denotes phenanthrene, and the abbreviations denote methylphenanthrenes.

These parameters have been measured in sediments bearing Type III kerogen (Radke *et al.* 1982*b*), Types II and III kerogens (Hall *et al.* 1985; Thompson & Lundell 1985; Radke *et al.* 1986, and the citations above), as well as in numerous coals (Radke *et al.* 1982*a* and 1984, also Kvalheim *et al.* 1987). Measured values were related to observed vitrinite reflectance values. The parameters are insensitive to increasing maturity in Type II kerogens at R_o levels below 0.65% (Radke *et al.* 1986) but are independent of kerogen type at higher levels (Radke *et al.* 1982*a*, as well as the combined data of Hall *et al.* 1985, and Thompson & Lundell 1985). Thompson (1990*b*) compiled data from the above sources to derive relationships between these phenanthrene-based parameters and vitrinite reflectance levels measured in sediments. Linear regressions were calculated. The abbreviations R_MPI1 and R_MPI3, were employed to designate calculated values of reflectance based upon phenanthrene indices. The derived regressions are:

$$R\_MPI1\% = 0.60 \,(MPI1) + 0.37$$

$$\{R_o < 1.4\%;\ n = 16,\ r = 0.96\}$$

and

$$R\_MPI3\% = 0.30(MPI3) + 0.68$$

$$\{0.80\% < R_o < 1.6\%;\ n = 31,\ r = 0.81\}.$$

MPI1 exhibits a maximum at $R_o = 1.4\%$, with considerable scatter at this level and beyond. MPI3 values scatter below $R_o = 0.8\%$, and again above $R_o = 1.6\%$. Nevertheless, a useful relationship exists between these limits. Most significantly, Type II, Type III, and coal data are inseparable in the interval.

Figure 7 is a generation curve representing progressive increase in methylphenanthrene indices in the Elmworth well (Alberta basin) based on the data of Radke *et al.* (1982*b*). The sediments principally bear Type III kerogen, with intercalations of Type II. MPI3 was calculated from digitized values of Radke *et al.*'s ratios MPR1, MPR2, MPR3 and MPR9 which equal, respectively, 1M/P, 2M/P, 3M/P and 9M/P (abbreviations as above). The curve is annotated with values of vitrinite reflectance calculated from the equation representing R_MPI3 given above. Petroleum data are compared with the curve in subsequent sections.

Non-thermal alteration processes in reservoired petroleums

The commonly reported major processes responsible for the alteration of petroleum are biodegradation and water washing, well reviewed by Connan (1984) and Palmer (1984), respectively. A further process was described by the writer, that of evaporative fractionation which significantly affects petroleum light ends. The concept of dismigration (Chiarelli & Du Rouchet 1977), is closely related. The latter processes are causally related to migration, involving the loss of gas-plus-condensate from oils. The first two processes can only be claimed to be adventiously related to petroleum migration, insofar as movement from depths and temperatures which lie outside those of the biosphere to those within it is a prerequisite. Nevertheless, because biodegradation is pervasive, though not severe, in the Tertiary-reservoired oils under consideration (Thompson & Kennicutt 1990), it is evaluated here.

Experimental Alteration

There is insufficient published data representing the effects of experimental biodegradation to determine the detailed behaviour of the ratios H and I. Although it is known that the ratio *n*-heptane/methylcyclohexane decreases (Winters & Williams 1969), causing predictable decrease in these parameters, a series of experiments was carried out to evaluate the behaviour of others, and their relative rates of change. Full details will be reported in a future publication. For the present, effects on the parameters H and I are shown in Fig. 8. The latter also shows previously unreported data obtained in the experiments described by Thompson (1987) representing the behaviour of H and I in evaporative fractionation.

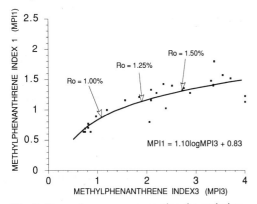

Fig. 7. Generation curve representing the evolution of MPI1 and MPI3 in Types II and III kerogens, based on the data of Radke *et al.* 1982*a*.

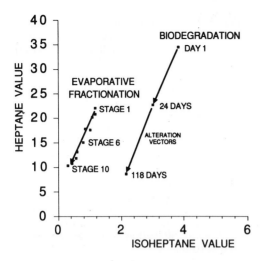

Fig. 8. Alteration vectors in terms of Heptane and Isoheptane Values as developed experimentally by evaporative fractionation and biodegradation.

Comparisons between petroleums in Mesozoic and Tertiary–Quaternary reservoirs

Molecular weight distributions

Figure 9 comprises gas chromatographic data representing 503 whole oil analyses. Horizontal axes indicate the carbon number of the dominant alkane in each. There is a striking difference in molecular weight distributions comparing the offshore Tertiary and the Mesozoic interior basin oils. The latter (Fig. 9a) represent normal, relatively pristine, fluids with their light ends intact. It seems feasible that limited losses took place in surface gas separating equipment, resulting in numerous cases in the recession of the dominant compound to n-hexane, or as far as n-nonane. Storage losses might also be represented. However, Fig. 9b presents an entirely different aspect. Oils in offshore Louisiana Tertiary reservoirs are generally depleted in light ends. The phenomenon is not due to storage loss, as has been demonstrated by the analysis of well-sealed samples within a short time after collection. There are characteristic differences in the compositional changes which accompany light end loss under reservoir pressure as opposed to normal pressure, as previously documented (Thompson 1987). Losses from the reservoirs represented in Fig. 9b are attributed to evaporative fractionation.

Three geological factors appear to influence the frequency and severity of evaporative frac-

tionation. The process is believed to be gas-driven. First, basins with extremely thick sedimentary successions which ensure complete gasification in the source rock, as well as of oils retained at depth, are prone to possess fractionated liquids. Examples have been described from the San Juan basin of New Mexico, and the Sacramento basin of California (Thompson 1990*b*). Secondly, evaporative fractionation appears to be facilitated by faulting which provides vertical migration pathways. This in turn is enhanced by gravitational instability, rapid sedimentation and abnormal fluid pressures. Lastly, lensoid sand units tend to mitigate against lateral migration.

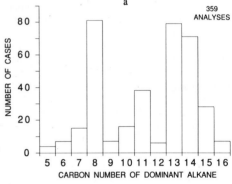

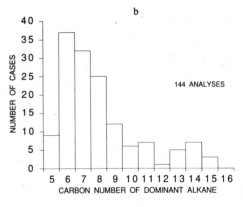

Fig. 9. Frequency distribution by carbon number of the dominant normal-alkane in 359 analyses of petroleums in (**a**) offshore Gulf Coast Tertiary and Quaternary reservoirs. and (**b**), in 144 analyses of petroleums from Mesozoic reservoirs.

Maturity parameters in oils in Tertiary and Quaternary reservoirs

Figure 10 shows the relationship between values of *H* and *I* in 202 oils occurring in Tertiary and

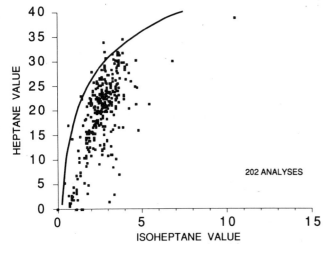

Fig. 10. Heptane and Isoheptane Values representing 202 analyses of oils in offshore Tertiary–Quaternary reservoirs. Extent of alteration is indicated by degree of departure from the generation curve.

Quaternary reservoirs in the offshore region. The broad band of data points directed away from the generation curve is parallel to experimentally generated alteration pathways (Fig. 8). Such a distribution is compatible with an explanation in terms of the secondary alteration of oils which originally possessed a limited range of initial values of *H*, between approximately 28 and 32. This suggests that original gasoline compositions represented reservoir petroleums which have experienced substantial thermal cracking. Figure 10 suggests that alteration, generally of substantial extent, is the rule in the gasolines in question. This lends further strength to the contention of Thompson & Kennicutt (1990) based on alkane/isoprenoid ratios and aromaticity relationships, that these gasoline compositions are, in virtually

all instances, modified by one or both of the processes of biodegradation and evaporative fractionation.

Figure 11 represents observed values of MPI1 and MPI3 in the suite of oils from offshore Gulf of Mexico reservoirs. Scatter increases at higher values, but the trend of the data tends to follow the generation curve. This suggests that secondary alteration of the distribution of phenanthrenes is limited, and that valid estimates of the maturity of the mid-range components of these oils can be made. It is perhaps significant that the data are asymmetrically disposed about the curve. The trend towards disproportionately high values of MPI1 at high values of MPI3 is much more apparent in the Mesozoic oils and is discussed further below.

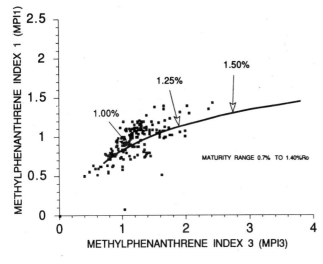

Fig. 11. Methylphenanthrene indices (MPI1 and MPI3) in the suite of oils represented in Fig. 10, evaluated in terms of a generation curve.

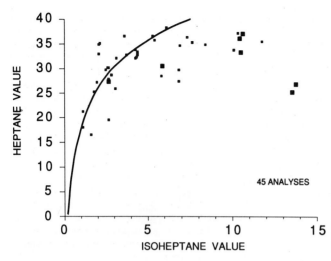

Fig. 12. Heptane and Isoheptane Values representing 45 analysed oils in Gulf Coast Mesozoic reservoirs. Alteration is apparently minimal except in cases of elevated Isoheptane Values, possibly due to aqueous fractionation.

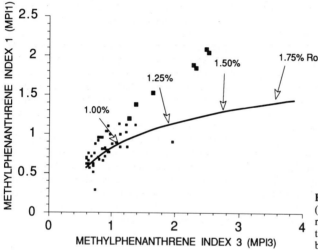

Fig. 13. Methylphenanthrene indices (MPI1 and MPI3) in the suite of oils represented in Fig. 12. Extreme departures from the generation curve are attributed to aqueous fractionation.

Maturity parameters in oils in Mesozoic reservoirs

Figure 12 shows the distribution of H and I in the suite of Mesozoic petroleums. Conformity to the curve is evidenced in the majority of cases wherein I is less than 5.0. This favours an interpretation in terms of valid representations of light-end maturity based upon indices of paraffinicity. However, those cases having values of I greater than 5, and those which do not conform to the curve, are problematical. In these, maturity in terms of R_o, assessed from values of I alone, reaches wholly improbable levels. Oils are rapidly cracked at reservoir temperatures sufficiently high to result in sediment vitrinite reflectances of the order of 1.6%, unless such tem-

peratures have been geologically short-lived. The cause of these aberrant Isoheptane Values is unknown. Seven of the anomalous cases, including some of the most extreme, comprise those oils which are suggested below to have suffered secondary alteration by aqueous fractionation (the set common to Figs 12 and 13 designated by the larger square symbols). This association does not uniquely establish a cause, although a contribution to the alteration by interaction with water, involving the preferential dissolution of dimethylcyclopentanes over branched heptanes, appears possible. The term 'aqueous fractionation' is suggested for the process, as opposed to 'water washing', because the movement of oil in regional carrier beds involves connate water merely as a passive stationary phase. The frequent expression of the movement of water past a stationary body of oil, biodegradation, is absent in these oils, as evidenced by the compositions of the gasoline fractions.

Figure 13 presents values of MPI1 and MPI3 in the Mesozoic oils. Once again, the data are conformable to the generation curve at lower maturities, but in this instance reveal a significant progressive departure at higher values, a trend emphasized by the generation curve. Seven cases are substantially affected in Fig. 13, 14% of the set. The experimental work of Lafargue & Barker (1988) suggests that the phenomenon could be attributed to interaction with water.

Aqueous fractionation

The hypothesis of selective removal of preferentially water-soluble compounds from oil during migration is traceable to several early petroleum geochemical studies. For example, Baker (1962), suggested that benzene was partially stripped from oils migrating into northern Oklahoma in the Arbuckle Formation. More recently, Palmer (1984) discussed the preferential removal of dibenzothiophenes. A comprehensive review of extraction hypotheses was given by Lafargue & Barker (1988). Radke et al. (1982a), investigating phenanthrene and its homologues in North German coals, noted deficiencies in phenanthrene in certain cases, compared to other coals of similar rank. At levels greater than 0.8% R_o the ratio of phenanthrene to each of the methylphenanthrenes increases systematically with maturity. Radke remarked on the greater solubility in water of phenanthrene compared to that of the methylphenanthrenes, and attributed anomalously high values of MPI1 in the phenanthrene-deficient coals to the extraction of that compound by ground water. Published aqueous solubility data substantiate this possibility for phenanthrene (9.94×10^{-5}g/100 g solution, {Andrews & Keefer 1949}) and 1-methylphenanthrene (2.69×10^{-5}g/100 g solution, {May et al. 1978}), both at 25°C. The foregoing data were compiled by Shaw (1989). Figure 14 depicts MPI1 and MPI3 data (Radke et al. 1982a) representing several of these coals from North Germany. Three cases, all possessing R_o values close to 0.99%, exhibit progressive departure from the generation curve at a constant value of MPI3. The coal of R_o 0.70% is relatively normal, while that at 1.70% is also inferred to be depleted in phenanthrene. Figure 14 also represents MPI data generated experimentally by Lafargue & Barker (1988) who treated a small sample of oil with increasingly large volumes of water. Gas chromatographic analyses of the oil were obtained at various stages of extraction. Figure 14 represents the initial condition of the oil, also an intermediate condition after elution with 20 litres of water, and the final condition after treatment with 45 litres. Substantial departures of the values of MPI1 from the generation curve were brought about, comparable to those observed in the Mesozoic oils.

Based on the observations of Radke and those of Lafargue & Barker, it is suggested that the progressive departure of values of MPI1 from the generation curve evidenced in certain Mesozoic oils (Fig. 13), and to a lesser extent in oils in Tertiary–Quaternary reservoirs (Fig. 11), is the result of aqueous fractionation. Any given frac-

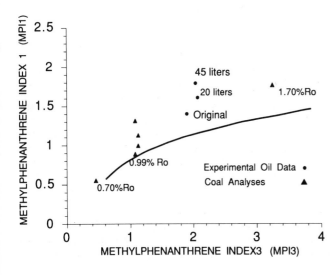

Fig. 14. Experimental data of Lafargue and Barker and observations on coals (Radke) indicative of the effect of aqueous fractionation on phenanthrene ratios.

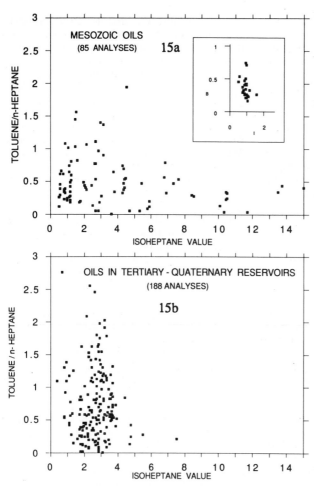

Fig. 15. (a) Toluene/n-heptane (B) versus Isopheptane Values (I) in Mesozoic oils. Low values of B are interpreted in terms of aqueous fractionation, which suggests the possibility that accompanying elevated values of (I) are due to the same cause. Inset shows clustered values in unaltered oils. (b) In offshore oils, elevated values of B are attributed to evaporative fractionation, low values to biodegradation.

tional loss of phenanthrene will result in progressively larger departures at increasing values of MPI1, without effect on MPI3.

Further evidence of aqueous fractionation in the Mesozoic oils is provided by the relationship therein between aromaticity (B, toluene/n-heptane) and associated Isoheptane Values. The ratio B has been evaluated for many sediment extracts, oils and gas-condensates (Thompson 1979, 1983, 1987). In unaltered oils of normal maturity, having R_o equivalences between 0.86% and 1.05%, B averages approximately 0.3. Values are relatively tightly clustered in 29 oils representing three basins in the western United States (inset, Fig. 15). In each basin these oils are remarkably uniform in composition over substantial geographic areas and are therefore considered to be unaltered (Thompson 1983). In the absence of alteration (a condition which is not commonly exhibited) H, I and B increase with maturation. An ideal illustration has been

observed in the oils and gas-condensates of the Smackover Formation of Alabama where paraffinicity, aromaticity and the proportion of light ends increase concomitantly (see above, and Thompson 1990b). In the Mesozoic oils described here, the highest values of both H and I are accompanied by the lowest values of B, contrary to expectation based on maturity. As illustrated in Fig. 15a, there is an inverse relationship between B and I at values of the latter greater than approximately 4. The extreme scatter of the data in Fig. 15a is interpreted in terms of alteration. The low observed levels of aromaticity are postulated to be due to aqueous fractionation, and it is feasible that the same process was responsible for the 'improbably' high levels of paraffinicity (particularly I) in numerous cases. In Tertiary-reservoired oils, on the other hand, substantial increases in B are accompanied by modest decreases in I (Fig. 15b), changes which are consistent evaporative fractionation.

Conclusions

In the Gulf Coast region, gasoline compositions of oils in post-Mesozoic clastic reservoirs are substantially altered, while in the mid-range phenanthrene distributions are relatively unaltered. Experimental evidence suggests that evaporative fractionation and biodegradation are the dominant alteration processes. Complements of light ends, and particularly normal-alkanes, are diminished. It appears that extent of alteration and original composition can be inferred from generation curves, defined in source rock extracts. Oils in Mesozoic reservoirs are relatively pristine. Lateral migration has been the rule; light end loss with evaporative fractionation, and biodegradation are uncommon. The competence and petroleum-retentiveness of their carbonate reservoirs, some with evaporite seals, facilitates maturation in situ and the eventual generation of thermal gas-condensates with light ends intact.

The foregoing data derive from a study of Gulf of Mexico petroleums conducted by the Geochemical and Environmental Research Group, (GERG), Texas A&M University. This four-year endeavour was supported by the following corporations: Amoco, ARCO, BP, Cities Service, Conoco, Elf-Aquitaine, Exxon, Gulf, Mobil, Pennzoil, Phillips, Tenneco, Shell, Sun, Transco and Unocal. Appreciation for their support and permission to publish is expressed, as well as thanks to J.M. Brooks, M.C. Kenicutt II, of GERG who initiated the study. The experimental and analytical assistance of T.J. McDonald is gratefully acknowledged. The findings and conclusions expressed are those of the writer, and do not necessarily reflect the views of the sponsoring companies.

References

ANDREWS, L. J. & KEEFER, R. M. 1949. Cation complexes of compounds containing carbon–carbon double bond. IV. The argentation of aromatic hydrocarbons. *Journal of the Amercian Chemical Society*, **71**, 3644–77.

BAKER, D. R. 1962. Organic geochemistry of Cherokee Group in southeastern Kansas and northeastern Oklahoma. *American Association of Petroleum Geologists Bulletin*, **46**, 1621–1642.

BERTRAND, R. & HÉROUX, Y. 1987. Chitinozoan, graptolite and scolecodont reflectance as an alternative to vitrinite and pyrobitumen reflectance in Ordovician and Silurian strata, Anticosti Island, Quebec, Canada. *American Association of Petroleum Geologists Bulletin*, **71**, 951–957.

BOREHAM, C. J., CRICK, I. H. & POWELL, T. G. 1988. Alternative calibration of Methylphenanthrene Index against vitrinite reflectance: application to maturity measurements in oils and sediments. *Organic Geochemistry*, **12**, 289–294.

BUFFLER, R. T. 1984. Early history and structure of the deep Gulf of Mexico basin. *In: Characteristics of Gulf Basin deep-water sediments and their exploration potential*. 5th Annual Research Conference, Gulf Coast Section, SEPM, Program and Abstracts, Austin, 1984, 31–34.

CHIARELLI, A. & DU ROUCHET, J. 1977. Importance des phenomenes de migration verticale des hydrocarbures. Revue de l' Institut Français du Pétrole, **32**, 189–208.

CLAYPOOL, G. E. & MANCINI, E. A. 1989. Geochemical relationships of petroleum in Mesozoic reservoirs to carbonate source rocks of Jurassic Smackover Formation, southwestern Alabama. *American Association of Petroleum Geologists Bulletin*, **73**, 904–924.

CONNAN, J. 1984. Biodegradation of crude oils in reservoirs. In: WELTE, D. H. & BROOKS, J. M. (eds) *Advances in Petroleum Geochemistry VI*, Academic Press, New York, 299–335.

COOK, T. D. & BALLY, A. W. 1975. *Stratigraphic Atlas of North America* (Shell Oil Company). Princeton Univ. Press, Princeton, NJ.

CURTIS, D. M. 1987. The northern Gulf of Mexico basin. *Episodes*, **10**, 267–270.

DOW, W. G. 1984. Oil source beds and oil prospect definition in the Upper Tertiary of the Gulf Coast. *Transactions of the Gulf Coast Association of Geological Societies*, **24**, 329–339.

ERDMAN, J. G. & MORRIS, D. A. 1974. Geochemical correlation of petroleum. *American Association of Petroleum Geologists Bulletin*, **58**, 2326–2337.

GARRIGUES, P. 1985. *Origine et évolution de series de composes aromatiques dans l'environnement sedimentaire*. Thése de doctorat d'Etat. UA 348 CNRS. Université de Bordeaux I.

GLASER, G. C. & JURASIN, A. C. 1971. Paleoecology, stratigraphy, production—getting it all together in offshore Louisiana. *Transactions of the Gulf Coast Association of Geological Societies*, **21**, 67–81.

HALL, P. B., SCHOU, L. & BJORØY, M. 1985. NPRA Alaska North Slope Oil-Source Rock Correlation Study. *In:* MAGOON, L. B. & CLAYPOOL, G. E. (eds) *Alaska North Slope Oil–Rock Correlation Study: Analysis of North Slope Crude*. American Association of Petroleum Geologists Studies in Geology, **20**, 509–556.

JACOB, H. 1983. *Disperse, feste Erdölbitumina als migrations und maturitätsindikatoren im rahmen der erdöl-/erdgas-Prospektion. Eine Modellstudie im NW-Deutschland*. Deutsche Gesellschaft für Mineralöwissenschaft und Kohlechemie e. U., Projekt 267.

KVALHEIM, O. M., CHRISTY, A. A., TELNAES, N. & BJØRSETH, A. 1987. Maturity determination of organic matter in coals using the methylphenanthrene distribution. Geochimica et Cosmochimica Acta, **51**, 1883–1888.

LAFARGUE, E. & BARKER, C. 1988. Effect of water washing on crude oil compositions. *American As-*

sociation of Petroleum Geologists Bulletin, **72**, 263–276.

LaPLANTE, R. E. 1974. Hydrocarbon generation in Gulf Coast Tertiary sediments. *American Association of Petroleum Geologists Bulletin*, **57**, 1272–1280.

LEYTHAEUSER, D. & RÜCKHEIM, J. 1989. Heterogeneity of oil composition within a reservoir as a reflection of accumulation history. *Geochimica et Cosmochimica Acta*, **53**, 2119–2123.

——, RADKE, D. & WILLSCH, H. 1988. Geochemical effects of primary migration of petroleum in Kimmeridge source rocks from Brae field area, North Sea. II: Molecular composition of alkylated naphthalenes, phenanthrenes, benzo- and dibenzothiophenes. *Geochimica et Cosmochimica Acta*, **52**, 2879–2891.

LIJMBACH, G. W. M. 1989. Geochemistry as an integral part of hydrocarbon exploration. *Abstracts, 14th International Meeting on Organic Geochemistry, Paris*.

MAY, W. E., WASIK, S. P. & FREEMAN, D. H. 1978. Determination of the solubility behavior of some polycyclic aromatic hydrocarbons in water. *Analytical Chemistry*, **50**, 997–1000.

MORTON, R. A., EWING, T. E. & TYLER, N. 1983. *Continuity and internal properties of Gulf Coast sandstones and their implications for geopressured fluid production*. Report Inv. 132, Bureau Economic Geology, University of Texas at Austin.

PALMER, S. E. 1984. Effect of water washing on C_{15+} hydrocarbon fraction of crude oils from northwest Palawan, Philippines. *American Association of Petroleum Geologists Bulletin*, **68**, 137–149.

PETERSON, J. A. 1983. *Petroleum geology and resources of southeastern Mexico, northern Guatemala, and Belize*. US Geological Survey Circular 760. US Government Printing Office, Washington, DC.

RADKE, M. 1988. Application of aromatic compounds as maturity indicators in source rocks and crude oils. *Marine and Petroleum Geology*, **5**, 224–236.

——, LEYTHAEUSER, D. & TEICHMÜLLER, M. 1984. Relationship between rank and composition of aromatic hydrocarbons for coals of different origins. *In*: LEYTHAEUSER, D. & RULLKÖTTER, J. (eds) *Advances in Organic Geochemistry, 1985. Organic Geochemistry*, **10**, 51–63.

——, WELTE, D. & WILLSCH, H. 1982b Geochemical study on a well in the Western Canada Basin: Relation of the aromatic distribution pattern to maturity of organic matter. *Geochimica et Cosmochimica Acta*, **46**, 1–10.

——, —— & —— 1986. Maturity parameters based on aromatic hydorcarbons: Influence of the organic matter type. *In*: LEYTHAUSER, D. & RULLKÖTTER, J. (eds) *Advances in Organic Geochemistry, 1985. Organic Geochemistry*, **10**, 51–63.

——, WILLSCH, H. & LEYTHAEUSER, D. 1982a. Aromatic components of coal: Relation of distribution pattern to rank. *Geochimica et Cosmochimica Acta*, **46**, 1831–1848.

SASSEN, R. 1990. Lower Tertiary and Upper Cretaceous source rocks in Louisiana and Mississippi:

implications to Gulf of Mexico crude oil. *American Association of Petroleum Geologists Bulletin*, in press.

——, MOORE, C. H. & MEENDSEN, F. C. 1987. Distribution of hydrocarbon source potential in the Jurassic Smackover Formation. *Organic Geochemistry*, **11**, 379–383.

SAWATZKY, H., GEORGE, A. E., BANERJEE, R. C., SMILEY, G. T. & MONTGOMERY, D. S. 1977. *Maturation studies of Canadian east coast oils*. CANMET Report 77–42, Dept. of Energy, Ottawa, Canada.

SHAW, D. G., (ed.) 1989. *Hydrocarbons with water and seawater. Pt. 1: Hydrocarbons C_5 to C_7, 528 p. Pt. 2: Hydrocarbons C_8 to C_{36}*. IUPAC Solubility Data Series. Pergamon, New York.

SILVERMAN, S. R. 1965. Reference to $^{13}C/^{12}C$ ratio of Challenger Knoll oil. In: Analyses of oil and caprock from Challenger (Sigsbee) Knoll. DAVIS, J. B. & BRAY, E. E. (Compilers). *Initial Reports of the Deep Sea Drilling Project*, **1**, U.S. Government Printing Office, Washington, DC, 456–467.

THOMPSON, K. F. M. 1976. Temperature and pressure of Cenozoic petroleum reservoirs in Louisiana. *Annual Meeting, South-Central Section, Geological Society of America*, Abstract 8, 1.

—— 1979. Light hydrocarbons in subsurface sediments. *Geochimica et Cosmochimica Acta*, **43**, 657–672.

—— 1983. Classification and thermal history of petroleum based on light hydrocarbons. *Geochimica et Cosmochimica Acta*, **47**, 303–316.

—— 1987. Fractionated aromatic petroleums and the generation of gas-condensates. *Organic Geochemistry*, **11**, 573–590.

—— 1988. Gas-condensate migration and oil fractionation in deltaic systems. *Marine and Petroleum Geology*, **5**, 237–246.

—— 1990a. Carbon and sulfur isotope ratios in aliphatic Cretaceous kerogens at DSDP Sites 535 and 540: inferences concerning Gulf of Mexico petroleums. In: SCHUMACHER, D. & KENNICUTT, M. C. (eds) *Geochemistry of Gulf Coast and Oils and Gases, Proceedings of the 9th Annual Gulf Coast Section*, SEPM Research Conference, December 1988. (in press).

—— 1990b. The generation of thermal and evaporative gas-condensates. *Advances in Petroleum Geochemistry*, **V3**. Academic Press, London, (in press).

—— & KENNICUTT, M. C. 1991. Correlations of Gulf Coast Petroleums on the basis of branched acyclic alkanes. *Organic Geochemistry*, (in press).

—— & —— 1990. Nature and frequency of occurrence of non-thermal alteration processes in offshore Gulf of Mexico platforms. *In*: SCHUMACHER, D. & PERKINS, B. F. (eds) *Geochemistry of Gulf Coast and Oils and Gases, Proceedings of the 9th Annual Gulf Coast Section, SEPM Research Conference, December 1988*. (in press).

—— & LUNDELL, L. L. 1985. Evaluation of selected sediments and petroleums, National Petroleum Reserve, Alaska. *In*: MAGOON, L. B. & CLAYPOOL,

G. (eds) *Alaska North Slope Oil/Rock Correlation Study*. American Association of Petroleum Geologists Studies in Geology, **20**, 459–480.

——, KENNICUTT, M. C. & BROOKS, J. M. 1990. Classification of offshore Gulf of Mexico oils and gas-condensates. *American Association of Petroleum Geologists Bulletin*, **74**, 187–198.

WALTERS, C. C. & CASA, M. R. 1985. Regional organic geochemistry of offshore Louisiana. *Transactions of the Gulf Coast Association of Geological Societies*, **35**, 277–286.

WINTERS, J. C. & WILLIAMS, J. A. 1969. Microbial alteration of crude oil in the reservoir *In: Petroleum Transformations in Geologic Environments, Symposium of the American Chemical Society, Petroleum Chemistry Division Preprints*, **14**, E22–E31.

A case study of migration from the West Canada Basin

N. PIGGOTT[1] & M. D. LINES[2]

[1] BP Exploration, 9401 Southwest Freeway, Houston, TX 77077, USA
[2] BP Exploration, 301 St Vincent Street, Glasgow G2 5DD, UK

Abstract: A regional model for hydrocarbon charge in the West Canada Basin is presented. Three discrete 'hydrocarbon cells': Devonian, Mississippian–Neocomian and mid Upper Cretaceous (Post-Mannville) host petroleum which was indigenously generated from five major source rocks. Source depositional environment and Columbian–Laramide basin geometry control this pattern. On a regional scale, lateral carrier bed migration dominates basin plumbing, with the mid Cretaceous Mannville Group (oil sands) acting as the ultimate gathering point for subcropping source–carrier systems.

At a play-specific level, an example of significant local vertical migration within the gas phase is presented to explain charge of Permo-Carboniferous sands in the Peace River Arch area.

The West Canada Basin has a generally NW–SE grain between the Canadian Shield to the east and the Western Cordillera to the West (Fig. 1). It extends from the Northwest Territories in the north to central Montana in the south where it is contiguous with the Williston Basin. From a petroleum geology standpoint it contains enormous (approximately $1\frac{3}{4}$ trillion barrels) oil reserves, mainly in tar deposits (Fig. 1) but significant conventional oil as well (approximately 40 billion barrels). Clearly, developing a perception of migration from source rock to trap, of such a significant resource is of major economic importance. Because the basin is in a mature state of exploration (approaching 200 000 wells) and thanks to an enlightened government policy on rock sample storage and access of this material to industry, it is possible to build a good understanding of trap charge for the numerous hydrocarbon plays.

The key to building this understanding lies in both oil and source geochemistry and in an appreciation of overall geometry. Principles of petroleum migration which can be established with some certainty in the West Canada Basin where control is relatively tight have application by analogy for play-making in frontier basins.

Geological setting

In gross terms, the stratigraphy of the West Canada Basin (Fig. 2) reflects two phases of its tectonic history; a Mid-Devonian–Mid-Jurassic 'rift-drift' phase and a Late Jurassic–Eocene flexural (Foreland Basin) phase.

Economic basement is commonly considered as pre-Devonian comprising Proterozoic (igneous rocks and metasediments) plus remnants of Lower Palaeozoic sediments. The Lower Palaeozoic (Sauk and Tippecanoe sequences of Sloss 1963) has yielded no significant petroleum reserves in the West Canada Basin but contains important productive plays in the Williston Basin where the succession is somewhat more complete.

A poorly understood extensional event of Early–Middle Devonian age started subsidence in a NW–SE-oriented intra-cratonic feature called the Elk Point Rift. It is unclear whether subsidence was truly thermally driven but a typical syn-rift assemblage of continental red beds and evaporites dominate the stratigraphy, deposited in a highly restricted basin. Basement Highs (rift shoulders?) such as the West Alberta Ridge, Tathlina High and Peace River Arch plus a major carbonate complex (the Presquile Barrier) sealed off the basin (Fig. 1). Small Pinnacle reef basins in NW Alberta acted as foci of subsidence. A regional unconformity 'the sub-Watt Mountain' ended this phase of sedimentation. The subsequent Devonian stratigraphy was dominated by four major carbonate/mud cycles (lithostratigraphically known as Beaverhill Lake, Woodbend, Winterburn and Wabamun). In this 'rift fill' period, the basement highs were progressively onlapped by successive transgressions. The basin was slowly evolving into an open-marine passive margin (the western flank of cratonic North America). The Elk Point Rift failed but possibly a successful, subparallel rift, in the present location of British Columbia, opened (? the Anvil Ocean of Tempelman–Kluit, 1979).

A regional unconformity (distal effect of the Ellesmerian/Antler Orogeny) closed off Devo-

From England, W. A. & Fleet, A. J. (eds), *Petroleum Migration*
Geological Society, Special Publication No. 59, pp. 207–225.

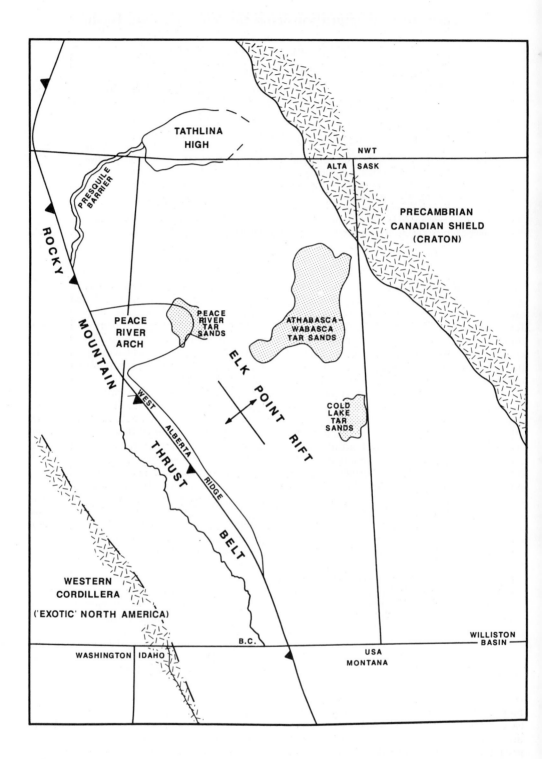

Fig. 1. Location map and tectonic elements of the West Canada Basin.

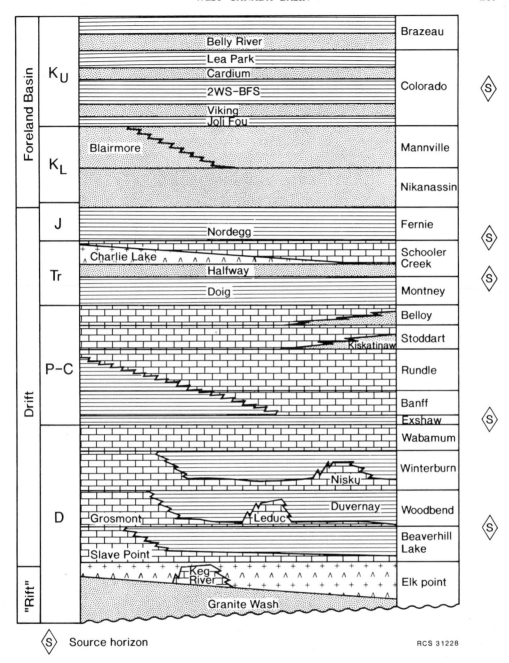

Fig. 2. Simplified lithostratigraphy for the West Canada Basin.

nian sedimentation. A major transgression in earliest Mississippian times was followed by predominantly carbonate sedimentation (westward prograding carbonate banks). In Late Mississippian–Permian times the palaeo-high Peace River Arch collapsed and became an intermit-tent local focus of clastic sedimentation (Kiskatinaw and Belloy sands) along a carbonate-dominated margin.

Clastic sedimentation becomes more important within the Triassic–Jurassic stratigraphy which is most complete in the northwest sector of the basin. Erosional unconformities are present in this interval (Base Montney, Doig/Halfway, Base Nordegg) but their precise tectonic significance is unclear.

The Late Jurassic Passage Beds passing up into the Nikanassin marks a radical switch in basin polarity. These and younger sediments are predominantly of westerly provenance, and mark the beginning of Foreland Basin deposition. The ocean to the west was closed as the exotic terranes which comprise British Columbia were accreted and the Rockies fold and thrust belt (of North American rocks) created by the associated compression. The precise accretionary history is not yet known but the Columbian and Laramide events record five major coarse clastic pulses: Nikanassin, Mannville, Viking, Cardium and Belly River interspersed with thick shales. This sedimentation infilled a long NW–SE oriented foredeep and since Eocene times (maximum burial) the basin has been undergoing uplift due to isostatic recovery, with drainage systems carrying Rockies derived sediments out to the continent margins (Mackenzie system–Beaufort Sea and Mississippi system–Gulf of Mexico)

Within this framework, oil and gas reserves occur in carbonate and clastic reservoirs of all ages from Mid-Devonian to Late Cretaceous. At first sight, such a distribution is, in itself, evidence of rich sources, widespread generation and expulsion and effective communication. While at a regional level, risk on source is clearly low, even in a mature basin an appreciation of how the oil gets from source to trap can give an explorer an edge in highlighting new plays, highgrading plays or major migration fairways and can downgrade prospects comprising good reservoir in trap configuration, if no charge could reach such a prospect.

Geochemical framework

The first serious geochemical investigation of the basin by Deroo *et al.* (1977) using basic analytical procedures, which were then current, delineated three oil families; Post-Mannville Cretaceous reservoired oils, Mississippian–Manville reservoired oils and Devonian oils. Oil–source correlations were not derived for these three families.

Since Deroo *et al.*, little geochemical work at the basin wide scale was available until Creaney & Allan (1990). In the intervening years many useful pieces of the jigsaw were assembled such as the Duvernay–Leduc story (Stoakes & Creaney 1985).

Using simple pyrolysis-TOC to detect rich source intervals and isotope-mass spectrometry and gas chromatography–mass spectrometry to characterize source extracts and oils and make oil–oil and oil–source correlations, we have evolved a geochemical framework consistent with Deroo *et al*'s three oil families and very similar to that of Creaney & Allan, (1990). On a basinwide scale, five significant source rocks (Fig. 2) have been detected and characterized: the Frasnian Duvernay, the earliest Mississippian Exshaw, the Middle Triassic 'Hot' 'Doig, the Sinemurian Nordegg and the Cenomanian–Turonian Second White Specks–Base Fish Scale zone (2WS-BFS).

A number of other source rocks have been detected notably off-reef Keg River/Ratner, Montney, Charlie Lake, Blairmore coals and Lea Park Shales. Although locally important (e.g., off-reef Keg River sources the Rainbow–Zama–Shekilie and Ratner the Winnepegosis pinnacles), these are minor contributory sources to the overall hydrocarbon flux.

The existence of the three families of Deroo *et al.* (1977) and the apparent lack of mixing of source products, apart from in the Mannville, allows the stratigraphy to be broken down into three 'hydrocarbon cells' (Fig. 3). The 'hydrocarbon cell' as defined here contains reservoired oils which are indigenously sourced from within the stratigraphic interval of that cell. Hence, the Devonian oils are sourced from the Duvernay (and its equivalent, the Muskwa, in Northern Alberta). Locally, Keg River/Ratner (NW Alberta) and possibly off-reef Nisku (West Pembina) have contributed, but both are intra-Devonian sources (Fig. 4).

Deroo *et al.*'s Mississippian–Mannville family in fact contains three oil families sourced by the Exshaw, Doig and Nordegg.

The Post-Mannville Cretaceous cell contains oils sourced from the 2WS-BFS source system.

On the east flank of the basin, the Mannville represents a zone of mixing where oils cannot be successfully tied to one source. This is, as Creaney & Allan (1990) showed and will be discussed later, a facet of migration.

Exceptions to this pattern exist but are few. As a result, the Duvernay although a prolific source has not generated any of the conventional oils in reservoirs younger than Devonian. Conversely, oils sourced from post-Devonian sources do not occur in the Devonian.

The existence of the hydrocarbon cells, defined by oil-source correlation, is a consequence of two fundamental controls; basin geometry and source rock fabric.

The wedge-shaped geometry of the Foreland

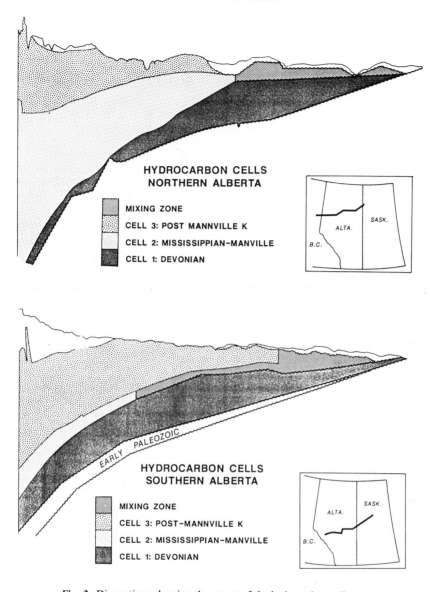

Fig. 3. Dip sections showing the extent of the hydrocarbon cells.

Basin exercises a control on both generation and migration. Insufficient burial by Middle Jurassic times means that no significant pre-Foreland Basin generation and migration of oil, even from the oldest and deepest Devonian source had occurred. As a result, the major cooking pot for the basin is in the foredeep of the west, the east flank being immature. At the same time regional dip is to the west-southwest and this provides for a steady continual gradient up which buoyancy driven migration moves oil and gas eastwards.

The second control is perhaps more subtle: source rock fabric. Of the five major source rocks, the Duvernay and 2WS-BFS systems are similar developments inasmuch as they were deposited in restricted basins. The prevailing shape of the basin at time of deposition meant deposition occurred in seaways effectively hemmed in by palaeotopography. The extensive Leduc reef chains (such as Rimbey–Meadowbrook and Bashaw), the Jefferson and Grosmont Banks plus the Peace River Arch during

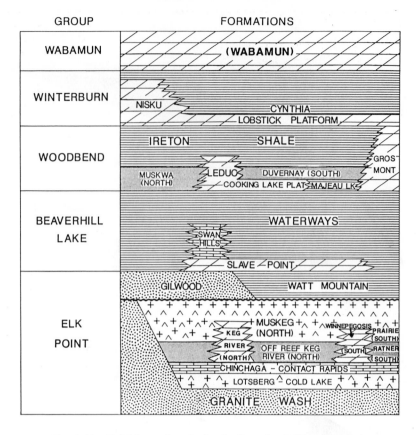

GROUP   FORMATIONS

Fig. 4. Detailed lithostratigraphy for the Devonian in West Canada.

depositon of the Duvernay and the Rockies during deposition of the 2WS-BFS, enhanced organic matter preservation, probably by limiting turn-over, producing a stratified water column and euxinic bottom waters. In contrast, no such topographic restriction existed during deposition of Exshaw, Doig or Nordegg source rocks. These were deposited on an open marine passive margin. The source development can best be modelled as organic productivity events. The phosphatic Doig and Nordegg can easily be rationalized. as upwelling events, while the Exshaw is part of a continent-wide Fammenian–Tournaisian Black shale event (Bakken/New Albany/Ohio/Chattanooga/Antrim being partial equivalents), possibly an oceanic anoxic event.

The ancient analogue for the Nordegg and Doig upwelling is the older Phosphoria of Wyoming. This unit was deposited on the same western continental margin during the Permian. Parrish (1982) shows how a predicted, dominant high pressure cell could have maintained shore parallel winds and created year-round upwelling in Kazanian times. Predictive models for Induan and Pliensbachian times (Parrish & Curtis 1982) show a similar situation persisted for the entire margin on which upwelling may be predicted during Doig and Nordegg deposition.

Since the Mississippian–Jurassic all source rocks were deposited during a period of transgression, whereby organic rich mudstones were developed over a large area of shelf in the northwest sector of the basin. In each case, the transgression was followed by regression in Upper Doig and Upper Nordegg–Rock Creek Formations. This imposes a coarsening-up fabric on the source unit. Such a fabric enhances upwards expulsion from a generating source rock. As a result little or no Exshaw-sourced oil was expelled downwards while the widespread nature of the transgression preserved the integrity of the Devonian cell. Similarly most phosphatic Doig oil was expelled up into Upper Doig/Halfway sands. The same reason explains why so little Nordegg-sourced conventional oil occurs in the basin; upward expelled oil finding its way into

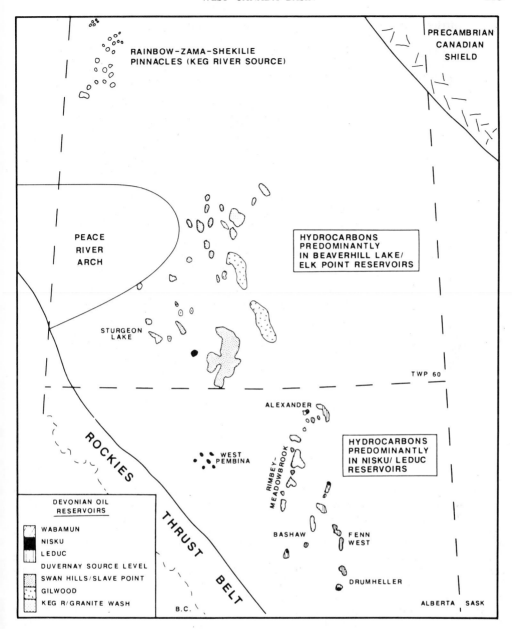

Fig. 5. Map showing stratigraphic positions of producing reservoirs in the Devonian.

the Foreland Basin sequence and ultimately into tar deposits (discussed later). The Nordegg is perhaps volumetrically the best source in the basin being consistently thickest and richest but it accounts for a disproportionately small volume of conventional pools.

Of course, exceptions occur and where highly porous rocks underlie the sources, some oil finds its way downwards. Examples include Wabamun oil at Alexander field (Exshaw-sourced) and Tangent oil at Rycroft field (Nordegg-sourced).

Lateral carrier bed migration

Given the geological and geochemical framework, it is possible to show that lateral carrier bed migration is the predominant process in source-trap communication, particularly for oil plays in the Devonian and Post-Mannville Cretaceous hydrocarbon cells and tar sand deposits (mixing zone, Fig. 3).

Devonian

Examining the stratigraphic occurrence of oil in Devonian reservoirs (Fig. 4) reveals a clear N–S dichotomy in reserves disposition. The Devonian oil in the south of the basin is reservoired in rocks the same age or younger than the Duvernay source (Fig. 5), mainly in Woodbend and Winterburn Group carbonate build-ups. To the north, the oils occur mainly in reservoirs older than the Duvernay source (mainly in Elk Point clastics and Beaverhill Lake carbonates). The Rainbow–Zama–Shekilie oils are off-reef Keg River sourced and not further discussed here.

The observed dichotomy is a result of migration control.

In the south, the Duvernay is frequently underlain by a partially dolomitized platform carbonate; the Cooking Lake. This acts as an obvious migration fairway conducting oil up dip, filling Leduc reefs as it migrates (Fig. 6). This led to the celebrated Gussow (1954) principle whereby the Rimbey–Meadowbrook trend is filled northeastwards by progressively spilled hydrocarbons, leaving gas in the southwest and moving oil to the northeast. In reality, there are anomalies within the phase distribution which Gussow accounted for by special pleadings. More importantly, many reefs are not full-to-spill and some are not rooted on the Cooking Lake meaning whilst Gussow's principle has general application, it does not strictly hold. Creaney & Stoakes (1990) explained these anomalies in terms of a 'leaky pipeline' whereby some reefs may be by-passed by the regional migrating hydrocarbon flux due to permeability barriers within the Cooking Lake/Leduc. Local subjacent mature Duvernay source rocks also play an important role particularly for isolated reef ac-

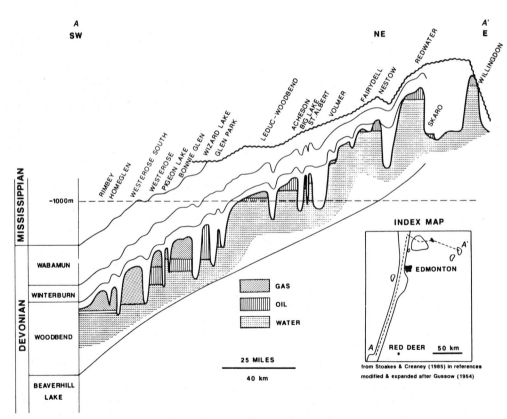

Fig. 6. Leduc reef–Cooking Lake Platform plumbing system.

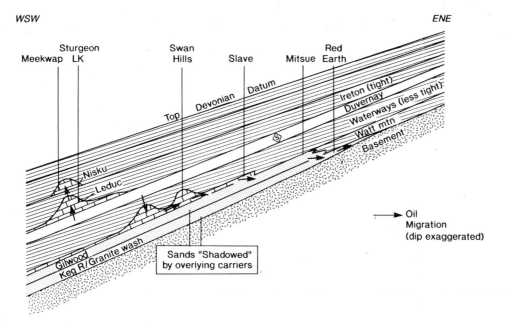

Fig. 7. Pattern of migration in Devonian of Northern Alberta.

cumulations. Hence, while long range carrier bed migration is the dominant process, in detail, vertical communication also plays a vital role in this play. This is further emphasised by the Nisku reservoired accumulations which commonly occur above Leduc reefs perhaps where Nisku build-ups root on Woodbend compaction highs. For instance, Fenn West and Drumheller fields contain geochemically identical oils at both Leduc and Nisku levels implying vertical charge of the latter. Certainly, no equivalent Nisku off reef source development (similar to the Duvernay) can be detected outside of the West Pembina Basin, which could set up an analogous Duvernay–Cooking Lake carrier bed system. The pinnacle reefs of West Pembina contain oils similar to Duvernay sourced oils but the reefs overlie a thick Ireton Shale development (sealed off from the Duvernay) and this argues strongly for some local off-reef Nisku source development. Unfortunately in such a westerly (mature) basin position it is difficult to detect and characterise the implied source due to generation effects reducing original potential.

In the northern sector of the basin (Fig. 5), excepting the Rainbow–Zama–Shekilie pinnacles (Keg River sourced), the major reserves reside within Swan Hills reefs, Slave Point dolomitized carbonate bank edge, Gilwood, Keg River and Granite Wash sands which are older

than their Duvernay source. The fundamental reason for this is the lack of a regionally developed dolomite platform facies beneath the Duvernay; the Cooking Lake passes northwards into the basinal Majeau Lake Formation (Fig. 4). Exceptions to this situation occur around the Peace River Arch where a Leduc fringing reef system is developed and also on isolated carbonate banks such as Sturgeon Lake where charge can be affected as in the southern sector of the basin. In general, however, charge of the older reservoirs is accomplished by short vertical migration through the Waterways Formation probably at the same time as downwards expulsion from the source. The Waterways forms a less effective barrier to migration than the tighter overlying Ireton Shale based on microporosimetry measurements on cores from the two units. Lateral migration then occurs with successive facies changes such as the Slave Point shale-out and Gilwood–Watt Mountain transition which force the oil into progressively stratigraphic deeper reservoirs as it moves up dip eastwards (Fig. 7). Closer examination of this migration model shows that deeper plays will be 'thieved' of oil by shallower carriers which effectively siphon it off. Hence, it is hard to charge Granite Wash sands in valid trap configuration where Slave Point dolomites or Gilwood sands intervene between the reservoir and its source

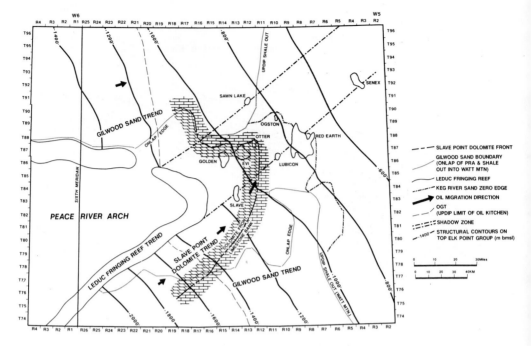

Fig. 8. Carrier bed facies changes and migration focusing, Peace River Arch area, Alberta.

(the Duvernay). The oil migration mechanism is thus imposing important constraints on the prospective extent of the various play systems.

Another communication issue of interest within the Devonian of Northern Alberta relates to the fact that the Duvernay was not deposited over the Peace River Arch, which should create a 'shadow zone' of no charge, given a simple up dip migration pattern. And yet accumulations such as Evi, Otter, Golden and Ogston occur in this zone. In reality, migration directions are not simply orthogonal up dip but instead are deflected by permeability barriers such as the Slave Point bank edge (Stoakes 1987) which moves oil oblique to dip, up plunge (Fig. 8).

Tar sands

All lateral carrier bed systems in the Devonian cell and younger Mississippian–Neocomian cell eventually subcrop the Mannville sand. This unit is thus a focus or gathering point for all the natural pipelines. The Mannville also underlies a regional shale seal; the Joli Fou Shale. This juxtaposition accounts for the massive Athabasca and Cold Lake Tar sand accumulations (Fig. 1) which are mixed oils at the terminus of the lateral carrier beds, having been sourced to varying degrees from all four major pre-Creta-

ceous sources. The Mannville itself is basically a non-source (Moshier & Waples 1985) except for Blairmore coals developed on the delta plain to the west, which are gas-prone only. Input from the Mesozoic (Doig and Nordegg) sources to the Cold Lake deposit cannot be seen by geochemical typing of the bitumen which is consistent with the absence of these source units downdip from Cold Lake in its drainage area in Southern Alberta. These sources are, however, the major precursors for the Peace River deposit which occurs west and downdip from the subcrop of Palaeozoic carriers for any Duvernay sourced oil (Fig. 9). The same is true to a large extent for the Exshaw which also subcrops to the East and locally lacks obvious carrier beds into which oil can be fed.

Upper Cretaceous

Lateral migration also dominates within the Upper Cretaceous cell. Here, the oils are *not* biodegraded as in the older Mannville reservoirs and are in fluid isolation from the rest of the basin due to the regional Joli Fou seal. All the oils, principally reservoired in Cardium, Viking and Belly River cannot be geochemically correlated with the older oils or pre-Mannville source rocks. A positive correlation can be made with

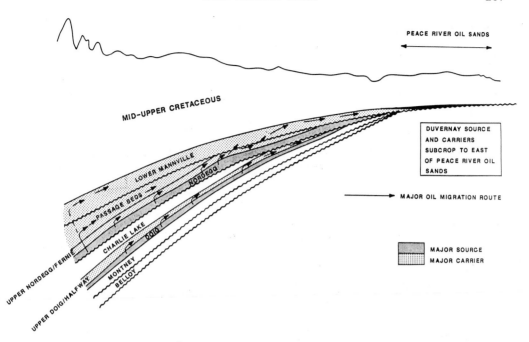

Fig. 9. Mesozoic sourcing for the Peace River Tar sands deposit.

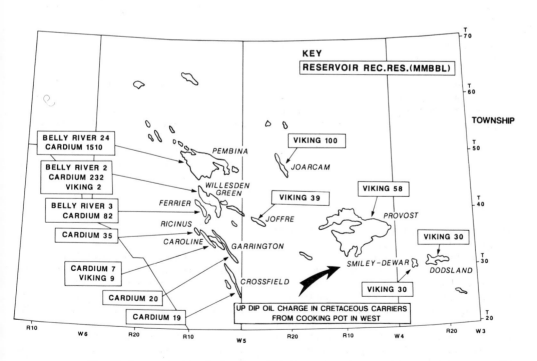

Fig. 10. Long range lateral migration within Post-Mannville Cretaceous sands.

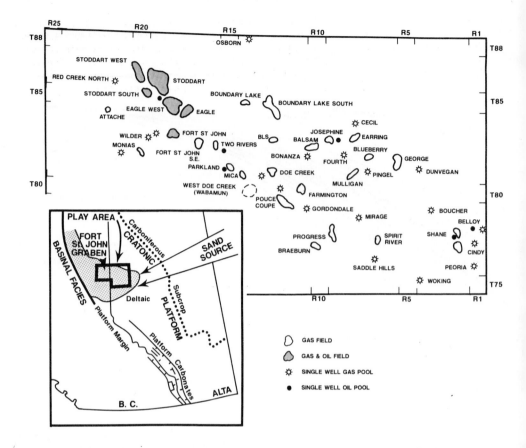

Fig. 11. Permo-Carboniferous gas play trend of west central Alberta and northeast British Columbia.

the 2WS-BFS source which is mature in the foredeep on the west side of the basin. Lateral migration has occurred over long distances, particularly within the Viking (viz. the Dodsland accumulation in Saskatchewan) where oil has moved 200–300 km up dip (Fig. 10).

Vertical migration

Within the Mississippian–Neocomian hydrocarbon cell there is an example of vertical hydrocarbon migration being the dominant process controlling both successful trap charge and phase type/distribution. The main play concerned is the Carboniferous Kiskatinaw sands gas trend in west central Alberta and adjacent parts of British Columbia (Fig. 11). Here, a much expanded Permo-Carboniferous section (Fig. 12) was deposited in the Fort St John Graben, a local clastic-filled sub-basin formed by inversion

of the Devonian Peace River Arch basement high. Gas is the predominant hydrocarbon type within the Stoddart and Belloy reservoirs although light oil also occurs in Stoddart and Eagle fields to the west. Condensate gas ratios (CGR) increase in pools on the west side of the playtrend. The presence of gas is anomalous inasmuch as the Carboniferous Rundle Formation produces oil from traps approximately on-strike to the south. The oil v. gas distribution within the Permo-Carboniferous is anomalous too, in that the increasing oil westwards is counter to regional maturity which increases in that direction suggesting the trend should become more gas prone.

The stratigraphic section in question lies significantly isolated from the major regional source systems of the Mississippian–Neocomian (Exshaw, Doig and Nordegg) with no obvious source-trap communication path. Four models could explain the gas charge:

(1) a new source rock system is working locally;
(2) local maturity increases compared to that observed along strike, leading to in-reservoir cracking of an oil charge to gas;
(3) the major Carboniferous source, the Exshaw has undergone a facies change from oil- to gas-prone;
(4) the phase is a consequence of the migration mechanism.

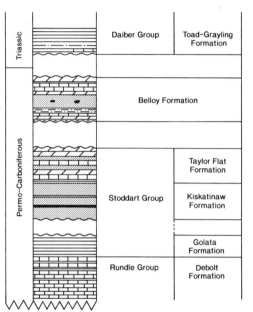

Fig. 12. Detailed Lithostratigraphy for the Permo-Carboniferous, Peach River Arch area.

Considering the first of these explanations, two likely candidates were investigated as local source rocks. The Golata Formation (Fig. 12) is a black shale, locally up to 75 m thick, has a high gamma ray log motif and is in excellent communication with the main Kiskatinaw Sands reservoir (frequently incised by channels). Geochemical data for the Golata however (Fig. 13) reveal it to be a non-source. The samples were collected from core along its subcrop edge and are only immature to early mature so do not represent low yield source rocks due to 'spent' potential. The only good yield was obtained from a thin volumetrically insignificant stylolite in the top of the Debolt Formation (boundary with Golata is transitional). The poor result for the Golata can be rationalized from inspection of a spectral gamma ray log through the unit. Although a 'hot' shale, the gamma response is

due to clays with high potassium and thorium levels. In contrast, the rich Nordegg and Doig source rocks give a high gamma response due to increased uranium content (Fig. 14) which is preferentially concentrated in many source rocks by high organic matter content.

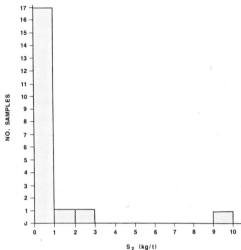

Fig. 13. Source potential of the Golata Formation (S_2 = Source Potential kg/tonne).

A second local source development considered was the possibility that coals within the Kiskatinaw were the source. Given the estuarine/deltaic depositional environment coal developments are possible. Examination of a large number of Kiskatinaw cores, however, showed the coal content of the section is very low and developments too thin to constitute a volumetrically significant source rock.

The second model advanced to explain the gas charge; in-reservoir cracking must also be rejected. Maturity data for the Exshaw (Fig. 15) show no on-strike deviation in isomaturity patterns.

A pyrogram for the major Exshaw shows it to be an excellent oil-prone source in the less mature eastern part of the study area (Fig. 16). It is no more gas-prone than core samples to the south which have sourced oil.

Hence, almost by default, the charge of the Permo-Carboniferous with gas must represent a facet of migration. There is a significant difference in the Banff–Debolt stratigraphy of the Fort St John Graben compared with areas to the south, in that this section is approximately 800 m thick between the Exshaw source and the reservoirs, and is mud-dominated wackestone and

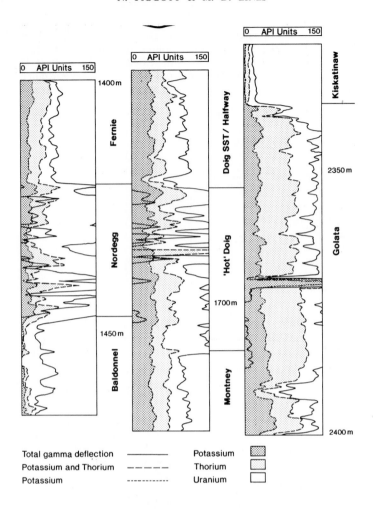

Fig. 14. Spectral Gamma-ray log response for Golata compared to rich Doig and Nordegg source rocks in BP Farmington 15-19-80-11W6.

mudstone rather than clean, partly dolomitized grainstones as it is to the south (Fig. 17). This lithology change occurred within a precursor of the Fort St John Graben called the Pekisko 'Shale' Basin, a feature which suggests this part of the Carboniferous shelf subsided more actively even during Banff–Rundle times. The tighter underlying Banff–Debolt section in the area of the Kiskatinaw play is postulated to act as a barrier to vertical oil-phase migration.

Figure 18 sums up the model we postulate to explain gas charge of the Kiskatinaw from the Exshaw. It shows oil generation commencing in the source but surrounding tight lithologies (the underlying Wabamun muddy carbonates were deposited in a distal ramp setting; Stoakes 1987)

inhibit effective expulsion and the oil is retained within the source or subjacent fractured Banff (Stage 1). Further burial (in Late Cretaceous–Palaeogene times) caused gas generation from the kerogen and cracking of unexpelled oil (Stage II), followed by gas-phase vertical migration through the Banff–Debolt up to Kiskatinaw level (Stage III). Within the Kiskatinaw, lateral carrier bed migration is possible (or within overlying Belloy dolomites or sands if there is no intervening seal) and the gas, saturated with condensate, at or close to dew point, would move up-dip eastwards (Stage IV). Moving up-dip involves moving to lower pressure and temperature regimes which will lead to progressive condensate stripping from the gas along

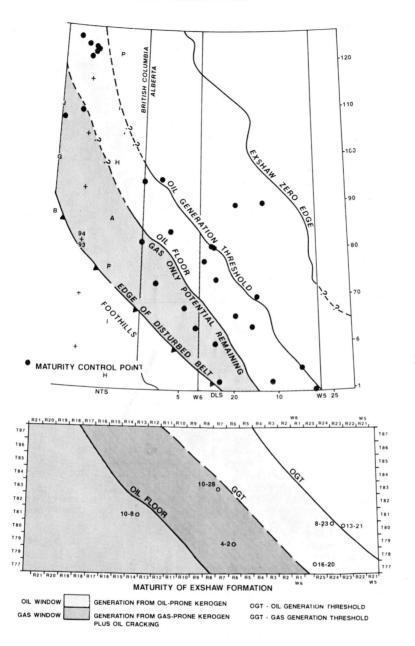

Fig. 15. Maturity maps for the Exshaw source rock (Top, regional; Bottom, local; Fort St John Graben).

the migration route. This would lead to wet gas rich in condensate in the west becoming increasingly dry eastwards.

A second process which will affect the reservoir hydrocarbon phase is the post-migration history. Since maximum burial in Eocene times the West Canada Basin has been undergoing isostatic uplift and erosion. Most uplift and erosion has been experienced in the Laramide foredeep in the west and this decreases eastwards. As a result, gas trapped in the west end of the Kiskatinaw/Belloy playtrends (which would

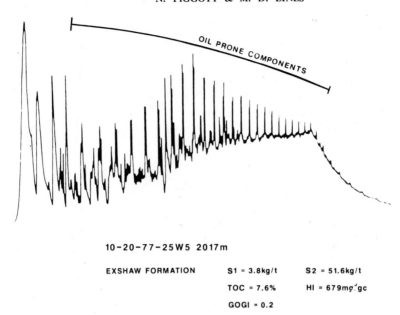

10-20-77-25W5 2017m

EXSHAW FORMATION   S1 = 3.8kg/t   S2 = 51.6kg/t

                         TOC = 7.6%   HI = 679mg/gc

                         GOGI = 0.2

Fig. 16. Pyrogram for the oil-prone Exshaw source from well 10-20-77-25W5 (2017M) (S1 = free hydrocarbons, S2 = source potential, TOC = Total Organic Carbon, HI = Hydrogen Index, GOGI = Gas Oil Generation Index where <0.2 = oil-prone, 0.2–0.5 = oil- and gas-prone and > 0.5 = gas-prone).

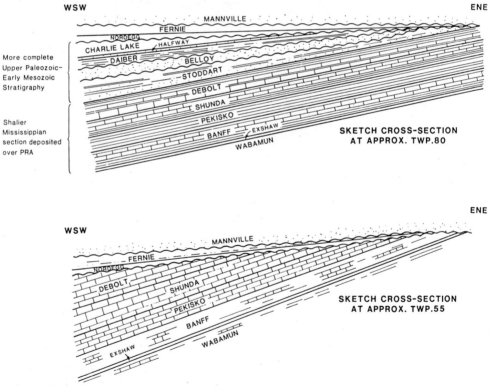

Fig. 17. Schematic dip cross-sections across West Canada Basin showing facies changes within Banff–Rundle carbonates.

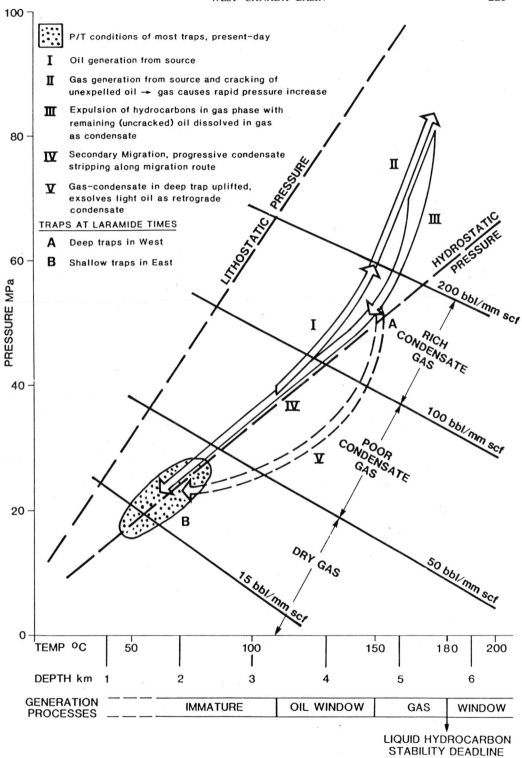

Fig. 18. Schematic showing generation–expulsion–migration–trap residence history for Exshaw source hydrocarbon.

have been richer in condensate/oil as it will have experienced less lateral migration/condensate stripping) will see a greater reduction in temperature and pressure because of greater uplift. The result would be exsolution of a light oil leg (retrograde condensation) beneath the gas (Stage V). Oils in the Eagle and Stoddart Fields are highly mature consistent with significant cracking and support the proposed model.

As well as explaining the observed anomalous oil v. gas distribution (oil in the west, drier gas in the east) within the playtrend and being consistent with the geochemistry of the oils the model satisfactorily explains two other observed phenomena. The first of these is the frequent pyrobitumen in fractures within the Exshaw and Banff cores. Pyrobitumen is also observed beneath the Exshaw in fractures and irregularly dolomitised Wabamun cores from the West Doe Creek area (Fig. 11). The pyrobitumen would represent the cracking residue as oil was thermally converted to gas. Secondly, gas shows are encountered in rare porosity streaks and fractures within the Pekisko and Banff when drilled beneath the Kiskatinaw trend. These would represent en route migration losses during vertical gas migration from the Exshaw source to the Kiskatinaw reservoir/carrier bed.

Conclusions

Three 'Hydrocarbon Cells' are defined: Devonian, Mississippian–Neocomian and Middle–Upper Cretaceous. Reservoired oils within each cell are indigenously sourced from five major source rocks. The cells exist as a result of basin geometry which controls maturity and migration patterns plus stratigraphy which controls both source depositional model (hence, dominant petroleum expulsion directions) and seal development.

At a regional level, charge of the major producing play systems in the basin is dominated by extensive lateral carrier bed migration. Good examples occur within the Devonian cell although in detail local vertical plumbing plays an important role, both in charge of Leduc and Nisku reefs and limiting the extent of prospectivity of various Middle Devonian plays south of the Peace River Arch. Communication within the Upper Cretaceous plays is totally governed by lateral carrier bed migration, which occurs over very long distances. At the same time, the lack of vertical fluid migration at this level has served to preserve oils from the biodegradation which affects the older Mannville oils. The Mannville sands are themselves charged by numerous subcropping carrier bed systems and, as such, represent a natural focus for migrant oils from multiple rich source rocks, which are the precursors of the massive tar deposits.

A model involving significant vertical migration in the gas phase is invoked to charge the Kiskatinaw sands play from late mature Exshaw source. This migration model is consistent with the observed hydrocarbon type distribution in the play and pervasive pyrobitumen plugging and gas shows on the migration route.

The authors would like to thank BP Resources Canada, BP Research Centre, Sunbury and BP Exploration, Houston for support in the preparation of this paper and permission to publish.

We would also like to thank the following individuals for useful discussions on the concepts and applications thereof, outlined here: B. Twombley, D. Mundy, S. du Toit, D. Mitchell, B. Clarke and B. Spring.

References

CREANEY, S. & ALLAN, J. 1990. Hydrocarbon generation and migration in the Western Canada Sedimentary Basin *In*: BROOKS, J. (ed.) *Classic Petroleum Provinces.* Geological Society, London, Special Publication, **50**, 189–202.

DEROO, G., POWELL, T. G., TISSOT, B. & McCROSSAN, R. G. 1977. The origin and migration of petroleum in the Western Canadian Sedimentary Basin, Alberta: A geochemical and thermal maturation study. *Geological Survey of Canada Bulletin*, **262**.

GUSSOW, W. C. 1954. Differential entrapment of oil and gas: a fundamental principle. *Bulletin of the American Association of Petroleum Geologists*, **38**, 816–853.

MOSHIER, S. O. & WAPLES, D. W. 1985. Quantitative evaluation of Lower Cretaceous Mannville Group as source rock for Alberta's Oil Sands. *Bulletin of the American Association of Petroleum Geologists*, **69**, 161–172.

PARRISH, J. T. 1982. Upwelling and petroleum source beds with reference to the Paleozoic. *Bulletin of the American Association of Petroleum Geologists*, **66**, 750–774.

—— & CURTIS, R. L. 1982. Atmospheric circulation, upwelling and organic-rich rocks in the Mesozoic and Cenozoic eras. *Palaeogeography, Palaeoclimatology, Palaeoecology*, **40**, 31–66.

SLOSS, L. L., 1963. Sequences in the Cratonic Interior of North America. *Geological Society of America*

Bulletin, **74**, 93–114.

STOAKES, F. A. 1987. Fault controlled dolomitisation of the Wabamun Group, Tangent Field, Peace River Arch, Alberta. *In*: KRAUSE & BURROWES (ed.) *Devonian lithofacies and reservoir styles in Alberta* C.S.P.G. publication for Second International Symposium on the Devonian System, 73–86.

—— & CREANEY, S. 1985. Controls on the accumu-lation and subsequent maturation and migration history of a carbonate source rock. *Society of Economic Paleontologists and Mineralogists Core Workshop Proceedings: Golden, Colorado.*

TEMPELMAN-KLUIT, D. J. 1979. *Transported cataclasite, ophiolite and granodiorite in Yukon: evidence of arc–continent collision.* Geological Survey Canada Paper **79–14**.

Hydrocarbon maturation, migration and tectonic loading in the Western Alpine foreland thrust belt

ROBERT W. H. BUTLER

Department of Earth Sciences, The Open University, Walton Hall, Milton Keynes MK7 6AA, UK

Present address: Department of Earth Sciences, The University of Leeds, Leeds LS2 9JT, UK

Abstract: The Western Alpine foreland fold and thrust belt developed in late Miocene times through the Neogene foredeep basin, superimposed on an early Mesozoic and local Palaeogene shelf–basin system. The relative timing between subsidence, tectonic burial and thermal maturation is examined along a section line through the Annecy district of the French Alps. Available stratigraphic data are used to define burial–time paths, using suitable decompaction parameters. The tectonic burial history is determined from regional structural sections. These burial histories are used with a simple thermal model (geothermal gradients: 28°C km^{-1} before thrusting, 30°C km^{-1} for the post-orogenic period) to establish hypothetical time–temperature histories. Both the Lopatin method and peak temperature model of source rock maturation suggest that the Jura anticlines are undermature while the Subalpine chains are overmature. However, Subalpine maturation occurred during foredeep subsidence with overmaturation during thrust sheet loading. Large-scale lateral migration of hydrocarbons is required to explain the distribution of bitumen accumulations in the foreland basin offering the possibility of a series of structural and stratigraphic traps in the palaeo-foreland which are not associated with the thrust belt. A complex secondary and remigration history would be needed if liquid hydrocarbons are to have accumulated in traps associated with compressional structures in this part of the thrust belt.

Foreland thrust belts are important regions for hydrocarbon exploration. Generally they have developed in sedimentary basins and passive margin sequences which experienced only moderate subsidence during their extensional histories. These types of basins may contain significant volumes of potential source rocks for hydrocarbon generation. If rift-related subsidence has been relatively weak these source rocks can reach maturity during the regional development of a thrust belt. This is important because large numbers of structural plays can be created during thrusting, either by entrapment of hydrocarbons beneath impermeable thrust sheets (se-called sub-thrust plays), by significant lateral migration to tilted stratigraphic and ancient structural traps or to more intricate structures within the thrust belt itself. Critically important is the relative timing between thrusting and hydrocarbon maturation in different parts of thrust belts. This problem is examined here, with specific reference to the foreland fold and thrust belt of the French Alps.

The outer parts of the western Alps are a classical example of a foreland fold and thrust belt. Structures involve Mesozoic shelf carbonates and shales together with regionally synorogenic sediment deposited during the later stages of Alpine compression, in Miocene times. The inspiration for the study described here comes from the documentation of bitumen seeps within the foreland basin, as shown on published geological maps (BRGM 1964, 1970, 1972) and discussed by Zweidtler (1985, see Fig. 1). In the Annecy area these accumulations have a history of exploitation for asphalt with most of the worked accumulations lying near the base of the Miocene molasse sands or in immediately underlying Barremian limestones. The aim here is to model hypothetically the thermal evolution of likely source regions for these accumulations using the classical Lopatin method as described by Waples (1980). This approach is strongly dependent on the various assumptions for the histories of subsidence and thermal structure for various parts of the basin. Therefore much of this paper is concerned with these assumptions, before moving on to discuss the implications of the modelling for the possible maturation and migration histories for hydrocarbons within the thrust belt. First, however, it is necessary to provide some background information.

An Alpine framework

The Alpine foreland thrust belt encompasses the regions of the Jura, the Swiss–Annecy molasse

From England, W. A. & Fleet, A. J. (eds), *Petroleum Migration*
Geological Society, Special Publication No. 59, pp. 227–244.

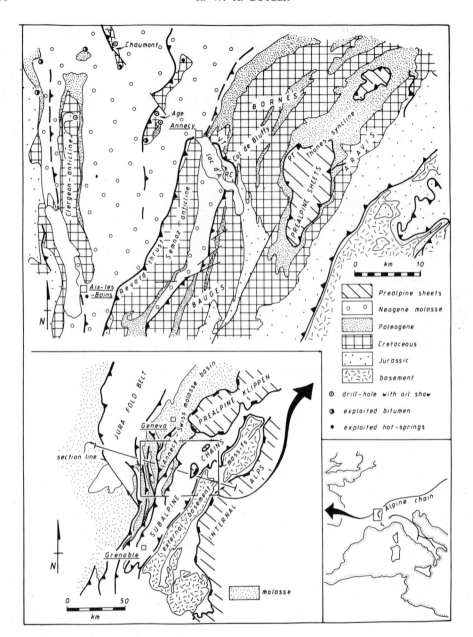

Fig. 1. Maps of the study area. (**a**) Simplified geological map of the Annecy district based on publications by BRGM. RC, Roc de Chère; PT, Prealpine thrust; VT, Veyrier thrust; lac d'A, Lac d'Annecy. (**b**) Location map showing the setting of the Annecy district in the Subalpine thrust belt. The section line of Fig. 2c is illustrated. Brick ornament along lines represents the Prealpine and Frontal Pennine thrusts. (**c**) Location of the western Alps.

basin and the Subalpine chains of France (Fig. 1). This region forms the study area of this contribution. I will consider the burial and inferred thermal history of a particular cross-section line (Fig. 2) which runs WNW–ESE, parallel to the direction of thrusting. The thrust belt involves basement structures further to the south and east (the so-called external basement

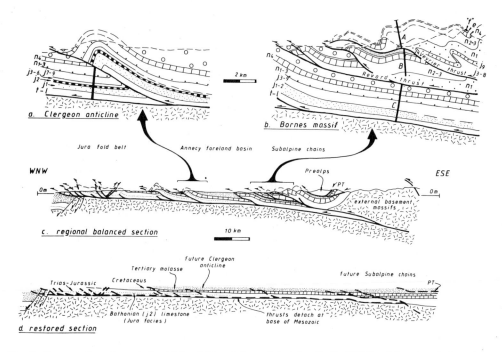

Fig. 2. Cross-sections through the Alpine fold and thrust belt (after Butler 1989*b*), all based on balanced sections constructed at 1 : 25 000. Thick black bars represent the profiles used for time–depth plots. (**a**) Through the Clergeon anticline (see Fig. 4 for time–depth plot). (**b**) The Bornes massif of the Subalpine chains (see Figs 7, 8 & 9 for the time-depth plots of A, B & C respectively). (**c**) Regional balanced cross-section through the Annecy sector of the thrust belt (PT, Prealpine thrust). (**d**) Regional restored section.

massifs) but these areas will not be considered here. Full descriptions of the regional structure are provided by Doudoux *et al.* (1982), Debelmas *et al.* (1983) and references therein. The region was interpreted in terms of conventional 'detachment-dominated' thrust tectonics by Boyer & Elliott (1982), based on published accounts of regional structure. More detailed interpretations of the thrust belt structure are provided by Butler (1989*a, b*) which incorporate new information and deep seismic reflection data (Bayer *et al.* 1987). We can consider the regional structure in three parts.

Structural styles

The outlying Jura forms the most westward expression of deformation directly associated with the Alps (but see Zeigler 1989 for other long-range effects). Structures include simple folds and thrusts, together with other fault arrays, formed within the Mesozoic sediments during regionally shortening which decoupled along

Triassic evaporites (Laubscher 1972). Hence basement is not considered to be involved in these outermost structures (Ménard 1980). To the SE of the Jura lies the Swiss–Annecy molasse basin. This too is considered to be underlain by thrust detachments (e.g. Rigassi 1977; Boyer & Elliott 1982) which pass out to the Jura but there has been relatively little folding or thrusting in comparison.

The third part of our structural triptych are the Subalpine chains. Recent seismic reflection profiling (Bayer *et al.* 1987) has shown these too to be detached from above the Palaeozoic basement and hence only involve Mesozoic and Tertiary sediments. In contrast to the Jura, the Subalpine chains along the section line of Fig. 2 were overlain by a major thrust sheet of internal Alpine units, now preserved as the Prealpine klippen. Boyer & Elliott (1982) simplified this description by considering the structures in the Jura to be directly emergent while the Subalpine chains formed as a duplex, capped by a roof thrust. This might be expected to cause radically different thermal structures in the thrust belt between the Jura and Subalpine chains.

Stratigraphic notes

Together with other parts of the Alps, the Subalpine chains and Jura represent the remnants of the old European continental margin and associated basins on the northern edge of Tethys (e.g. Aubouin et al. 1986). As such they have experienced a complex history of subsidence and uplift. In recent years stratigraphic research has been directed at re-appraising the Mesozoic sediments of the Alps in terms of this tectonic setting (Lemoine et al. 1986). A compilation of important papers can be found following the review by Lemoine & de Graciansky (1988). Along the segment of the thrust belt to be discussed here, traditional stratigraphic information is provided by the 'notices explicatives' accompanying published maps (BRGM 1963, 1964, 1966, 1969, 1970, 1972, 1980a, b, 1981 & 1987). These show that a major period of subsidence affected the European continent during the Mesozoic, with the Jurassic now widely seen as being syn-rift with the lower Cretaceous being the post-rift thermal subsidence fill (e.g. Gillcrist et al. 1987).

Additional active faulting occurred in the middle Cretaceous (Barremian) in some localities (e.g. Arnaud 1988) and these basins were in turn overlain during the Upper Cretaceous worldwide highstand in sealevel (Vail et al. 1977). All these Mesozoic rocks are dominantly shales and carbonates which are assumed to contain varying amounts of organically-lean marine source rocks but there are no widely available geochemical analyses to support this contention.

Throughout much of the French Alpine foreland thrust belt a thin unit of presumed Eocene sands unconformably overlie a karstic surface on Barremian limestones, suggesting significant Palaeocene uplift (Debelmas & Kerckhove 1980). These units are overlain by sequence of submarine siliciclastics long-recognized as forming a foredeep fill. The style of deposition varies in both space and time although there is evidence in some localities that the initial subsidence was created by active faulting in the floor of the basin (e.g. Villars et al. 1988). Otherwise the Alpine foreland basin is widely believed to have developed by the downward flexure of the European

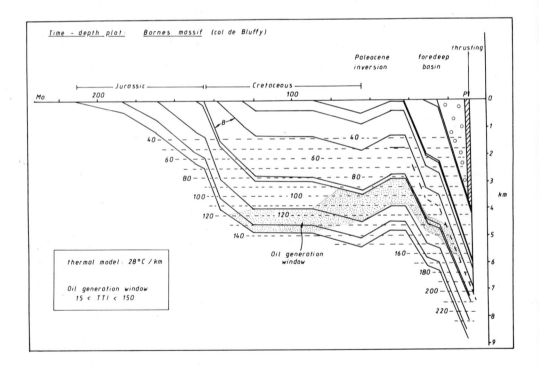

Fig. 3. Time–depth plot for the Subalpine stratigraphy, based on measured sections in the Bornes massif (Cretaceous) and projected from the restored section (Fig. 2d) and the Aravis chain (Jurassic). B, Berriasian shales, the assumed principal source rock. Open circles denote molasse, P, emplacement of Prealps (shaded). Pecked line within Berriasian denotes level of hanging-wall of Veyrier thrust.

lithosphere by loading from the adjacent, and developing, Alpine orogenic belt (e.g. Karner & Watts 1983). The foredeep fill is dominated by Miocene shallow-marine sandstones. These stratigraphic details are variable between localities and their implications for subsidence and burial histories will be discussed later.

Timing of thrusting

Although subsidence due to lithospheric flexure and thrust sheet loading began as early as Priabonian times (latest Eocene, c. 39 Ma) on the Alpine foreland, the emplacement of thrust sheets onto the Mesozoic shelf sediments in this region, together with dismembering of the shelf by thrusting, did not occur until somewhat later. A full discussion is provided by Butler (1989a), based on the time-averaged thrusting rate of about 1 cm a^{-1}, a figure which incorporates available stratigraphic and geochronological data. Using this value, and the notion that the thrusts to the east of the study area roof into the base of the Prealpine thrust sheet, implies that the Prealps finally were emplaced onto their

immediate footwall by about 12 Ma. Significant stratal shortening is generally considered to have ended in the Jura by about 6 Ma following which time the region has experienced uplift, driven primarily by isostatic rebound from erosion of the orogenic interior. Between these times thrusting in the Subalpine chains and Jura appears to have migrated outwards, so that higher thrusts are folded by lower, more outlying structures (Butler 1989a).

Subsidence history in the Subalpine chains

A generalized subsidence history for the Subalpine chains can be built up from measured stratigraphic sections in the Bornes–Aravis massif (Fig. 1). In this and all other stratigraphic models the absolute dates are based on the time scale of Harland et al. (1982). The sections are decompacted by a lithology-dependent factor, as listed on Table 1. The thickness of Upper Cretaceous sediments before the Palaeocene inversion is determined by comparison with the well-preserved successions on the east of the Aravis area (Villars et al. 1988) and is likely to be a minimum

Table 1. *Stratigraphic data for the Bornes massif*

Ma		Lithology	DF	Tp	T$_0$
	Miocene	Molasse	1	3500	3500
25					
	Oligocene	Conglomerates	1	250	250
32					
	Eocene	Flysch	1.33	1500	2000
42					
		Uplift, erosion, non-deposition			
65					
	Senonian	Shales, limestones	–	50	500
89					
	Alb-Apt.	Sandstones, conglomerates	1	5	5
119					
	Barremian (n$_4$)	Limestone	1.17	300	350
125					
	Haut-Val. (n$_{2-3}$)	Limestone, shales	1.29	700	900
138					
	Berriasian (n$_1$)	Shales	2	700	1400
144					
	Tithonian (j$_9$)	Limestone	1	150	150
150					
	Cal-Oxf-Kim (j$_{3-8}$)	Shales, limestones	1.77	650	1150
169					
	Aal-Bath (j$_{1-2}$)	Limestones, shales	1.4	500	700
188					
	Liassic (l)	Shales	2	200	400
213					

Stratigraphic abbreviations coincide with groupings used in Fig. 2b. DF, decompaction factor; Tp, present-day stratigraphic thickness; T$_0$, decompacted original thickness (both in metres). Time scale from Harland et al. (1982).

estimate. The other poorly constrained estimates involves the original thickness of Miocene molasse sediments, arbitrarily defined as being 3500 m for the Bornes massif. Most of this section is no longer preserved, being cut-out by the overriding Prealpine thrust sheets. This value will be discussed later.

Using the stratigraphic thickness illustrated in Table 1, a time–depth plot was constructed (Fig. 3). A simple linear decompaction routine was adopted using the ratios (DF) shown on Table 1. However, Fig. 3 takes no account of the sea water depth of deposition so the pathways can only be used qualitatively to determine tectonic controls on subsidence. However, this simple plot is more useful when thermal histories come to be calculated since the depth below the sea bed is more important than depth below sea level in determining the temperature of a given rock at a given time.

Figure 3 shows important periods of sediment accumulation, in the Early Cretaceous and in the Tertiary. The first of these may be the simple infilling of basin topography created during Jurassic rifting (Lemoine *et al.* 1986). Likewise the Eocene sediments may also be infilling a basin created by late Palaeocene–Early Eocene extension. However, the Eocene basin fill and the subsequent Miocene units were deposited during flexural subsidence of the European foreland, driven by thrust sheet loading. This subsidence history is appropriate only for the Bornes area. Lemoine *et al.* (1986) point out that further to the west, towards 'stable' Europe, there was far less subsidence. This can be recognized on the restored template through the thrust belt (Fig. 2b). The Mesozoic shelf of the Jura probably never exceeded a thickness of more than 2 km, in contrast to over 5 km in the Bornes area (Fig. 3).

The Clergeon anticline

A number of exploited bitumen deposits lie around the crest of the Clergeon anticline, on the eastern edge of the Jura (Fig. 1). I assume these to be the degraded remnants of thermally matured and migrated hydrocarbon liquids which, because of their volume, represent an exhumed palaeo-accumulation rather than remnants of a short-lived migration conduit. Thus a model of the Clergeon structure is the first to be discussed. Table 2 illustrates the stratigraphic model used for this structure, using measured sections taken from maps (BRGM 1972) and local drillhole data (BRGM 1970). The thickness of molasse deposits is known from map data to exceed 1 km. This is a gross underestimate, since

only the lower half of the time-record is preserved. The total thickness, if the average deposition rate of the preserved stratigraphic section was maintained, would have been nearly 2 km. It is likely that thrusting reached this area by about 7 Ma.

Burial history

Figure 4 illustrates the burial-time path for the last 30 Ma at the Clergeon anticline. The model is for the footwall of the thrust which uplifts the anticline. It is likely that this region has experienced the higher temperatures for the longest time and hence the model represents a thermally extreme case. If possible source rocks in this footwall have not reached maturity then it is unlikely that the bitumen seeps at Clergeon had a local source. We will see shortly that maturity cannot have been achieved prior to the generation of the molasse basin (initiated *c.* 23 Ma) without invoking abnormally high geothermal gradients in Mesozoic times. Therefore this modelling only considers the Neogene history.

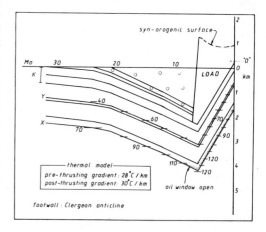

Fig. 4. Time–depth plot for the footwall to Clergeon thrust (see Fig. 2a). X, top of basement; Y, Callovo-Oxfordian; K, Cretaceous; open circles denote molasse. Oil window defined for X only (Y is under-mature), using the time–temperature indices of Waples (1980).

The burial–time plot (Fig. 4) is conventionally constructed for the period before thrust sheet emplacement. The datum during Miocene times will have been within about 50 m of sea level since the molasse sandstones of this time were deposited in a tidally dominated shallow-marine environment (Allen & Homewood 1984). This datum level is fixed during and after thrust sheet

Table 2. *Stratigraphic and compaction data*

Ma	Lithology	DF	Tp	T_0
(a) footwall to the Clergeon thrust				
Barremian (n_4)	Cut out by Clergeon thrust			
125				
Haut-Val-Berr (n_{1-3})	Limestones, shales	1.4	350	490
144				
Upper Jurassic (j_{7-9})	Marly limestones	1.25	400	500
156				
Call-Oxf (j_{3-6})	Shales, marls	2	100	200
169				
Bathonian (j_2)	Limestones	1.1	200	220
177				
Trias-L.Jurassic ($t-l-j_1$)	Limestones	1.28	390	500
200				
(b) Semnoz sheet				
Miocene	Molasse, cut out by Veyrier thrust			
25				
Eoc-Oligocene	Flysch	1.32	250	330
40				
	unconformity			
119				
Barremian (n_4)	Limestones	1	250	250
125				
Haut-Val. (n_{2-3})	Limestones, shales	1.33	700	930
138				
Berriasian (n_1)	Shales	1.66	800	1330
	Revard thrust.			
(c) Lowest subalpine thrust sheet				
	Revard thrust-cuts out most of molasse section			
Miocene	Molasse	1.2	400	500
25				
	unconformity			
119				
Barremian (n_4)	Limestones	1	250	250
125				
Haut-Val (n_{2-3})	Shales, limestones	1.35	250	400
138				
Berriasian (n_1)	Limestones	1.2	100	125
144				
Upper Jurassic (j_{3-9})	Limestone, shales	1.2	600	750
169				
Middle Jurassic (j_{1-2})	Limestones	1.1	400	450
188				
Lias-Triassic ($t-l$)	Limestones	1	250	250
200				

DF, Decompaction factor; Tp, present-day stratigraphic thickness; T_0, decompacted original thickness; both in metres
Table 1 contains the information for the Bornes sheet, with the Veyrier thrust running within Berriasian shales (see Fig. 3). Timescale from Harland *et al.* (1982).

emplacement so the load top represents the height above sea level of the syn-orogenic surface. The top of basement is not considered to show significant subsidence due to the emplacement of the Clergeon sheet. Karner & Watts (1983) have shown the European continental lithosphere to display significant flexure strengths so that the loading of individual thrust sheets is swamped by the long range flexure driven by the main mountain belt (cf. Mugnier & Ménard 1987). Uplift of the Clergeon section after 6 Ma is driven by the gradual removal of

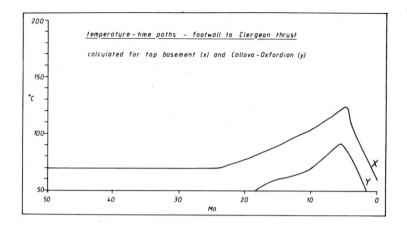

Fig. 5. Time–temperature paths for the Clergeon anticline, calculated from Fig. 4.

the main orogenic load by erosion in the Alpine interior, causing isostatic rebound of the foreland.

Thermal model

Although the burial model for Clergeon is fairly sophisticated, the thermal model is very simple, as there are no ages on given temperatures as might be achieved by fission track analysis, for example. In the absence of suitable data, the prethrusting geothermal gradient is modelled at $28°C \, km^{-1}$. This implies the palaeotemperature at the basement top, *c.* 24 Ma, to have been 70°C. Just prior to thrust sheet emplacement the temperature for this level has risen to 120°C.

Unfortunately there are no suitable models of thermal structure following thrusting which may be applied to the Clergeon thrust sheet. The simple thermal diffusion models of Angevine & Turcotte (1983) amongst others are not considered appropriate since thermal re-equilibration is likely to be dominated by advective rather than conductive processes. The derivative model of Furlong & Edman (1984) has the added requirement of lateral persistence of the emplaced thrust sheet so that thermal structure can be modelled in one dimension. Clearly this is inappropriate for the Clergeon structure.

The present-day geothermal gradient in the Alpine foreland is assumed to be $30°C \, km^{-1}$. In reality the thermal structure of the foreland basin is likely to be rather complex with, by analogy with the Uinta basin, western USA (Chapman *et al.* this volume), significant lateral components of heat flow. There are as yet no regional-scale measurements of sufficient

number to quantify the thermal structure more precisely. It is likely, with thermal re-equilibration during thrusting driven by advective means, that the regional thermal structure will dominate over local, transient effects. The distribution of low grade metamorphic minerals in the Subalpine chains (Thones syncline area, Fig. 1) suggest crude geothermal gradients after maximum burial of about $30°C \, km^{-1}$ (Sawatzki 1975). In the thermal model for Clergeon the present-day assumed geothermal gradient is projected back to 5 Ma. The undefined period between 7 and 5 Ma can then be simply projected in.

Time–temperature indices

Using the rather simple thermal model a series of time-temperature paths have been plotted (Fig. 5) for horizons in the footwall to the Clergeon thrust. Of these, the top of basement reaches a thermal high of *c.* 130°C at 5 Ma (X on Fig. 5). In contrast Callovo-Oxfordian rocks (Y on Figs 4 and 5) only reach a high of 82°C. The thermal histories can then be used to calculate time-temperature indices (TTIs), using the Lopatin method popularised by Waples (1980). These are plotted cumulatively against time on Fig. 6. The top-basement horizon (X) achieves a TTI of 28.6 while the Callovo-Oxfordian peaks at a TTI of just 1.9. Waples (1980) considers oil generation to occur in rocks which achieve a TTI in the range between 15 and 150. This is achieved by the top-basement, just after maximum burial by the thrust sheet and the model predicts that hydrocarbons are still being generated at this horizon today. However, this theoretical approach has taken no account of the distribution of source rocks. The Jura stratigraphy which applies to the Clergeon area is dominated by limestones with relatively low availabilities of organic carbon for hydrocarbon generation. Certainly there are no records of potential source rocks at the base of the Mesozoic succession. Therefore, although the thermal history of the top-basement may be suitable for the partial generation of hydrocarbons there are no suitable sources which have experienced this history.

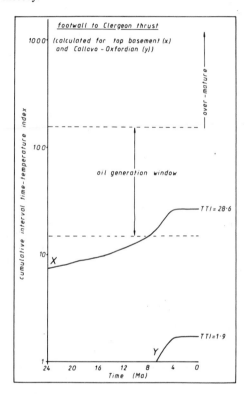

Fig. 6. Evolution of TTIs for the Clergeon anticline, calculated from Fig. 5 using the method of Waples (1980).

The implication of the above reasoning is that the Clergeon structure is unlikely to have generated any significant volume of hydrocarbons. The likely source rock of Callovo-Oxfordian organic-rich limestones has not reached maturity. If shear heating along the thrust plane was important it may have been possible to achieve sufficient temperatures for maturation. However, in sedimentary rocks it is likely that this provides a minor component to the thermal budget of a sedimentary basin. Bustin (1983)

shows that shear heating is unimportant in the Rocky mountain thrust belt. The long periods of quiescence along seismically slipping faults would generally allow thermal anomalies generated during the slip events to dissipate so that long term thermal anomalies would not develop. Rapid thermal re-equilibration would be greatly enhanced by the high degrees of fluid flow in thrust belts (e.g. Engelder 1984; Rye & Bradbury 1988). The rarely recorded thermal anomalies at thrust faults (e.g. Underwood *et al.* 1988) are probably generated by the migration of deep basinal fluids rather than shear heating. This may be the case for the Clergeon thrust, since hot springs are exploited along strike at Aix les Bains (Fig. 1), but they are not considered here to be important in driving hydrocarbon maturation.

An alternative approach to predict source rock maturity, involving merely the peak temperatures (e.g. Quigley & Mackenzie 1988), is examined later. The results of TTI modelling of the Clergeon structure suggest that the accumulations of bitumen in the foreland basin are not derived locally but have migrated in from elsewhere in the area. The most likely sources lie in the main Subalpine thrust belt where the rocks reached thermal maturity (Sawatzki 1975).

Thermal modelling in the Subalpine chains

The structure of the Subalpine chains in the Annecy area (Fig. 2) is dominated by three thrust sheets, beneath the nappe of internal Alpine rocks of the Prealps. The highest Subalpine sheet outcrops in the Bornes–Aravis massif (Fig. 1). Its lower contact, the Veyrier thrust, can be mapped around the north side of the Lac d'Annecy (Fig. 1) and traced southwards into the Bauges hills. The next thrust sheet down is represented by the Semnoz anticline (Fig. 1) which plunges gently northwards beneath the Bornes massif. The basal, or Revard thrust is poorly exposed although splays from it are known from near Annecy (Doudoux 1967) and from further south towards the town of Chambéry (Fig. 1). The main evidence for the thrust, together with the presence of a long panel of Miocene molasse in its footwall comes from section balancing (Butler 1989a). The lowest thrust sheet lies beneath the Revard thrust and directly above the crystalline basement. Its basal thrust, a structure with no stratigraphic separation, is inferred from the model of Jura detachment outlined earlier. This slip surface is considered to pass back eastwards from the Jura beneath all the subalpine structures before finally cutting back into basement beneath the external crystalline massifs (Butler 1989a, b). Regional folding relationships suggest that these sheets were emplaced in top to bottom, or piggyback, sequence (Butler 1989a).

Before discussing the time-burial paths for these three thrust sheets we must establish the stratigraphic framework for the Subalpine chains. The model presented as Fig. 3 refers to the upper thrust sheet of the Bornes massif. There are differences in the Mesozoic thickness between the Bornes and Jura, as discussed earlier. However, there are critical differences in the thicknesses of Tertiary rocks also. In the Bornes there lies an important flysch basin of late Eocene, early Oligocene age (Sawatzki 1975). The thickness of these units is constrained from the geological maps (BRGM 1963, 1980). It did not continue out towards the west where, during this time, the Mesozoic shelf was emergent (e.g. Debelmas & Kerckhove 1980). Rather less easy to constrain is the thickness of Miocene molasse deposits in the Subalpine chains. Substantial uplift, both by thrusting and post-orogenic isostatic rebound has removed much of the sedimentary record through erosion.

The thickness of molasse most probably increased eastwards during the development of the foredeep basin. Typically, subsidence in foredeep basins increases towards the mountain belt and the basin itself progrades away as the locus of thrusting migrates into the foreland (Beaumont 1981). Hence the stratigraphic thickness of Miocene molasse sandstones should have been significantly greater over the Bornes than the Jura.

Additional information comes from the unroofing history of the Subalpine thrust belt. Along the length of the molasse basin in the NW Alps there is evidence of local thrust structures controlling sediment deposition. There are no influxes of clasts derived from the Mesozoic sediments of the Subalps within the Miocene molasse. The clasts are almost entirely derived from the internal Alps, either in the metamorphic interior of the mountain belt or from the Prealpine sheets. In contrast the post-orogenic sediments of Pliocene and younger ages are dominated by Subalpine clasts. These observations suggest that during thrusting the Subalpine Mesozoic rocks were not uplifted sufficiently to have been eroded.

Present-day topography, and the regional cross-section (Fig. 2) shows the Mesozoic rocks of the Subalps to have been uplifted by several kilometres during thrusting. Yet they did not source the molasse basin at that time. Therefore these rocks in the thrust belt must have been protected by a mantle of higher units which eroded first. In the Bornes this carapace was provided by the Prealpine sheets. However, in the other units it must have been the Miocene molasse. The depth–time plots presented here (Figs 7, 8 and 9) show subsidence due to isostatic flexure of the foreland lithosphere to approach 5 km. Notice that this is not excessive, the present depth to the top-of-basement beneath the Subalpine chains can reach almost 7 km. At the end of Mesozoic times the basement would have been overlain by perhaps 1500 m of Mesozoic rocks so that the net subsidence is now 5 km. There would have been additional subsidence during the later part of the Miocene and uplift since then.

Depth–time plots

With some confidence in the pre-thrusting stratigraphy of the Subalps we can begin to model the burial history of the three thrust sheets. The models refer to a palaeo-vertical profile beneath the Col de Bluffy (Fig. 1). Figure 7 shows the bural–time path for that stratigraphy preserved in the Bornes sheet. The Veyrier thrust runs within the Berriasian shales so this is the oldest unit modelled here. The load, emplaced at 11 Ma is the Prealpine nappe complex. Map patterns show this to lie directly on Oligocene sediments along strike (Fig. 1) suggesting that the Prealps were emplaced within the molasse basin, rather than across the top of it. It is likely that the leading edge of the Prealps were only a few kilometres thick hence the topographic expression of this thrusting would have been subdued.

Uplift of the Bornes and Prealpine sheets began with the displacement on the Veyrier thrust, carrying its hanging-wall up onto the future Semnoz anticline. Further thrusting and isostatic rebound continue the uplift process so that the base of the Bornes sheet, the Veyrier thrust, now outcrops about 300 m above sea level (Fig. 7), on the flanks of the Col de Bluffy (Fig. 1).

The next sheet down to be modelled (Fig. 8) is part of the Semnoz structure, a small imbricate slice of Cretaceous and younger rocks which outcrop on the shores of the Lac d'Annecy (Fig. 1). The subsidence history prior to thrusting is broadly comparable with the Bornes (Fig. 7) although the Palaeogene basin is much more weakly developed. At about 9 Ma the thrust load was emplaced. This consists of the Prealpine and Bornes complexes. The emplacement event coincides with uplift of the Bornes profile (Fig. 7). Immediately following this emplace-

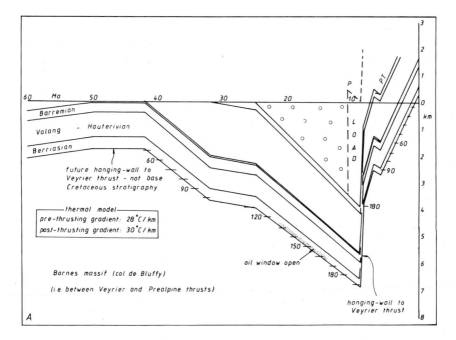

Fig. 7. Time–depth plot for the Bornes massif (A on Fig. 2b). Oil window defined for the preserved base of the Berriasian only (but see Fig. 3). P, Prealpine emplacement; PT, Prealpine thrust.

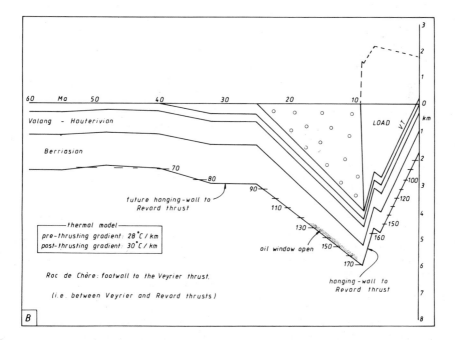

Fig. 8. Time–depth plot for the Semnoz sheet, based on the Roc de Chère area (B on Fig. 2b). Oil window defined for base of Berriasian only. VT, Veyrier thrust.

ment the Semnoz structure itself was uplifted along the Revard thrust.

The lowest thrust sheet is the footwall to the Revard thrust (Fig. 9). This still preserves its full Mesozoic stratigraphy. This far to the west in the Alps the Paleogene flysch basin was not developed (e.g. Debelmas & Kerckhove 1980). The stratigraphy is essentially the same as for the Clergeon sheet in the eastern Jura (Fig. 4) but with a greater thickness of Miocene molasse. Some of this molasse is preserved in the footwall to the Revard thrust. Forming the load in the hanging-wall lie the Semnoz, Bornes and Prealpine sheets. At this stage Prealpine sheets must still form the syn-orogenic surface so that the Mesozoic rocks of the Bornes and lower sheets are protected from erosion. It is unlikely that the topography exceeded 2 km above sea level at this time hence the thrust sheets must sit deeply in the foredeep basin. It is this feature which determines the thickness of the Miocene foredeep in these models.

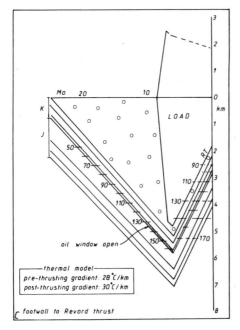

Fig. 9. Time–depth plot for the lowest Subalpine sheet, the footwall to the Revard thrust (RT). Oil window defined for the Berriasian only. K, Cretaceous; J, Jurassic; molasse denoted by open circles. Note that at time = 0 Ma this plot fits beneath those of Figs 7 and 8.

Thermal model and TTI

A simple thermal model can now be overlain on the burial–time plots for the three Subalpine

thrust sheets. In all cases the assumed geothermal gradients are 28°C km^{-1} before the emplacement of thrust loads and 30°C km^{-1} for the period between 6 Ma and present. We can follow the thermal history of one horizon in each of these sheets. The horizon chosen is the preserved base of the Berriasian, a shale formation which probably represents the best hydrocarbon source rock in the region. The results of the model are displayed at temperature time-paths (Fig. 10).

During the Mesozoic extensional basin history, and continuing during flysch and molasse foredeep basin evolution there was more subsidence in the east than the west. Therefore, we might expect that the palaeotemperature for a given horizon to be greater for parts of the basin derived from further to the east. This simple pattern is not shown on Fig. 10 because the Berriasian horizon traced for the Bornes sheet is not the stratigraphic base but the base of the thrust sheet. The Veyrier thrust cut part-way through the formation. Hence the Bornes pathway represents a higher part of the Subalpine basin. After 40 Ma the Bornes overtakes the Semnoz pathway because of the stronger development of the Palaeogene flysch basin. Regardless of the starting temperatures the three thrust sheets are shown to reach their peak temperatures at different times. This illustrates the westwards progradation of thrusting and the foredeep basin.

As with the Clergeon structure, the temperature-time paths can be used to generate time–temperature indices, using Lopatin's method (Waples 1980). Figure 11 shows how the TTI builds up for each of the three thrust sheets. The thermal input for the sheets is different, with the Bornes having a longer and hotter history than the Semnoz which in turn was hotter for longer than the lowest unit. Hence the final TTIs vary from 1393, through 589 to 385 for the Bornes, Semnoz and lowest sheets respectively. The implication is that, although the Clergeon structure was undermature, all three Subalpine sheets have passed through the oil generation window defined by Waples (1980: TTI between 15 and 150). Therefore, the Subalpine chains certainly provide a possible source for the migrated hydrocarbons now out in the foreland basin.

Timing of maturation and local structural trap development

The next question to address is the timing of oil generation in the Subalpine chains. Following Waples (1980) assumption that this occurs when TTI has reached a value between 15 and 150, the

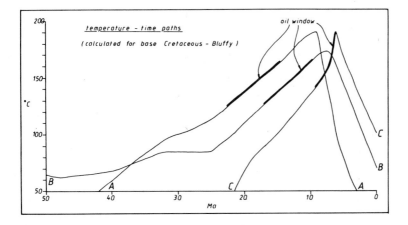

Fig. 10. Temperature–time paths calculated for the base of the Berriasian rocks found within the three Subalpine thrust sheets, using the burial histories shown in Figs 7, 8 and 9 (paths A, B & C respectively). The oil windows are defined using the method of Waples (1980; see Fig. 11).

cumulative plots of TTI against time (Fig. 11) will provide an answer. Not surprisingly, given the subsidence history, the oil generation window is encountered diachronously for the three thrust sheets. The first to mature was the Bornes, followed by the Semnoz and finally the lowest sheet. These sheets also reached peak TTI in the same sequence.

Figure 12 illustrates the absolute timing of oil generation, as modelled above, as lasting from about 22 Ma to the present day. However, the site of maturation has changed. More critically, Fig. 12 shows the relative timing between maturation and tectonics. In all three Subalpine examples the maturation occurs not during thrust sheet emplacement but before, during the burial of the Berriasian source beneath the fore-deep basin. Both the Bornes and Semnoz sheets have passed out through the oil generation window before thrusting while the lowest Subalpine sheet passes through the window just as thrusting occurs.

The modelling implies that most of the oil generated from source rocks in the Subalpine chains had undergone maturation before thrust-related structures had developed either above or ahead of the source region. If there had been no secondary migration away from the source rocks these hydrocarbons would pass into the gas-generation field during the later part of foredeep subsidence and certainly during thrusting. There is little direct evidence for the maximum temperatures or even cumulative TTI reached by most of the rocks being considered here. They remain buried and, on this section line, have not been drilled.

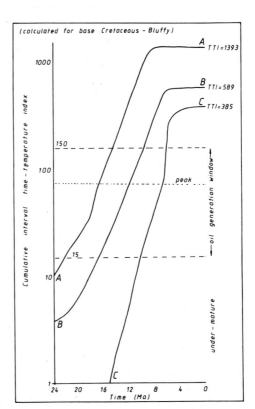

Fig. 11. Evolution of TTI's for the Subalpine thrust sheets illustrated in Figs 7, 8 & 9 (paths A, B & C respectively).

In a comprehensive review on the geology of the Prealpine flysch now preserved in the Thones syncline (Fig. 1), Sawatzki (1975) describes the

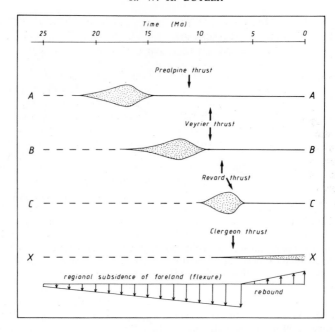

Fig. 12. The relative timing of oil generation (stippled 'blobs', peak generation where 'blob' is widest, see Fig. 11) and thrusting for each of the sites along the section line of Fig. 2. X, top basement in the footwall to the Clergeon thrust; A, base Berriasian in the Bornes sheet; B, base Berriasian in the Semnoz sheet; C, base Berriasian in the lowest Subalpine sheet. The local emplacement of thrust sheet loads are arrowed. Based on constructions shown in previous figures.

presence of low grade metamorphic minerals including the zeolite laumonite together with corrensite. He suggests maximum temperatures of about 170–200°C and pressures of 1.5–2 kbar. On Fig. 7 the maximum temperature during burial for the base of the flysch was estimated at 180°C, achieved following the emplacement of the Prealpine sheets as a tectonic load. These peak temperatures are supported by Sawatzki's reports of an illite crystallinity value of 0.42 from the same Thones area.

The modelling strategy used here clearly involves many uncertainties particularly relating to thermal aspects. We have assume a geothermal gradient of 28°C km^{-1} for the sedimentary basin before thrusting. This seems a reasonable figure but if the gradient was lower it would be possible for the source rocks to avoid maturation during foredeep development. In the case of the Bornes sheet the time available for heating during sedimentary burial (31 Ma from the onset of the Palaeogene flysch basin to the emplacement of the Prealps) is much greater than between tectonic burial and uplift (*c.* 3 Ma). We know from Sawatzki's (1975) studies that rocks within the Bornes sheet have reached temperatures of 170–200°C. This supports the contention that most of the thermal input occurred

before loading by the Subalpine thrust sheets.

The post-thrusting thermal structure is very poorly known, particularly during thermal re-equilibration of the tectonically disturbed geotherms. Rigorous models should incorporate the effects of advective cooling of deep and therefore hotter basinal rocks emplaced onto younger units over a significant time period. The final structures are not laterally persistent and so the thermal structure would need three dimensional modelling. Yet this does not alter the conclusion that maturation must have occurred in the Bornes and Semnoz sheets before thrust loading.

The modelling here has referred to the Berriasian formations. There are few potential source rocks from younger, shallower and hence cooler parts of the basin. Deeper source rocks in the Bornes, such as may have been deposited in Kimmeridgian and Callovo-Oxfordian times will have reached maturity even before foredeep development (Fig. 3).

Peak temperature and maturation

During the late 1980s the Lopatin method's (Waples 1980) dependence on TTIs to predict maturation has been seriously questioned. Kinetic studies by Quigley & Mackenzie (1988) suggest that a simple adoption of the Arrhenius equation central to the Lopatin method is not well-founded as it 'overestimates the effect of time and underestimates the effect of temperature on source rock maturation' (Quigley & Mackenzie 1988, p. 551). They argue that peak temperature is of prime importance, with rich oil-prone source rocks reaching maturity between 100 and 150°C. The leaner marine source rocks would mature over the narrower 120–150°C field. Above 150°C oil generated from both initial kerogen types would begin to crack to gas, with rapid cracking above 170°C. We can apply this somewhat simpler image of hydrocarbon maturation to the time-temperature plots. In all cases peak temperature would be reached after tectonic loading but before exhumation on the back of the next active thrust.

The Callovo-Oxfordian of the Clergeon structure (Fig. 5) is predicted to have experienced peak temperatures of about 90°C and so did not reach maturity. Thus the principle conclusion of this study, that the bitumen deposits on the Clergeon anticline were not sourced from the immediate area, is substantiated. Modelling the three Subalpine structures shows that the base Berriasian for each reached the temperature-dependent oil window (100–150°C) before that predicted by the Lopatin method (TTI values between 15 and 150). So the Lopatin method gave us the better chance of predicting maturation during the local emplacement of thrust loads. Both Lopatin and 'peak temperature' methods suggest maturation during the precursor foreland basin subsidence. The Berriasian of all three sites is predicted to have been buried to 6 km ahead of thrusting, so complete maturation would have occurred for oil-prone sources (rich or lean) with palaeo-geothermal gradients of 25°C km^{-1} and over.

Discussion

We can summarize the results of the modelling of hydrocarbon maturation in the external French Alps along the section line of Fig. 2. It appears that there are no suitable source rocks which have reached maturity in the Clergeon structure of the eastern Jura, unless we have seriously underestimated the thickness of foreland basin sediments. In contrast all the potential source rocks are either mature or, more likely, overmature in the Subalpine chains. The bitumen deposits in the Annecy molasse basin probably record a significant amount of lateral migration of oil from sources now within the main Subalpine thrust belt. This process has been documented from many foreland basins with several hundred kilometres lateral migration, most spectacularly in the Western Canadian basin (e.g. du Rochet 1984).

Lateral migration during foredeep development will be enhanced by the general hinterland-dip of stratigraphy. Since thrust-related structures will not yet have developed up dip of the maturing source rocks simple stratigraphic migration pathways should still exist (Fig. 13a). Nevertheless, important normal faults which dip back towards the mountain belt have been described throughout the Subalpine chains (Lemoine & de Graciansky 1988; Butler 1989b) which could have provided suitable structural traps within the Mesozoic sediments. In the case of the Alpine foreland basin an important hydrocarbon carrier probably lies at the contact between Cretaceous limestones and the overlying molasse sandstones. This is a faulted karstic surface and unconformity, locally marked by thin clean and well-sorted sandstones. The overlying Miocene molasse sandstones have a surprisingly low porosity, due to the high original clay content and the diagenetic breakdown of metamorphic and igneous materials derived from the Alps. So the possibility exists for combined stratigraphic and structural traps within the foreland basin which are not thrust-related (Fig. 13b) although later deformation has incorporated such structures into the thrust belt.

The fate of hydrocarbons which migrate out onto the flanks of the foreland basin is to move into a rather cooler environment. Indeed, if these fluids migrated to the very edge they would have emerged on the outer shore of the narrow Miocene seaway. This type of pathway greatly increases the chance of substantial biodegradation, a possible explanation for the heavy asphalts and bitumen found in the basin today. Unfortunately there are, as yet, no published accounts of the organic geochemistry of these deposits combined with those of basin fluids and the potential source rocks in this part of the Alps. It is hoped that current studies of clay mineral structure, spore colour and vitrinite reflectance underway throughout the Subalpine chains (S. Moss, Durham University) will constrain the thermal modelling reported here.

The peak temperatures predicted by the modelling presented in this paper suggest that most Subalpine source rocks should have passed well beyond the oil window. If there was no large-scale secondary migration from the source kitchens during foredeep development the subsequent thrust loading would have generated

(a)

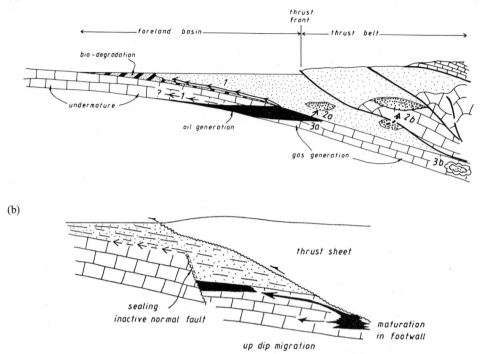

(b)

Fig. 13. (a) Idealized view of the structural setting of petroleum generation and migration pathways in the Subalpine thrust belt. Oil generation occurs during molasse deposition through much of the area, ahead of the thrust sheet loads. Pathway 1 involves large-scale lateral migration up stratigraphically defined conduits which dip towards the orogen. The foreland basin hydrocarbon accumulations of the Alps probably followed this type of pathway, migrating along the base of the molasse. Pathway 2 involves an episode of secondary migration into a higher level reservoir (2a) and then remigration (2b) into structural traps in the thrust belt. This circuitous route enables the hydrocarbons to avoid intense burial during the later stages of molasse deposition and the subsequent tectonic loading. Pathway 3 involves no secondary migration (3a) but shows the overthrusting of overmatured source rocks at depth (3b). Both this and derivations of pathway 2 could generate substantial volumes of gas within the thrust belt.
(b) The development of a potential hydrocarbon play in the footwall to a thrust sheet over-riding a pre-existing extensional fault which seals a suitable reservoir formation (based on a section by Butler 1989*b*). In this simple case maturation is driven by the combined tectonic and sediment burial causing up-dip migration of liquid hydrocarbons (Pathway 1 in **a**). If the pre-existing normal fault is sealing it will prevent the continued up-dip migration (pecked arrows) to shallow levels near the feather edge of the foreland basin, thereby creating a suitable closure in the footwall to the emplaced thrust sheet. The role of the thrust sheet as a migration pathway has not been considered here (see Roberts, this volume).

substantial volumes of gas (Fig. 13). The consequences of this on influencing the rheological and hence geometric evolution of the thrust belt have yet to be investigated. The Jurassic units which originally underlay the Bornes and Aravis section are predicted on Fig. 2 to now lie deeply buried below and east the Subalpine chains. They lie buried at over 5 km, continuing beneath the leading edge of the basement thrust sheets. In this environment the TTIs will have increased to beyond the preservation of any liquid oil and

will be approaching the upper threshold for preserving dry gas, assuming the values of Waples (1980). Likewise peak temperatures here are predicted to have exceeded the 170°C threshold for rapid cracking of oil to gas proposed by Quigley & Mackenzie (1988).

The above reasoning suggests that most of the sub-thrust plays are very poor targets for liquid hydrocarbons in this sector of the Alps. However, they may exist if the fluids had a complex migration history (Fig. 13). It would be possible for oil to escape high burial temperatures if it was able to migrate to higher stratigraphic positions, perhaps at the base of, or within, the molasse. The karst-reservoir and carrier of the outer parts of the basin do not exist in the east and there are few other sedimentary reservoirs available. Active thrusting might be expected to change the permeability and hence reservoir capacity in the section but the absence of a suitable stratigraphic holding area in this example probably precludes this model.

However, there have been minor oil shows in drill holes from the Annecy district, namely from the Age anticline (BRGM 1972, Fig. 1). This lies between the Subalpine chains and Clergeon. Perhaps subsidence during Miocene times was at a critical magnitude to permit the Mesozoic source rocks to just reach sufficient temperatures for maturation but not exceed it. Regrettably the Age structure has not been penetrated by many drill holes and there have been no modern geochemical studies published on those materials recovered. Hence we know neither the source of this petroleum nor its migration pathway.

The modelling strategy adopted here has been applied to a single cross-section line. Hence the conclusions reached are only applicable to adjacent parts of the Subalpine chains if there are no lateral variations in the Mesozoic through Tertiary basin history and the structure. In reality all of these aspects vary. The Jura and Subalpine chains converge to the south near Grenoble (Fig. 1). More critically the thickness of the foredeep basin decreases to the south so that the critical period of source rock heating may occur during or following thrust loading. Arguably the best place to look for suitable sub-thrust plays would be beneath the western edge of the Chartreuse, if suitable reservoir formations are present. In contrast it seems decreasingly likely that liquid hydrocarbons will be found northwards around the Subalpine chains from Annecy. In this direction the foredeep basin becomes deeper and the Prealpine thrust load increases too. The exception might lie at the base of the Prealps themselves, if a migration pathway developed during thrusting which allowed the hydrocarbons sourced in the buried Subalpine units to escape. These comments could move away from being entirely speculative if additional palaeothermal, fission track or present-day geothermal data become available.

I thank S. Bowler and G. Roberts for discussions on the geology of the Alpine foreland thrust belt. I also thank the convenors of the Petroleum Migration meeting, particularly A. Fleet, for inviting this contribution thereby forcing my thoughts into a more coherent form and both he and S. Hay for useful comments on the manuscript. Research in the French Subalpine chains was supported through a NERC research grant (GR3/6172).

References

ALLEN, P. A. & HOMEWOOD, P. 1984. Evolution and mechanics of a Miocene tidal sand wave. *Sedimentology*, **31**, 63–81.

ANGEVINE, C. L. & TURCOTTE, D. L. 1983. Oil generation in overthrust belts. *Bulletin of the American Association of Petroleum Geologists*, **67**, 235–241.

ARNAUD, H. 1988. Subsidence in certain domains of southeastern France during the Ligurian Tethys opening and spreading stages. *Bulletin de la Société Géologique de la France*, **8**, 725–735.

AUBOUIN J., LE PICHON, X. & MONIN, A. S. (eds) 1986. Evolution of the Tethys. Part II: Maps. *Tectonophysics*, **123**, plates I–X.

BAYER, R., CAZES, M., DAL PIAZ, G. V., DAMOTTE, B., ELTER, G., GOSSO, G., HIRN, A., LANZA, R., LOMBARDO, B., MUGNIER, J-L., NICOLAS, A., THOUVENOT, F., TORREILLES, G. & VILLIEN, A. 1987. Premiers resultats de la traversee des Alpes occidentales par sismique reflexion verticale (Programme ECORS-CROP). *Comptes rendus de l'Academie des Sciences, Paris*, **305**, 1461–1470.

BEAUMONT, C. 1981. Foreland basins. *Geophysical Journal of the Royal Astronomical Society*, **65**, 291–329.

BOYER, S. J. & ELLIOTT, D. 1982. Thrust systems. *Bulletin of the American Association of Petroleum Geologists*, **66**, 1196–1230.

BUSTIN, R. M. 1983. Heating during thrust faulting in the Rocky Mountains: friction or fiction? *Tectonophysics*, **95**, 309–328.

BUTLER, R. W. H. 1989a. The geometry of crustal shortening in the Western Alps. *In*: SENGÖR, A. M. C. (ed.) *Tectonic evolution of Tethyan Regions* Proceedings of NATO Advanced Study Institute **259**, Kluwer, Dordrecht, 43–76.

—— 1989b. The influence of pre-existing basin structure on thrust system evolution in the Western Alps. *In*: COOPER, M. A. & WILLIAMS, G. D. (eds) *Inversion Tectonics* Geological Society, London, Special Publications, **44**, 105–122.

BRGM 1963. *Carte géologique de la France à 1: 80,000. Feuille 160bis, Annecy.* Bureau des Récherches Géologiques et Minières, Orleans.

—— 1964. *Carte géologique de la France à 1: 80,000. Feuille 160, Nantua.* Bureau des Récherches Géologiques et Minières, Orleans.

—— 1966. *Carte géologique de la France à 1: 80,000. Feuille 169bis, Albertville.* Bureau des Récherches Géologiques et Minières, Orleans.

—— 1969. *Carte géologique de la France à 1: 50,000. Feuille 702, Chambéry.* Bureau des Récherches Géologiques et Minières, Orleans.

—— 1970. *Carte géologique de la France à 1 : 50,000. Feuille 701, Rumilly.* Bureau des Récherches Géologiques et Minières, Orleans.

—— 1972. *Carte géologique de la France à 1 : 50,000. Feuille 697, Seyssel.* Bureau des Récherches Géologiques et Minières, Orleans.

—— 1980a. *Carte géologique de la France à 1 : 250,000. Feuille 29, Lyon.* Bureau des Récherches Géologiques et Minières, Orleans.

—— 1980b. *Carte géologique de la France à 1 : 250,000. Feuille 30, Annecy.* Bureau des Récherches Géologiques et Minières, Orleans.

—— 1981. *Carte géologique de la France à 1 : 50,000. Feuille 696, St-Ramberg en Bugey.* Bureau des Récherches Géologiques et Minières, Orleans.

—— 1987. *Carte géologique de la France à 1 : 250,000. Feuille 24, Chalon sur Sâone.* Bureau des Récherches Géologiques et Minières, Orleans.

CHAPMAN, D. S., WILLETT, S. D. & CLAUSER, C. 1991. Using thermal fields to estimate basin-scale permeabilities. *This volume.*

DEBELMAS, J. & KERCKHOVE, C. 1980. Les Alpes Franco-Italiennes. *Géologie Alpine,* **56,** 21–58.

——, ESCHER, A. & TRUMPY, R. 1983. Profiles through the Western Alps. *In:* RAST, N. & DELANEY F. M. (eds) *Profiles through Orogenic Belts* Geodynamics Series of the American Geophysical Union and the Geological Society of America **10,** 83–96.

DOUDOUX, B. 1967. Nouvelle etude de la montagne du Semnoz pres d'Annecy. *Annales du Centre Universitaire, Savoie,* **5,** 121–143.

——, DE LEPINAY, B. M. & TARDY, M. 1982. Une interprétation nouvelle de la structure des massifs subalpins savoyards (Alpes occidentales): Nappes de charriage oligocènes et deformations superposées. *Comptes rendus de l'Academie des Science, Paris,* **295,** 63–68.

ENGELDER, T. 1984. The role of pore water circulation during the deformation of foreland fold and thrust belts. *Journal of Geophysical Research,* **89,** 4319–4325.

FURLONG, K. P. & EDMAN, J. D. 1984. Graphic approach to determination of hydrocarbon maturation in overthrust terrains. *Bulletin of the American Association of Petroleum Geologists,* **68,** 1818–1824.

GILLCRIST, R., COWARD, M. P. & MUGNIER, J. L. 1987. Structural inversion: examples from the Alpine foreland and French Alps. *Geodynamica Acta,* **1,** 5–34.

HARLAND, W. B., COX, A. V., LLEWELLYN, P. G., PICKTON, C. A. G., SMITH, A. G. & WALTERS, R. 1982. *A geologic time scale.* Cambridge University Press.

KARNER, G. D. & WATTS, A. D. 1983. Gravity anomalies and flexure of the lithosphere at mountain ranges. *Journal of Geophysical Research,* **88,** 10449–10477.

LAUBSCHER, H. P. 1972. Some overall aspects of Jura dynamics. *American Journal of Science,* **272,** 293–304.

LEMOINE, M. & DE GRACIANSKY, P. C. 1988. Histoire d'une marge continentale passive: les Alpes occidentales au Mésozoïque. Introduction. *Bulletin de*

la *Société géologique de la France,* **8,** 597–600.

LEMOINE, M., BAS, T., ARNAUD-VANNEAU, A., ARNAUD, H., DUMONT, T., GIDON, M., BOURBON, M., DE GRACIANSKY, P. C., RUDKIEWICZ, J.-L., MEGARD-GALLI, J. & TRICART, P. 1986. The continental margin of the Mesozoic Tethys in the Western Alps. *Marine and Petroleum Geology,* **3,** 179–199.

MÉNARD, G. 1980. Profondeur du socle antetriassique dans le sud-est de la France. *Comptes rendus de l'Academie des Sciences, Paris,* **290,** 299–302.

MUGNIER, J. L. & MÉNARD, G. 1986. Le developpement du bassin molassique suisse et l'evolution des Alpes externes: une modele cinematique. *Publications of Elf Aquitaine,* 167–180.

QUIGLEY, T. M. & MACKENZIE, A. S. 1988. The temperature of oil and gas formation in the subsurface. *Nature,* 333, 549–552.

RIGASSI, D. 1977. Genèse tectonique du Jura: une nouvelle hypothèse. *Palaeolab News,* **2,** 1–27.

ROBERTS, G. 1991. Structural controls on fluid migration through the Rencurel Thrust Zone, Vercors, French Sub-Alpine chains. *This volume.*

ROCHET, J. DU. 1984. Migration in fracture networks; an alternative interpretation of the supply of the "giant" tar accumulations in S Alberta, Canada—1. *Journal of Petroleum Geology,* **7,** 381–402.

RYE, D. M. & BRADBURY, H. J. 1988. Fluid flow in the crust: an example from a Pyrenean thrust ramp. *American Journal of Science,* **243,** 197–235.

SAWATZKI, G. G. 1975. Etude géologique et minéralogique des flyschs a grauwackes volcaniques du synclinal de Thones (Haute-Savoie, France). *Archives des Sciences, Genève,* **28,** 18–368.

UNDERWOOD, M. B., FULTON, D. A. & McDONALD, W. W. 1988. Thrust control on thermal maturity of the frontal Ouachita mountains, central Arkansas, USA. *Journal of Petroleum Geology,* **11,** 325–340.

VAIL, P. R., MITCHUM, R. M. Jr. & THOMPSON, S. III 1977. Seismic stratigraphy and global changes of sea-level. *Memoirs of the American Association of Petroleum Geologists,* **26,** 49–212.

VILLARS, F., MÜLLER, D. & LATELTIN, O. 1988. Analyse de la structure du Mont Charvin (Haute Savoie) en termes de tectonique synsédimentaire paléogène. Conséquences pour l'interpretation structurale des chaînes subalpines septentrionales. *Comptes rendus de l'Academie des Sciences, Paris,* **307,** 1087–1090.

WAPLES, D. W. 1980. Time and temperature in petroleum formation: application of Lopatin's method to petroleum exploration. *Bulletin of the American Association of Petroleum Geologists,* 64, 916–926.

ZEIGLER, P. 1989. Geodynamic model for Alpine intraplate compressional deformation in Western and Central Europe. *In:* COOPER, M. A. & WILLIAMS, G. D. (eds) *Inversion Tectonics* Geological Society, London, Special Publications, **44,** 63–85.

ZWEIDTLER, D. 1985. *Genèse de gissements d'asphalte des formations Pierre Jaune de Neuchatel et des calcaires Urgoniennes de Jura (Jura Neuchatelois et Nord-Vaudois).* Thèse Université Neuchatel (unpublished).

Structural controls on fluid migration through the Rencurel thrust zone, Vercors, French Sub-Alpine Chains

GERALD ROBERTS

Department of Geological Sciences, Science Laboratories, South Road, Durham DH1 3LE, UK

Present address: Department of Geology, The University of Manchester, Manchester M13 9PL, UK

Abstract: The Vercors, French Sub-Alpine Chains, is part of a classic foreland thrust belt which accommodates the last few kilometres of WNW-directed shortening in the Alps. The Rencurel thrust zone locally emplaces Barremian limestones onto Miocene sandstones. The thrust zone is well exposed allowing detailed structural logging and sample collection to be carried out. Calcite veins occur within gouge zones along minor faults occurring pervasively throughout the fault zone. However, at the base of the exposed fault zone, where a large displacement thrust has developed, the dominant vein fill is ferroan calcite, locally accompanied by bitumen. Cross-cutting relationships indicate that the pervasive set of minor faults developed early in the deformation sequence. The lack of ferroan calcite and bitumen within the early set of faults suggests that they were impermeable during later faulting. Fluid migration during deformation along the large displacement thrust was confined to underneath the impermeable zone of early minor faults. This study indicates that the style and location of deformation changes during the incremental development of a foreland thrust zone. Distributed deformation by mesoscale faulting or folding may be characteristic of deformation which occurs before the localization of displacement onto a major fault zone. Deformation probably occurs episodically with diffusive mass transfer and cataclasis as the dominant deformation mechanisms. Microstructures evolve during progressive deformation and this may induce changes in the permeability along a fault zone through time. These changes in the nature of fault zones during incremental deformation must be considered in order to assess fully the role of faulting in controlling syn-tectonic fluid migration.

Fault zones are often considered to be conduits to fluid migration (Porter & Weimer 1982; Jourdan *et al.* 1987; Lee *et al.* 1989; Burley *et al.* 1989). However, studies which have concentrated on the finite state of permeability along fossil fault zones suggest that in some cases fault zones may have a high sealing potential (Smith 1966, 1980; Weber & Daukoru 1975; Weber *et al.* 1978; Harding & Tuminas 1989). This paper suggests that the porosity and permeability of fault rocks change during incremental deformation due to microtextural evolution. Also, the microtexture of individual portions of a fault zone may evolve independently, such that the sealing potential within a fault zone at any one time, may be spatially variable. Thus, fault zones are not simply a seals or a conduits to subsurface fluid migration.

Sibson *et al.* (1975) suggested that significant volumes of pore water could possibly be mobilised by dilatancy associated with the build up and release of stress during earthquakes along active fault zones. Evidence for fluid migration during active faulting include several instances when changes in stream discharge, ground-water levels and other hydrological phenomena have been observed during large earthquake episodes (Briggs & Troxell 1955; Stermitz 1964; Swensen 1964; Wood 1985). Thus, hydrocarbons which are being generated or are trapped in the subsurface in the vicinity of active faults, may also be mobilised along with migrating pore waters. Active fault zones are among the prime candidates for fluid migration pathways. An investigation of the sealing potential of fault zones during active deformation can aid assessment of the role of faulting in fluid migration. However, indirect means must be used to consider the likelihood of migration along a particular ancient fault zone.

First, the study of microstructures from inactive fault zones can give valuable insights into the nature of the porosity and permeability within fault zones during active deformation.

From England, W. A. & Fleet, A. J. (eds), *Petroleum Migration*
Geological Society, Special Publication No. 59, pp. 245–262.

Mitra (1988), in a study of the Appalachians thrust belt, has shown that the microstructure present in a deformed rock can control its porosity and permeability, with high strain zones being less permeable than low strain zones. This approach can be used to assess the permeability of fault zones during deformation if we also use microstructural information to assess whether the deformation style, and hence the permeability changed during the accumulation of strain. Sibson (1989), in his review paper on seismic faulting, has shown that active faults in the upper 12–15 km of continental crust may be seismogenic. Seismogenic faults accumulate displacement during a seismic cycle where large variations in strain rate occur. Knipe (1989), in his review paper on the recognition of deformation mechanisms from microstructures, has shown that at the relatively low temperatures which occur in the upper part of the continental crust, cataclasis is the dominant deformation mechanism at high strain rates with diffusive mass transfer (DMT) dominant at low strain rates. Seismogenic faults are therefore likely to experience cyclicity between diffusive mass transfer and cataclasis during the seismic cycle. Fault rock textures will show cross cutting relationships between microstructures produced by cataclasis and diffusive mass transfer. The permeability of a fault zone will be controlled by the microstructures which already exist within the fault rock. However, the permeability will be altered during the next increment of fault slip when the strain rate changes. This will activate new deformation mechanisms and produce overprinting microstructures. An evolving microtexture will have important effects on the permeability of a fault zone and any incremental syn-tectonic fluid migration.

Secondly, fault zones have been shown to exhibit complex internal geometries. Complex arrays of minor faults are found together with gouge zones which can be metres thick (Woodward et al. 1988; Wojtal & Mitra 1986). Sibson (1989), showed that fault zones do not deform at a constant strain rate across their whole length and width during one increment of deformation. For example earthquake ruptures show changes in their propagation velocities at fault bends and offsets. Also, only certain strands of a wide fault zone are known to be activated during a single earthquake. This all suggests that the deformation mechanisms together with the microstructures they produce, will vary both across and along a fault zone, and the situation will change with time.

To assess the role of faulting in controlling syn-tectonic hydrocarbon migration we must consider several aspects of the nature of fault zones. Individual faults slip episodically, inducing complex microstructural evolution which must be recognized as it will alter the permeability of the fault rocks with time. The relative permeabilities and sealing potentials of actively deforming and abandoned parts of a single fault zone must also be considered. This will allow the size and distribution of potential fault zone fluid migration pathways to be assessed.

This paper presents a case study where this line of enquiry has been followed. The study is from the Vercors in the French Sub-Alpine Chains where a classic foreland thrust belt is developed. Structural mapping and microstructural assessment of possible migration pathways in well exposed structures such as those in the Vercors, may aid the assessment of migration pathways in provinces where hydrocarbn accumulations exist in thrust related structural traps.

Geological background

The Vercors, French Sub-Alpine Chains forms part of a classic foreland thrust belt (Goguel 1948; Gidon 1981; Butler 1988, 1989; Mugnier et al. 1987. The area, shown in Fig. 1, accommodates the last few kilometres of WNW-directed shortening in the Alps (Butler 1988). Thrusting occurred in post middle Miocene times, as revealed by the existence of middle Miocene sediments in the footwall to thrusts in the Vercors. Basement is probably not involved in the thrusting within these thin-skinned structures which are thought to detach along the basement-cover contact facilitated by Triassic evaporites (Butler in press; Bayer et al. 1987; Mugnier et al. 1987).

Triassic times saw the deposition of sabkha environment evaporites and platform carbonates. The rest of the Mesozoic saw the deposition of extensive carbonate sequences during subsidence and sea level fluctuations associated with the opening of Tethys (Graciansky et al. 1979; Lemoine et al. 1986; Arnaud-Vanneau & Arnaud 1990). The Mesozoic sequence is overlain by Tertiary foredeep clastic sediments. Bitumen seeps along strike in the area around Annecy indicate the maturation of source rocks and migration of hydrocarbons within the thrust belt (Zweidtler 1985). The thrust belt has been exhumed by uplift and erosion related to isostatic rebound of the lithosphere after the cessation of thrusting. This allows examination deep into the thrust structures which may have contained migration pathways and structural traps before uplift occurred.

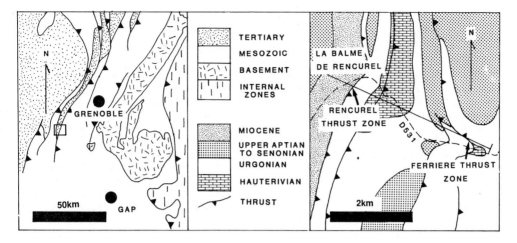

Fig. 1. Location map of the Rencurel thrust zone.

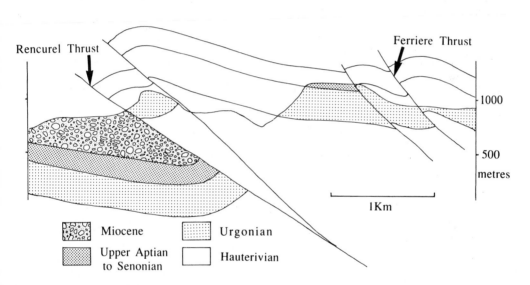

Fig. 2. Cross Section through the Gorges de la Bourne in the Vercors, showing the main structures within the Rencurel Thrust Sheet. The line of section is indicated on Fig. 1.

Deformation within the Rencurel thrust sheet

Several deeply incised gorges occur within the Vercors which allow examination deep within thrust structures. The Gorges de la Bourne, in its upper section between La Balme de Rencurel and Villard de Lans contains excellent exposures showing the deformation within the Rencurel Thrust Sheet (Fig. 2). Particularly well exposed are the Barremian platform carbonates which are locally termed the Urgonian. This carbonate formation forms part of a regionally extensive platform sequence which contains lateral thickness and facies changes (Arnaud-Vaneau & Arnaud 1990). Within the Gorges de la Bourne the Urgonian consists of massively bedded bioclastic grainstones which locally show an overprinting coarse dolomite mosaic, which was produced during burial diagenesis (Vieban 1983). The secondary inter-crystalline porosity, produced during dolomitization has given the Urgonian potential as a reservoir for hydrocarbons (see Fig. 3). Deformation within the Urgonian consists of bedding-parallel slip horizons cut by

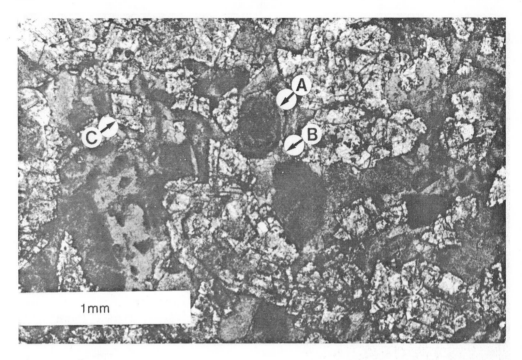

Fig. 3. Photomicrograph of the Urgonian Limestone from the Gorges de la Bourne. Skeletal grainstones with an early calcite cement are overprinted by a coarse dolomite mosaic. A, carbonate grain; B, early calcite cement; C, overprinting dolomite.

small displacement imbricate thrusts which have localized folds associated with them. The thrust sheet is carried on a localized cataclastic thrust named the Rencurel thrust, which has a displacement of around 1.5 km. The internal deformation to the thrust sheet displayed by the Urgonian limestones is described first, followed by a description of the basal thrust zone.

Deformation in the Rencurel thrust sheet

The bedding-parallel slip horizons consist of zones of cataclastic fault gouge which are generally around 1–5 cm thick, but one example 10 km to the south along strike in the Grande Goulets gorge is around 30 cm thick. These layer-parallel faults are thought to form early in the deformation history before folding related to imbricate thrusting. Kinematic indicators such as reidel shears, fault plane steps, and gouge fabrics indicate constant movement direction of top to the WNW for these slip horizons.

Imbricate thrusts developed locally after the slip horizons. Figure 2 shows the locations of 3 of these thrusts which developed significant displacements, although it must be stressed that

examples with smaller displacements occur commonly throughout the Urgonian.

The Ferriere thrust zone has been the subject of special study by Roberts (1990). The thrust locally emplaces Hauterivian limestones and shales on top of Urgonian limestones. The displacement on this thrust can be demonstrated to be around 300 metres. The fault zone experienced both localized faulting and distributed folding during its displacement history, indicated by the cross cutting relationships which exist between elements of its fault zone geometry shown in Fig. 4. Deformation began with the development of a distributed array of minor thrust faults which have gouge zones less than 2 cm thick. The faults exhibit ramp-flat geometries within the Hauterivian limestones and shales. Folding then became the dominant deformation process so that the early thrusts have become locally downward facing. Folding probably occurred ahead of a growing large-displacement thrust zone which can now be seen to cut the steep western limb of the fold developed in the Hauterivian hanging-wall rocks. This thrust zone, which has a 10 cm thickness of carbonate gouge developed along it, emplaces the older Hauterivian on top of the younger Urgonian.

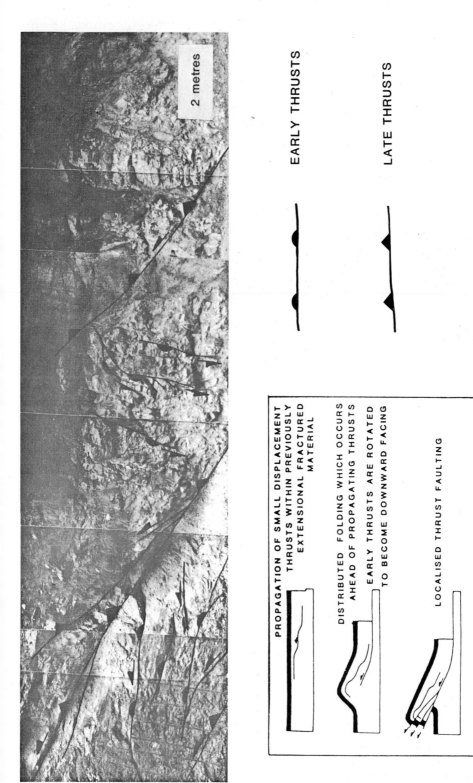

2 metres

EARLY THRUSTS

LATE THRUSTS

PROPAGATION OF SMALL DISPLACEMENT
THRUSTS WITHIN PREVIOUSLY
EXTENSIONAL FRACTURED
MATERIAL

DISTRIBUTED FOLDING WHICH OCCURS
AHEAD OF PROPAGATING THRUSTS

EARLY THRUSTS ARE ROTATED
TO BECOME DOWNWARD FACING

LOCALISED THRUST FAULTING

Fig. 4. Interpreted structural section and model for the development of the geometry of the Ferriere thrust zone (adapted from Roberts 1990).

Fig. 5. View looking south onto the Rencurel thrust zone one kilometre to the east of La Balme de Rencurel. A, two imbricate slices of Urgonian limestones; B, Miocene sandstones and conglomerates exposed within the woods.

This structural study demonstrates that the style and position of deformation changed during the development of thrusts within the Rencurel thrust sheet. The implications for structurally controlled fluid migration through this structure are discussed elsewhere (Roberts 1990).

All the faults studied within the Gorges de le Bourne were dominated by frictional slip. All the lineations found on fault planes consist of striations on polished gouge coated surfaces. No mineral fibre lineations have been found. X-ray diffraction and petrological studies of the gouge have shown that it consists of finely comminuted carbonate material which may include fragments of wall rock and cement precipitated within the fault zone. Also, faulting does not involve pervasive deformation of the wall rocks by intense veining. Veins are mostly restricted to within gouge zones. Thrusting did not produce a pervasive fracture porosity.

Deformation within the Rencurel thrust zone

1 km to the east of La Balme de Rencurel the Rencurel thrust zone is exposed (shown in Fig. 5 and Fig. 2). This thrust zone forms the base to the thrust sheet and from regional considerations it can be shown that the fault accommodates around 1.5 km of displacement. At the present erosion level the fault emplaces Urgonian limestones onto Miocene molasse sediments with the thrust contact well exposed on the D103 road between Pont de Goule Noire and St Julien en Vercors.

The fault zone can be divided into two parts. A basal thrust, containing at least 30 cm of lithified carbonate gouge derived from the overlying Urgonian limestones; and at least 2 m of gouge, derived from both the Urgonian hanging-wall rocks and the molasse footwall rocks. Structurally overlying the basal thrust are the Urgonian limestones which contain a densely spaced array of minor faults in a zone at least 50 m thick.

Figure 6 shows a field photo of a characteristic fault from the Urgonian limestones involved in the Rencurel thrust zone. Deformation is localized into a zone around 3 cm across while the wall rocks are relatively undeformed. Figure 7

Fig. 6. Field photo of a minor fault within the Urgonian limestones of the Rencurel Thrust Zone. Cataclastic deformation is confined to a thin zone approx. 5 cm across. Grain-size reduction has produced a fault gouge within which textural features of the wall rocks are not recognizable. The wall-rocks to this fault zone are not pervasively deformed by fracturing with features such as ooids, skeletal grains and early cements still recognisable. Lens cap is 5 cm in diameter.

shows the microstructure of the fault zone. Petrological and X-ray diffraction studies indicate that the finely-comminuted carbonate material which makes up the fault gouge is probably derived from the crushing of wall rocks and cement precipitated within the fault zone. The gouge fabric is cut across by extensional fractures which are filled with calcite cement which exhibits a drusy fabric.

Figure 8 shows a field photo of the faulted contact between the Urgonian limestones and the Miocene molasse sediments within the Rencurel Thrust Zone. 30 cm of lithifield carbonate gouge separates the relatively undeformed Urgonian wall rocks from the gouge derived from both the Urgonian and the Miocene molasse sediments which is at least 2 m thick. As the footwall to the high strain zone is not exposed the style of deformation within the molasse wall-rocks is not discussed here.

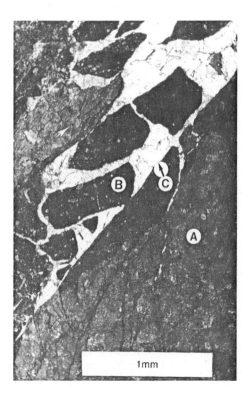

Fig. 7. Microstructures developed along a minor fault within the Urgonian, which has been locally dolomitized before deformation took place. A, Dolomite wall-rocks to the fault zone which are not pervasively deformed by fracturing; B, carbonate gouge; C, drusy calcite cement infilling extensional fractures.

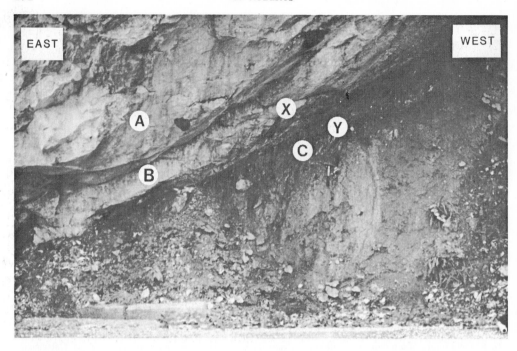

Fig. 8. Field photo of the faulted contact between the Urgonian and the Molasse in the high strain part of the Rencurel thrust zone. A, Urgonian limestones deformed by spaced minor faults (See Fig. 14); B, lithified carbonate gouge; c, gouge zone composed of both Urgonian and molasse material. Hammer is 40 cm long. X and Y indicate the positions of the close up views shown in Fig. 9.

Figure 9 shows close up views of the high strain zone, the position of which is indicated on Fig. 8. These views indicate that within both the carbonate gouge and the carbonate/molasse gouge, deformation and gouge production occurred by displacement on minor slip surfaces during the last phases of faulting. It may be that the large thickness of gouge which exists within the high strain zone was produced by many increments of slip on such minor fault surfaces.

In thin section, cataclastic gouge can be seen to cut and be cut by extensional fractures which are filled with carbonate cements. Also stylolites produced by the action of pressure dissolution can be seen.

The discrete slip surfaces described above, from within the high strain zone, separate fault rocks which have different chemical compositions. Figure 10 shows two photomicrographs of samples taken from the top and the bottom of the carbonate gouge shown in Fig. 9. Sample A, taken from the hanging-wall to the slip surface is composed of a calcite and dolomite rich gouge which is cut by calcite filled extensional fractures. All the elements of the fault rock fabric stain pink when tested with a combined stain of Potassium Ferricyanide and Alizarin Red S.

Sample B from the footwall to the slip surface is also composed of calcite and dolomite rich gouge. Clasts of an early-formed gouge fabric, which stains pink occur within a later iron rich gouge fabric which stains blue. Extensional fractures which are filled with ferroan calcite cut across the whole fault rock fabric as shown in Fig. 11. Some extensional fractures are filled with bitumen (Fig. 12). In double polished wafers, bitumen-filled extensional fractures are associated with fractures which appear light blue under ultra violet fluorescence. When the wafers were immersed in an organic solvent, the fluorescence disappeared. This suggests that the fluorescence is caused by the presence of live oil in the fractures (Robert 1988). The hydrocarbon bearing fractures cross-cut early gouge fabrics but are in some cases themselves cut by later gouge zones and extensional fractures. This suggests that the hydrocarbons entered the fractures during the displacement history of the fault zone, and not after the cessation of deformation. Also, if the mobilization of hydrocarbon was not controlled by the structural development of the high strain part of the fault zone, but entered fault zone after the cessation of deformation due to processes occurring in the water table after

Fig. 9. Close up views of the high strain zone, the position of which are indicated in Fig. 8. X, Discrete slip horizons within the lithified carbonate gouge. Pencil is 16 cm long. A and B indicate the positions of the samples shown in Fig. 10. Y, Discrete slip horizons within the foliated gouge composed of both Urgonian and molasse material. Knife is 9 cm long.

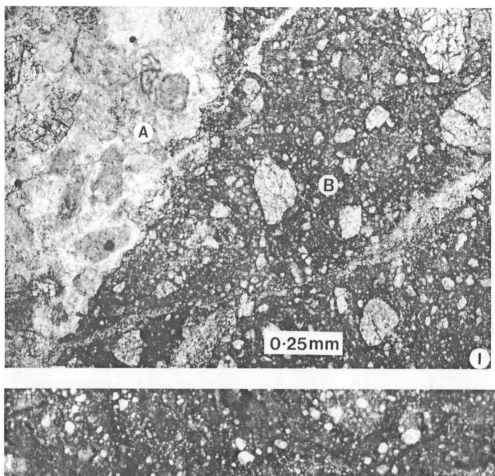

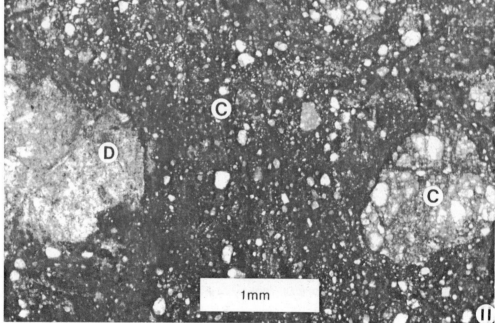

Fig. 10. Photomicrographs of samples taken from above (I) and below (II) the slip horizons shown in Fig. 9 (**X**). A, Partially dolomitized grainstone; B, gouge composed of calcite and dolomite; C, iron-rich gouge composed of calcite and dolomite; D, clast of gouge composed of calcite and dolomite within the iron-rich gouge.

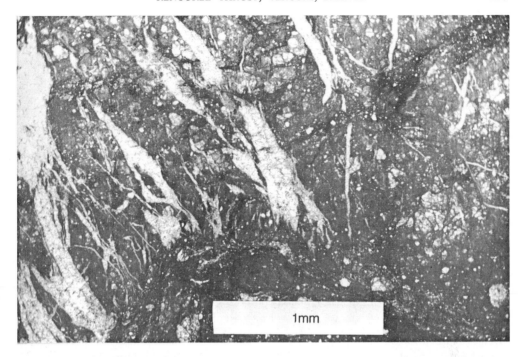

Fig. 11. Extensional fractures filled with ferroan calcite cement which cut across the iron rich gouge.

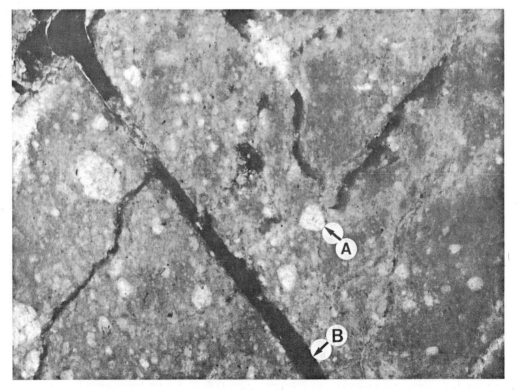

Fig. 12. Fracture filled with bitumen. A, Clast within iron rich gouge; B, bitumen.

the cessation of deformation, then it is likely that bitumen would be distributed throughout the fault zone infilling fractures and intercrystalline secondary porosity within the dolomites. Over 100 samples were collected from the Rencurel thrust zone and the overlying thrust sheet within the Gorges de la Bourne. Bitumen is not found infilling secondary inter-crystalline porosity within the Urgonian and is not found associated with fractures elsewhere within the thrust sheet. Hydrocarbon and iron rich samples were only found within the 2.5 m thick area within the high strain part of the Rencurel thrust zone. Thus, the evidence cited above suggests that hydrocarbons entered the fault zone during its displacement history, and that the migration pathway was controlled by structural features. Butler (this volume) suggests that along strike in the thrust belt to the north, maturation of possible source rocks occurred prior to development of the Sub-Alpine structures. A complex hydrocarbon migration history involving entrapment in palaeo-foreland structures followed by re-migration during Sub-Alpine deformation is suggested. Hydrocarbons involved in migration along thrust zones such as those described in this paper may have been sourced from existing accumulations. The fluids may have undergone up-dip migration in response to fluid pressure variations with depth along the thrust zone. In addition, the dilatant nature of cataclastic deformation which dominates the thrust zone may have produced fluid pressure gradients and forced fluid migration. It not clear whether the hydrocarbons dominated the fluid composition or whether mainly aqueous phases existed with only a minor charge of hydrocarbons.

Microstructural controls on fluid migration within the Rencurel thrust sheet

A characteristic feature of the gouges found within the Rencurel thrust sheet is that they contain evidence for early cataclastic gouge fabrics which now occur within clasts in later gouges as shown in Fig. 13. The early gouges underwent lithification and became indurated before renewed cataclasis broke them into clasts during the production of later gouges. It is known that cataclasis becomes dominant over diffusive mass transfer as strain rate increases (Knipe 1989). High strain-rate events where cataclasis is activated as a deformation mechanism must therefore occur episodically. The low strain rates between these periods would elevate diffusive mass transfer as a deformation mechanism which would induce lithification of the gouge. Within the gouge, pressure dissolution would cause chemical compaction, liberating material which may be precipitated as cements elsewhere in the vicinity of the fault zone. Within gouge zones in the Rencurel thrust sheet, induration of gouge fabrics during the action of diffusive mass transfer is sufficient such that cataclastic brittle failure can occur. This is suggested by the observation that cataclasis which post-dates induration, produces a new gouge fabric which contains clasts of the early gouge fabric (see Fig. 13). If the gouge was not well indurated, further high strain rate deformation may have been accommodated at least in part by flow due to sliding of the grains. This deformation mechanism termed independent particulate flow (Borradaille 1981), is difficult to recognize during microstructural studies as it accommodates strain by re-packing the grains. It is therefore not clear whether the gouge became indurated prior to all the episodes of cataclasis and the importance of independent particulate flow is difficult to assess. Clearly independent particulate flow can only occur in gouge after cataclastic grain-size reduction has produced grains. The onset of independent particulate flow is governed by the strength of the contacts between grains. Processes which control the strength of these contacts include cementation and chemical compaction by pressure dissolution which are part of the diffusive mass transfer deformation mechanism. Thus, the onset of independent particulate flow in the gouges of Rencurel Thrust is controlled by the extent to which diffusive mass transfer has been a dominant deformation mechanism.

The microstructures found within the Rencurel thrust sheet indicate that cataclasis and diffusive mass transfer (Knipe 1989; Engelder 1974; Rutter 1983) were the dominant deformation mechanisms. Cross-cutting relationships which exist between the microstructures indicate that deformation accumulated incrementally with faulting conditions varying over a great enough range to activate both diffusive mass transfer and cataclasis. Periods when diffusive mass transfer was the dominant deformation mechanism were punctuated by times when cataclasis was dominant. The change of deformation mechanism with time and the resulting microtextural evolution will have important effects on the porosity and permeability of the fault rocks.

Lithified and indurated gouge would have low values of porosity and permeability due to the very fine grain sizes and poor sorting, and also due to the diagenetic effects of diffusive mass transfer such as cementation and chemical

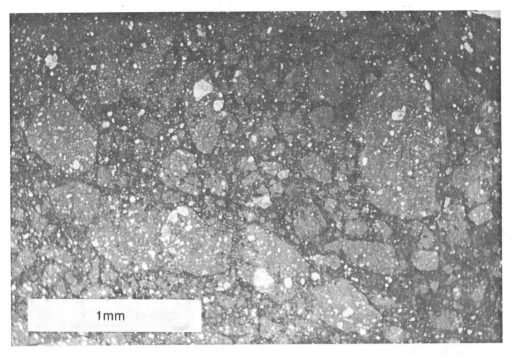

Fig. 13. Early formed gouge fabrics within clasts which make up a later gouge fabric. This suggests that periods of cataclasis and grain size reduction are interspersed with phases of gouge induration probably due to the action of cementation and pressure dissolution.

compaction. Carbonate gouge zones which are not undergoing cataclastic deformation are low permeability barriers to fluid migration (Roberts 1990).

The largest volumes of cement within the fault zone exist within extensional fractures. This suggests that the process of cataclastic extensional fracturing, which is known to involve dilatancy (Knipe 1989), concentrates the influx of cement-precipitating pore waters through the porosity contained within the fractures. Also, the production of a fault gouge is a dilatant process and involves extensional fracturing on a small-scale (Engelder 1974; White 1976; Knipe 1989). Cataclastic gouge formation involving fracturing within an impermeable medium will thus cause an increase in the porosity. Thus, fluid flow may occur during the formation of a gouge fabric. However, if cataclasis occurs within a material of high porosity and permeability such as a reservoir sandstone, grain-size reduction may reduce the porosity and permeability of the fault zone relative to the medium. Thus, whether fluid flow is concentrated through the fault rocks within a gouge zone undergoing dilatant deformation will be governed by the porosity and permeability of the wall-rocks to the fault zone.

Episodes of cataclasis within indurated gouge will episodically increase the permeability of the cataclastically-deforming gouge relative to inactive indurated gouge. Thus, fluid flow may become localized within a gouge which is actively undergoing cataclasis.

Independent particulate flow also produces complex volume changes involving dilatancy due to sliding and re-packing of grains. The inter-granular porosity will vary during such deformation.

The permeability of the gouge relative to the wall-rocks will also vary in a complex manner. At times grain re-packing may reduce the size of inter-granular pore spaces so that permeability within the gouge may approach the low values of the indurated gouge. Fluid flow would not become localised within the actively-deforming zone. However, dilatant grain repacking may at times increase the size of the inter-granular pore spaces so that the permeability of the actively-deforming zone is increased, localizing fluid flow.

The actual values of porosity and permeability will be governed by the textures of the fault rocks. However, grain-size reduction during cataclastic gouge formation will destroy textures formed early in the deformation history of an individual gouge zone. The permeability of the early textures of the gouge cannot be directly measured or estimated using microtextural studies. Thus, qualitative assessment of the relationship between fault rock textures, permeability and fluid flow must be used to infer the history of fluid flow during incremental deformation along a gouge zone.

The geometry and sequence of faulting within the Rencurel thrust sheet: implication for fluid flow

In foreland thrust belts the position and style of active deformation is known to change through time. Studies of fault sequences have shown that individual faults are abandoned with time and faulting switches to a position elsewhere within the thrust belt. Piggy-back and break-back fault sequences as well as combinations of the two styles, have all been recognized (Butler 1987). Also, individual volumes of rock are known to experience different styles of deformation during the growth of an individual structure. An example is the change between distributed deformation in the form of folding which occurs before the localization of displacement onto a thrust (Elliott 1976; Coward & Potts 1983; Williams & Chapman 1983; Roberts 1990). During the growth of thrust structures, the style and position of deformation changes with time. This implies that the style and position of fluid migration through such structures will change accordingly.

The Rencurel thrust sheet contains both folds and faults. To document the migration pathways which were important during its structural development, both the geometry and sequence of deformation must be ascertained. Detailed structural analysis was used to determine the geometry and deformation sequence within the Rencurel thrust sheet.

Deformation within the Urgonian limestones of Rencurel thrust sheet occurred through the development of bedding-parallel slip horizons which later became folded and cut by the growth of small imbricate thrusts such as the Ferriere thrust zone. Away from these localised deformation zones the Urgonian limestones are relatively undeformed. The deformation was accompanied by the migration of pore waters which were over saturated with respect to calcite. This is evidenced by the existence of calcite filled extensional fractures within the fault rocks of the imbricate thrusts. The intensification of this style of deformation produced a zone of closely spaced cataclastic faults around 50 m thick

forming the precursor to the Rencurel thrust zone. At this stage the thrust zone may still have had a small stratigraphic separation across it, perhaps with the Urgonian limestones still in contact across the fault zone.

The fault rocks to this precursor fault zone now occur as clasts within the iron and hydrocarbon rich gouge along the high strain part of the Rencurel thrust zone. This later gouge must have formed after the early cataclastic faulting within the Urgonian had ceased. This later deformation was localized structurally below the preserved part of the precursor fault zone. It was accompanied by the migration of an iron rich and carbonate saturated pore water which contained hydrocarbons. The early cataclastic faults were not active and were relatively impermeable during this time because they do not show evidence for the through migration of iron and hydrocarbon rich pore waters. The gouge along these faults had become lithified and impermeable due to the action of diffusive mass transfer after the cessation of high strain rate fault slip. This later stage of fluid migration was channelized along the active late gouge zone beneath the now inactive and impermeable precursor fault zone which was being carried in the hanging-wall to this later deformation. The high strains recorded within this late gouge zone and the existence of molasse in the footwall and Urgonian in the hanging-wall suggest that the Rencurel thrust zone accumulated a large proportion of its 1.5 km of displacement during this late deformation.

The change in the fluid composition may have occurred as a direct result of the growth of the fault zone. Faults that accommodate large displacements are generally longer along strike and down dip than faults which have small displacements (Elliott 1976; Walsh & Watterson 1988). The small displacement precursor fault zones within the Rencurel thrust zone may only have been able to access fluids from an area approaching the length of an individual fault. If the precursor fault zone was confined to within the Urgonian at this time then the fluid composition may reflect the composition of the Urgonian. This may be the case as the fluid composition did not include hydrocarbons or significant amounts of iron which is true of the composition of the Urgonian. The large displacement accommodated by the later faulting implies that the fault zone may have had a greater length at this time. The fluid composition could reflect the compositions of other formations within the sequence, as suggested by the presence of hydrocarbons and significant amounts of iron within the pore water migrating during this deformation. The Hauterivian limestones and shales which lie stratigraphically beneath the Urgonian limestones have a vitrinite reflectance value of R.E. 0.35%. This indicates that the thermal maturity of organic matter was low, and significant amounts of hydrocarbons were not generated within the rocks in the immediate vicinity of the exposed portion of the Rencurel thrust zone. Displacements on individual thrusts within the Vercors are small being in the order of a few kilometres (Butler in press). This indicates that the Rencurel thrust sheet has not been overthrust by thick thrust sheets containing rocks with high thermal maturities. This suggests that the hydrocarbons found within the fault rocks of the high strain part of the Rencurel thrust zone underwent maturation structurally beneath the Hauterivian rocks. Possible source rocks may be the Berriasian or Oxfordian shales which exist deeper in the stratigraphy and might be expected to have higher maturity values.

This suggests that the pore waters and hydrocarbons moved up dip through the structures in the thrust belt. At the level exposed by erosion within the Rencurel thrust sheet, migration of these hydrocarbon-bearing pore waters occurred through the Rencurel thrust zone. At deeper structural levels, the hydrocarbon-bearing pore waters may have migrated up-dip along carrier beds. The question arises as to whether any structural traps existed during this hydrocarbon migration. An obvious candidate is the anticline which is carried in the hanging-wall to the Rencurel thrust zone shown in Fig. 2. This structure contains possible Urgonian reservoir rocks which contain secondary porosity produced during burial dolomitization. Also the clay-rich molasse formations which would have structurally overlain the anticline before uplift and erosion, may have provided a good top seal. The geometry of the thrust-related hanging-wall anticline suggests that it can be interpreted as a tip line fold (Elliott 1976; Coward & Potts 1983; Williams & Chapman 1983). This implies that the fold was already formed by the time that large displacements had developed and the migration of hydrocarbons had begun along the Rencurel thrust zone. The anticline was available as a structural trap during the migration of hydrocarbons along the Rencurel thrust zone. However, the existence of a densely-packed impermeable zone of cataclastic faults between the trap and the hydrocarbon migration pathway within the Rencurel thrust zone suggest that the hydrocarbons may not have been able to enter the trap upwards across the precursor fault zone. Also, the hydrocarbons may have encountered difficulties in migrating through potential carrier

beds such as the Urgonian limestones. In the well-exposed gorge sections in the hanging-wall of the Rencurel thrust zone, the Urgonian limestones can be seen to contain a high density of small-scale gouge coated faults which would have a low permeability (Roberts 1991). These faults accommodate a range of displacements which are in general less than 50 m. The Urgonian is around 300 m thick and contains a variety of carbonate platform facies which in places are pervasively dolomitized (Arnaud & Arnaud-Vanneau 1990; Vieban 1983). The fluid flow within the Urgonian would be in part controlled by the complex heterogeneities within the formation due to facies variation and textural alteration produced by diagenetic processes. Faults which accommodate displacements approaching 50 m cross-cut the whole formation. Faults accommodating smaller displacements cross-cut individual carrier horizons within the formation. Thus, the presence of a large number of impermeable fault zones within the Urgonian would compartmentalize the formation in a complex manner. Cross-formational flow may be reduced to insignificant levels by the features described above.

Thus, the pore waters containing hydrocarbons may have encountered difficulties in entering trap in the hanging-wall anticline by migrating across the higher parts of the Rencurel thrust zone, or up-dip along Urgonian carrier beds in the limbs of the anticline. Figure 14 shows a cartoon of this migration model. Hydrocarbons may however have accumulated in smaller traps in the footwall if the hydrocarbon migration pathway identified is truly at the base of the fault zone. Accumulations could then occur below the footwall cut off of the base of the molasse sequence where a small closure exists formed by the interaction of the fault zone seal and the seal at the base of the molasse sediments. The footwall to the Rencurel thrust zone is not well exposed, so the existence of such an accumulation cannot be discussed further here.

Discussion and conclusions

This is an example where the fault sequence and fault zone geometry may have combined to channelize hydrocarbon migration away from a structural trap. However, if the high strain part of the Rencurel thrust zone had localized along the top of the impermeable zone of early minor faults, then hydrocarbons would not be impeded from migrating into the trap. Also, if thrusting had been accompanied by pervasive fracturing and brecciation of the wall-rocks, then migration

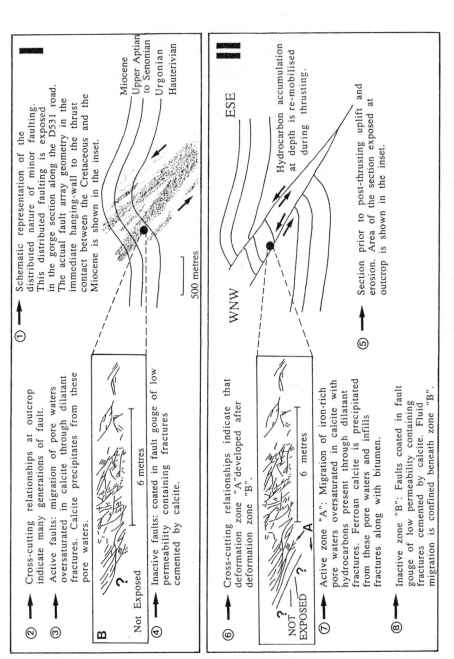

Fig. 14. Model of fluid migration through the Rencurel Thrust Zone. I, Early deformation and fluid flow. II, Late deformation and fluid flow. The fault patterns shown in the insets were mapped by marking the position of faults with fault rocks visible at outcrop onto a photo-montage of the fault zone. This methodology produces a minimum estimate of the fault density as some faults cutting massive limestones may be difficult to see at outcrop. The minor faults have displacements in the order of a few metres or less.

① Schematic representation of the distributed nature of minor faulting. This distributed faulting is exposed in the gorge section along the D531 road. The actual fault array geometry in the immediate hanging-wall to the thrust contact between the Cretaceous and the Miocene is shown in the inset.

Miocene
Upper Aptian to Senonian
Urgonian
Hauterivian

500 metres

② Cross-cutting relationships at outcrop indicate many generations of fault.

③ Active faults: migration of pore waters oversaturated in calcite through dilatant fractures. Calcite precipitates from these pore waters.

B

Not Exposed

6 metres

④ Inactive faults: coated in fault gouge of low permeability containing fractures cemented by calcite.

⑤ Section prior to post-thrusting uplift and erosion. Area of the section exposed at outcrop is shown in the inset.

Hydrocarbon accumulation at depth is re-mobilised during thrusting.

WNW

ESE

⑥ Cross-cutting relationships indicate that deformation zone "A" developed after deformation zone "B".

NOT EXPOSED

—A—

6 metres

⑦ Active zone "A": Migration of iron-rich pore waters oversaturated in calcite with hydrocarbons present through dilatant fractures. Ferroan calcite is precipitated from these pore waters and infills fractures along with bitumen.

⑧ Inactive zone "B": Faults coated in fault gouge of low permeability containing fractures cemented by calcite. Fluid migration is confined beneath zone "B".

may have been possible across early formed impermeable zones through secondary fracture porosity. These alternative styles of fault zone may exist in areas where different lithologies exist or where faulting is occurring with different bulk strain rates, pressures, temperatures and fluid availabilities. Clearly more case studies from other fault zones are needed before we can begin to predict migration pathways through deformed rock. This style of investigation will be of equal importance when assessing normal and strike slip fault zones as migration pathways.

This contribution forms part of a study of Sub-Alpine thrust system geometry, 3D evolution, fault rock evolution, fluid migration and thermal/diagenetic history. I thank R. Butler, M. Tucker, S. Bowler and S. Moss who are involved in this research effort. I am most grateful to D. Hunt and J. Henton for their comments during the preparation of this paper and C. Milsom and J. Porter for their assistance during fieldwork. I would also like to thank A. Fleet and S. Hay for their constructive reviews. This study was undertaken during the tenure of a BP studentship. Further studies of deformation and diagenetic histories around fault zones are funded by a NERC Research Fellowship (GTS/F/90/GS/8).

References

ARNAUD-VANNEAU, A. & ARNAUD, H. 1990. Hauterivian to Lower Aptian carbonate shelf sedimentation and sequence stratigraphy in the Jura and northern Sub-Alpine Chains (south eastern France and Swiss Jura). *In*: TUCKER, M. E., WILSON, J. L., CREVELLO, P. D., SARG, J. R. & READ, J. F. (eds) *Carbonate Platforms: Facies, Sequences and Evolution*. Special Publication of the International Association of Sedimentologists, **9**, 203–233.

BAYER, R., CAZES, M., DAL PIAZ, G. V., DAMOTTE, B., ELTER, G., GOSSO, G., HIRN, A., LANZA, A., LOMBARDO, B., MUGNIER, J-L., NICOLAS, A., THOUVENOT, F., TORRIELLES, G. & VILIEN, A. 1987. Premiers resultants de la traversee des Alpes Occidentales par sismique reflexion verticale (Programme ECORS-CROP). *Compte Rendus de l' Academie des Sciences Paris*, **305**, 1461–1470.

BORRADAILE, G. J. 1981. Particulate flow and the generation of cleavage. *Tectonophysics*, **72**, 306–321.

BRIGGS, R. C. & TROXELL, H. C. 1955. Effects of the Arvin-Tehachipi Earthquake on spring and stream flows. *In*: OAKESHOFF, G. B. (ed.). Earthquakes in Kern County, California during 1952. *California Division of Mines and Geology Bulletin*, **171**, 81–97.

BURLEY, S. D., MULLIS, J. & MATTER, A. 1989. Timing diagenesis in the Tartan Reservoir (U.K. North Sea): constraints from combined cathodoluminescence microscopy and fluid inclusion studies. *Marine and Petroleum Geology*, **6**, 98–120.

BUTLER, R. W. H. 1987. Thrust Sequences. *Journal of the Geological Society, London*, **144**, 619–634.

—— 1988. The geometry of crustal shortening in the Western Alps. *In*: SENGOR, A. M. C. (ed.) *Tectonic evolution of the Tethyan regions*. Proceedings of NATO Advanced Study Institute, **259** Kluwer, Dordrecht, 43–76.

—— 1989. The influence of pre-existing basin structure on thrust system evolution in the Western Alps. *In*: COOPER, M. A. & WILLIAMS, G. D. (eds) *Inversion Tectonics*. Geological Society, London, Special Publication, **44**, 105–122.

—— 1991. Hydrocarbon maturation, migration and tectonic loading in the Western Alpine foreland thrust belt. *This volume*.

BUTLER, R. W. H. in press. Thrust evolution within previously rifted regions: an example from the Vercors, French Sub-Alpine Chains. *Memoria Societa Geologica Italiana*, **389**.

COWARD, M. P. & POTTS, G. J. 1983. Complex strain patterns developed at the frontal and lateral tips to shear zones and thrust zones. *Journal of Structural Geology*, **5**, 383–399.

ELLIOTT, D. 1976. The energy balance and deformation mechanisms of thrust sheets. *Philosophical Transactions of the Royal Society, London*, **A283**, 289–312.

ENGELDER, J. T. 1974. Cataclasis and the generation of fault gouge. *Geological Society of America Bulletin*, **85**, 1515–1522.

GIDON, M., 1981. Les deformations de la couverture des Alpes Occidentales Externes dans la region de Grenoble: Leurs rapports avec celles du socle. *Compte Rendus de l' Academie des Sciences, Paris*, **292**, 1057–1060.

GOGUEL, J., 1948. Le Role des Failles de Dechrochement dans le massif de la Grande Chartreuse. *Bulletin de la Societe Geologique de la France*, **18**, 277–235.

GRACIANSKY, P. C., BOURBON, P., CHENET, P. Y., DE CHARPAL, O. & LEMOINE, M. 1979. Genese et evolution comparee de deux marges continentales passives: Marge Iberique de l' Ocean Atlantique et Marge Europeene de la Tethys dans les Alpes Occidentales, *Bulletin de la Societe Geologique de la France*, **21**, 663–674.

HARDING, T. P. & TUMINAS, A. C. 1989. Interpretation of footwall (lowside) fault traps by reverse faults and convergent wrench faults. *Bulletin of the American Association of Petroleum Geologists*, **72**, 738–757.

JOURDAN, A., THOMAS, M., BREVART, O., ROBSON, P., SOMMER, F., & SULLIVAN, M. 1987. Diagenesis as the control of Brent sandstone reservoir properties in the Greater Alwyn area, E. Shetland Basin. *In*: BROOKS, J., & GLENNIE, K. (eds) *Petroleum*

Geology of North-West Europe. Graham & Trotman, London, 951–961.

KNIPE, R. J. 1989. Deformation Mechanisms—Recognition from natural tectonites. *Journal of Structural Geology*, **11**, 127–146.

LEE, M., ARONSON, J. L., & SAVIN, S. M. 1989. Timing and conditions of Permian Rotliegendes sandstone diagenesis, Southern North Sea: K/Ar and oxygen isotopic data. *Bulletin of the American Association of Petroleum Geologists*, **73**, 195–215.

LEMOINE, M., BAS, T., ARNAUD-VANNEAU, A., ARNAUD, H., DUMONT, T., GIDON, M., GRACIANSKY, D. E., RUDKIEWICZ, J. L., MEGARD-GALLI, J. & TRICART, P. 1986. The continental margin of the Mesozoic Tethys in the Western Alps. *Marine and Petroleum Geology*, **3**, 179–199.

MITRA, S. 1988. The effects of deformation mechanisms on reservoir potential in the Central Appalachians Overthrust Belt. *Bulletin of the American Association of Petroleum Geologists*, **72**, 536–554.

MUGNIER, J. L., ARPIN, R. & THOUVENOT, F. 1987. Coupes Equilibrees a travers le massif sub alpine de la Chartreuse. *Geodinamica Acta, Paris*, **1**, 125–137.

PORTER, K. W. & WEIMER, R. J. 1982. Diagenetic sequence related to structural history and petroleum accumulation: Spindle Field, Colorado. *Bulletin of the American Association of Petroleum Geologists*, **66**, 2543–2560.

ROBERT, P. 1988. *Organic metamorphism and geothermal history. Microscopic study of organic matter and thermal evolution of sedimentary basins.* D. Reidel Publishing Company, Dordrecht, Holland.

ROBERTS, G. 1990. Structural controls on fluid migration in Foreland Thrust Belts. *In*: LETOUZEY, J. (ed.) *Petroleum and tectonics in mobile belts.* Editions Technip, Paris, 193–210.

—— 1991. *Deformation and diagenetic histories around foreland thrust faults.* PhD Thesis, University of Durham.

RUTTER, E. H. 1983. Pressure solution in nature, theory and experiment. *Journal of the Geological Society, London*, **140**, 725–740.

SIBSON, R. H. 1989. Earthquake faulting as a structural process. *Journal of Structural Geology*, **11**, 1–14.

——, MOORE, J. McM., RANKIN, A. H., 1975. Seismic pumping—A hydrothermal fluid transport mechanism. *Journal of the Geological Society, London*, **131**, 653–659.

SMITH, D. A. 1966. Theoretical consideration of sealing and non-sealing faults. *Bulletin of the American Association of Petroleum Geologists*, **50**,
363–374.

—— 1980. Sealing and non-sealing faults in Louisiana Gulf Coast Salt Basin. *Bulletin of the American Association of Petroleum Geologists*, **64**, 145–172.

STERMITZ, F. 1964. Effects of the Hebgen Lake earthquake on surface water. *US Geological Survey Professional Paper*, **435**, 139–150.

SWENSEN, F. A. 1964. Ground-water phenomena associated with the Hebgen Lake earthquake. *US Geological Survey Professional Paper*, **435**, 159–165.

VIEBAN, F. 1983. *Installation et evolution de la plateform Urgoniene (Hauterivian a Bedoulien) du Jura meridional aux chaines sub alpines (Ain) Savoie, Haute Savoie.* These 3e cycle, Grenoble.

WALSH, J. & WATTERSON, J. 1988. Analysis of the relationship between displacement and dimensions of faults. *Journal of Structural Geology*, **10**, 437–462.

WEBER, K. J. & DAUKORU, E. 1975. Petroleum Geology of the Niger Delta. *Proceedings of the Ninth World Petroleum Congress*, **2**, 209–221.

——, MANDL, G., PILAAR, W. F., LEHNER, F. & PRECIOUS, G. 1978. The role of faults in hydrocarbon migration and trapping in Nigerian growth fault structures. *Proceedings of the 10th Annual Offshore Technology Conference, Houston, Texas*, **4**, 2643–2653.

WHITE, S. H. 1976. The effects of strain on the microstructures, fabrics and deformation mechanisms in quartzites. *Philosophical Transactions of the Royal Society, London*, **A283**, 69–86.

WILLIAMS, G. D. & CHAPMAN, T. 1983. Strains developed in the hanging-walls to thrusts due to their slip/propagation rate: a dislocation model. *Journal of Structural Geology*, **5**, 563–571.

WOJTAL, S. & MITRA, G. 1986. Strain hardening and strain softening in fault zones from foreland thrusts. Geological Society of America Bulletin, **97**, 674-687.

WOOD, S. H. 1985. The Borah Peak, Idaho earthquake of October 28, 1983—Hydrologie Effects. *Earthquake Spectra*, **2**, 127–150.

WOODWARD, N. B., WOJTAL, S., PAUL, J. B., ZADINS, Z. Z., 1988. Partitioning of deformation within several external thrust zones of the Appalachian Orogen. *Journal of Geology*, **96**, 351–361.

ZWEIDTLER, D., 1985. *Genese des gisements d' asphalte des formations de la Pierre Jaune de Neuchatel et des calcaires Urgoniens du Jura (Jura Neuchatelois et nord-vaudois).* These 3e cycle, Universite Neuchatel.

PART IV:

TRAP LEAKAGE AND SUBSEQUENT MIGRATION

Petroleum seepage and post-accumulation migration

R. H. CLARKE[1] & R. W. CLEVERLY[2]

BP Exploration, Britannic House, Moor Lane, London EC2Y 9BU, UK

Present addresses: [1] *BP Research, Chertsey Road, Sunbury-on-Thames, Middlesex TW16 7LN, UK*

[2] *BP Exploration Operating Company Ltd, 4/5 Long Walk, Stockley Park, Uxbridge, Middlesex UB11 1BP, UK*

Abstract: The processes governing the loss of petroleum from traps and its migration to the surface are poorly understood. Leakage from traps does not necessarily result in surface seeps, nor are all occurrences of surface petroleum the result of present-day trap leakage. Three types of seep are recognized: unaltered gassy seeps, seeps affected by surface dispersal processes, and seeps altered by subsurface processes. Unaltered seeps form about 30% of the recorded population and are characterized by the presence of a free gas component. The buoyancy of the gas phase appears to drive most seeps to the surface. Non-associated oil seeps, often sharing their point of efflux with spring water, are altered by near-surface processes, especially the segregation of the oil and gas phases at the water table. Seeps altered by subsurface processes are frequently sour, containing free hydrogen sulphide, and are often associated with leakage pathway mineralisation. The geological context and characteristics of seepage deserve attention, leading to more realistic models for oilfield growth and decay, and to a better understanding of the environmental and mineralising effects of seepage.

Trap leakage appears to be so widespread that it should form an integral part of the material balance relationships used to model oilfield growth and decay. Out of 370 basins with recorded petroleum reserves at least 126 are leaking to the surface at the present day, according to our records. Little attention has been paid to seepage in some major provinces. It has only recently become apparent, for example, that active seeps occur in the North Sea (Hovland & Judd 1988). This seepage, regardless of the fact that much of it may be biogenic gas, was effectively hidden from the pioneer explorers of this basin by the North Sea itself. Similar concealment of seepage may take place elsewhere, both onshore and offshore and may hinder appreciation of its importance and the development of a more dynamic theory of petroleum entrapment.

This paper incorporates a simple classification of seeps which draws attention to some of the ways in which they become altered or obscured. Systematic study of seepage taking account of environmental effects is likely to yield significant information. Seeps are, after all, the ends of migration pathways and the only parts of them which can readily be used to determine flux rates and composition. Most reported information comes from onshore areas although the offshore environment is in many ways less hostile to petroleum seeps.

Descriptions of about 6000 petroleum seeps obtained from unpublished BP reports and a scattered literature have been reviewed as part of an effort to understand more about petroleum leakage and its distribution. Recorded seeps occur throughout the world, mostly onshore, but the database does not include information from most of the United States and parts of Western Europe (Fig. 1).

Classification

Discussion

Link (1952) published the most recent detailed review of seeps. He recognized five types (Fig. 2):

(1) seeps emerging from homoclinally dipping beds;
(2) petroleum emerging from mature source rock outcrops;
(3) petroleum leaking from breached traps;
(4) seeps and impregnations associated with unconformities;
(5) seeps associated with intrusives and diapirs.

His classification emphasized the geological context of leakage but his discussion was phrased in terms of certainties which cannot be supported without more precise analysis. Many

From England, W. A. & Fleet, A. J. (eds), *Petroleum Migration*
Geological Society, Special Publication No. 59, pp. 265–271.

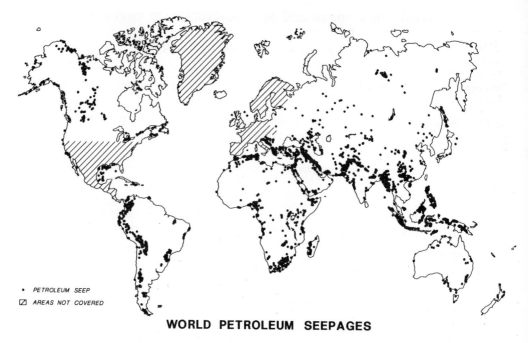

PETROLEUM SEEP
AREAS NOT COVERED

WORLD PETROLEUM SEEPAGES

Fig. 1. World petroleum seep map. All categories of seep are represented. Data are derived from published reports and from BP field surveys over the last 90 years.

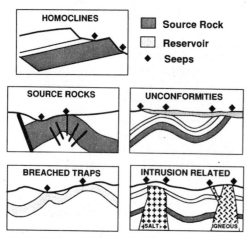

Fig. 2. Geological classification of seepage, simplified from Link (1952).

aspects of petroleum geology have changed dramatically since his report, especially the definition of structure and the understanding of fluid movement. This paper recognizes the need for an understanding of geological context and emphasizes the part alteration has to play in changing the composition of the petroleum and obscuring seepage.

The environment in which trap leakage occurs is substantially different from the conditions in which source to reservoir migration takes place. Trap leakage pathways are generally cooler and usually have gentle, buoyancy dominated fluid potential gradients. Permeable intervals are often active aquifers, allowing microorganisms to enter the system and feed on seep petroleum (iron-reducers, sulphate-reducers, fermenters). Trap leakage can continue for as long as there are oilfields in the subsurface whereas the source to reservoir migration system is a relatively ephemeral product of source rock maturation.

Most seeps appear to define the locations of fracture pathways along which trap leakage is substantially isolated from the matrix. Gas at shallow depths has such a low specific gravity that it is barely susceptible to displacement by moving aquifer water. However, if either oil or gas leaks into the matrix of an aquifer containing moving water then it is likely to be destroyed by dissolution and oxidation.

Seeps have been grouped as shown in Table 1, which also lists other phenomena with which seeps are often confused. Three types of seep or seep related geological phenomena are

(1) unaltered seeps;
(2) seeps altered by surface dispersal;
(3) seeps altered by subsurface processes.

A brief description of each type follows.

Table 1. *Seepage classification.*

Seepage type	Similar but unrelated
Unaltered	
Flowing gas	Volcanic gas seeps
Flowing gas and oil (either with/without salty spring)	Landfill gas
Mud volcano	
Altered by surface dispersal	
Flowing gas-free oil	Detrital petroleum in clasts
Oil at water table spring	Oil pollution
Free soil gas above water table	Soil fermentation products
Oil impregnation in soil or rock (overlying seep, above water table)	Mature source rock outcrop
	Fossil matrix impregnation
	Fossil fracture injection
	Fossil oilfield reservoir
Altered by subsurface processes	
Sour gas seeps	Volcanic exhalations
Sulphur springs	
Biogenic gas seep or seep component	
Seep pathway mineralization	Source to trap migration pathway mineralization
Superficial mineralization, e.g. *gach-i-turush*	

Definition of 'seep'

A petroleum seep is defined as the surface expression of a leakage pathway along which petroleum is currently flowing, driven by buoyancy, from a subsurface source.

The petroleum may be biogenic or thermogenic gas with, or without a liquid oil component. Unaltered seeps of gas-free oil are rare or non-existent, partly because of the rarity of gas-free oil pools, and partly because oil by itself has little buoyancy. Most seeps appear to originate from petroleum accumulations. If they do originate from maturing source rocks directly, considerable flow focussing is implied, given the low rates at which source rocks are supposed to expel petroleum and the rates at which seeps typically flow (see below).

Unaltered seeps

Unaltered seeps constitute about 30% of the database (Fig. 3) and include seeps whose composition resembles unaltered petroleum, with or without associated formation water. Over half of this group are unassociated gas seeps which have apparently arisen directly from a subsurface gas pool or gas cap. About a quarter of the unaltered seeps are gassy oil seeps.

SEEP TYPE PROPORTIONS

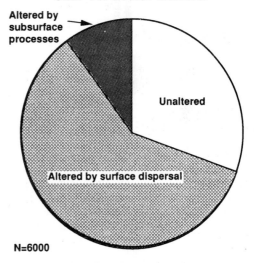

Fig. 3. World seep data classified according to their surface alteration characteristics.

The oil in such seeps could be exsolved from gas cap gas or could be oil from which the solution gas has exsolved. There are few oilfields in which the oil leg petroleum is not predominantly gas by volume at surface conditions. An oilfield with a typical gas/oil ratio of 500 scf/bbl (standard cubic feet per barrel, 1 bbl = 5.6 scf) will produce (Fig. 4) about a hundred volumes of gas at surface conditions for every volume of oil by progressive exsolution and expansion of the gas phase as the petroleum approaches the surface. The low density of gas, especially under near-surface conditions, provides a strong buoyancy force which drives seepage.

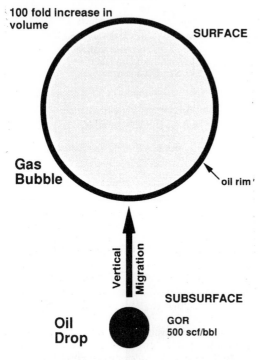

Fig. 4. Schematic phase behaviour of an oil drop as it rises to the surface.

Gas seeps and gassy oil seeps are associated with trapped petroleum in both normally pressured and overpressured environments. It is unclear whether they can arise directly from maturing source rock sequences. However, in basins where source rocks have been 'switched off' by uplift and cooling seeps can only arise from trapped oil and gas. Sometimes, as in the case of the Karoo Basin, southern Africa, the seeps may record the last stages of the escape of the petroleum charge following inversion.

Gassy seeps are occasionally (less than 10% of total) associated with salt water springs. The salt water is apparently representative of sub-surface formation fluids and appears to be brought to the surface by a 'gas lift' process, for example in Papua New Guinea (BP proprietary data).

These gassy springs are quite different from the more catastrophic mud volcanoes which are even less common (about 5% of total) and have a specific association with young compressive terrains (Trinidad, Burma, Makran) and are often associated with mud diapirism. Mud volcano gas seems to be of secondary importance to processes which appear to involve the mobilisation of underconsolidated sequences in response to rapidly applied stresses.

Flow rates have been estimated (in mostly unpublished field notes) for 28 gassy seeps in the list, including a few offshore seeps where measurement is easier (Fig. 5). The rates vary from ten cubic metres to three million cubic metres per year, with a median value of about two hundred thousand cubic metres per year. Associated oil flow rates are volumetrically small, in general.

Volcanic gas seeps, for example in Italy and Sicily, can be mistaken for gassy petroleum seeps. These, however, are usually predominantly carbon dioxide and an analysis is usually sufficient to resolve uncertainties.

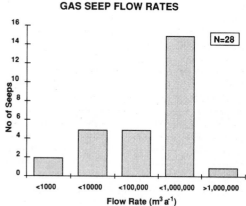

Fig. 5. Gas seep flow rates reported for seeps worldwide.

Seeps altered by surface dispersal

Flowing oil seeps constitute about 40% of the database. Their easy visibility and commercial significance has probably caused them to be over-represented. Many flowing oil seeps may in fact be gassy and unaltered but the gas, especially in oils which are extremely viscous at surface conditions, may go unnoticed.

Gas-free oils are usually altered. Analyses usually show that they are more or less biodegraded. Early BP field geologists referred

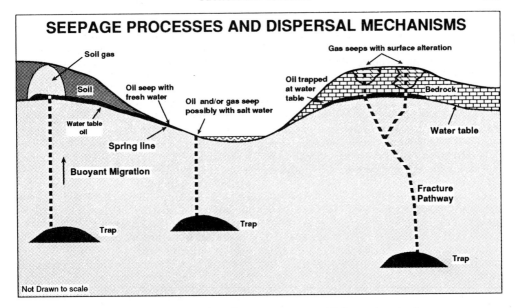

Fig. 6. Diagrammatic illustration of seepage processes and dispersal mechanisms.

to such petroleum as 'water table oil' in unpublished reports, and recognized that a segregation process typically allowed the seep gas to escape into the soil atmosphere, while the oil became trapped at the water table, no longer mobilized by the gas's buoyancy. Such water table oil can either become completely biodegraded, or it might flow downslope to emerge with groundwater at spring lines. The oil can form residual saturations ('impregnations') anywhere between the spring line and the intersection of the seep with the water table. Consequently such water table oil is displaced some distance from the source of the seep (Fig. 6). The gas component of seeps is not subject to such lateral displacement at the water table, hence Lees & Richardson's (1940) observation that 'gas escapes are associated with all the existing fields [in SW Iran and Iraq], and oil seepages with four of them'.

Residues in the soil may be associated with seepage directly, as in the case of free gas concentrations immediately above the places where seeps intersect the water table. Many soil alkane concentrations measured by conventional methods are, however, so small that they could easily be accounted for by the products of fermented vegetation, primarily methane, and by the occurrence of detrital petroleum in clasts. It appears that the low concentrations of petroleum which can occur in the soil as a result of the near-surface seep dispersal processes described in the previous paragraph, together

with methane from fermentation, and detrital petroleum, have given rise to the idea that it can escape towards the surface in low concentrations and over a wide area by an ill-defined process referred to as 'microseepage'. However, oil and gas cannot flow in the subsurface except under well defined conditions, governed by the viscosity, capillarity, density, and miscibility of the various phases involved. These conditions preclude the flow of trace quantities (typically a few ppm) often associated with 'microseepage'.

Oil impregnations constitute about 20% of the database but are deliberately under-represented as most are not true flowing seeps. In addition to the soil impregnations caused by real seepage there are many other phenomena which have been mistaken for seeps, including:

petroleum impregnations in mature outcropping source rocks;

oil impregnations of outcropping fossil migration pathways, both fractures and matrix;

impregnated outcropping fossil oilfield reservoir rocks.

Seeps altered by subsurface processes

The formation water which accompanies gassy seeps moves down pressure and temperature gradients as it is gas lifted towards the surface and may precipitate minerals as vein fillings or matrix cements. Both the water and the petroleum may react with the host rock surrounding the seep pathway, affecting the composition of both the fluids and the rock. Iron mobilisation and reduction reactions are associated with seepage, as are bacterially mediated reactions between seep gas and anhydrite-bearing evaporites. The latter reaction produces sour gas and secondary aragonitic limestones in both the stratified evaporites of the Middle East and in salt dome cap rocks in the USA Gulf Coastal region (Thomas 1952; Kyle & Price 1986).

The most obvious products of such subsurface alteration processes are sulphur springs and *gach-i-turush*. The latter is a local Persian word introduced by Thomas (1952) to describe the surface effects of seeps made sour by subsurface reactions with evaporites. *Gach-i-turush* is an association of rapidly oxidizing seep petroleum, grey powdery gypsum, jarosite, sulphuric acid and sulphur, and has been described from numerous Middle Eastern localities. Similar associations have been recognized in Sicily and elsewhere. Sulphur-related seepage phenomena constitute about 10% of the database. Other types of altered seeps may occur, especially those associated with iron mobilization, but they do not appear to constitute a group with readily recognizable surface characteristics.

The mineralization of source to reservoir migration pathways may resemble the products of sulphur-associated seepage alteration. Several of the Gulf Coast salt domes affected by seepage through their caprocks contain commercial orebodies. Sulphur may also have a volcanic origin and the geographical association between seep and volcanic sulphur in Sicily is particularly close and potentially confusing.

Where leakage pathways intersect intervals which are sustaining active fermenting micro-organisms biogenic gas, typically isotopically light methane, may be released. This either produces biogenic gas seeps at the surface or adds biogenic component to seep petroleum.

Discussion

Abundance and detectability of seeps

Seeps are most abundant in petroliferous basins that have suffered most deformation, especially those such as foldbelts and active transpressional margins which lack unfaulted cover sequences. Over intracratonic rifts, passive margins, and some foreland basins where arable land and pasture, forests, standing water, and dune-fields

tend to inhibit surface observation the proportion of seeps which are readily accessible to field observers may be quite small. It is also unlikely that more than a small proportion of seeps reach the surface in places where the water table is perennially deep, especially in deserts and karst terrain. Even where the water table occasionally reaches the surface during heavy precipitation, the field worker has to be there to see the seeps which are thereby flushed out. Nevertheless, the general impression gained from the database is that seeps are not particularly widespread, even in environments which favour their visibility, and that they tend to have an obvious connection to subsurface structure, where it is known.

Flowing seeps are localized and associated with permeable pathways through the rock, nearly always fractures. Their permeability, which may be affected by stress regime changes, plus an adequate supply of petroleum, determine a seep's capacity to flow and favour its continuing occurrence. Estimates of the flow rates of 32 oil seeps, including a few offshore seeps, vary between 6 and 120 000 barrels/year (1 to 20 000 cubic metres/year) with a median value of about 300 barrels/year (50 cubic metres/year) (Fig. 7). Such rates are a compelling indication of the significance of flowing seeps, since few oilfields are large enough to sustain such rates through geological time even with a low density of seeps per unit area.

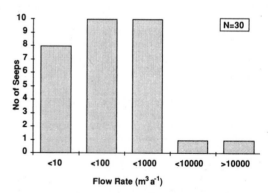

OIL SEEP FLOW RATES

Fig. 7. Oil seep flow rates reported for seeps worldwide.

Recorded oil seeps are most abundant in major petroleum provinces such as those in the Middle East where big fields are indeed available to sustain the rates observed. The total number of seeps in our database, a few thousand, may not be more than an order of magnitude less than the actual worldwide total, unless global petroleum reserves are much larger than current estimates. For example, 10 000 seeps flowing at 300 barrels/year would drain the world's total oil and gas resources (this is not a definite number but may be conservatively estimated to be about 3 trillion, 3×10^{12}, barrels of oil or oil equivalent) in about one million years.

Conditions favouring seeps

The exact geological context of seepage has seldom been recorded. In the Zagros fold-belt of Iran and Iraq, lines of seeps tend to occur along tip-line folds or thrust fault outcrops on the foreland (southwest) side of the oilfields which source them. This lateral displacement of seepage indicates that the vertically directed buoyancy of the gassy seep petroleum acts in a rock which has predominantly basinward-dipping fracturing. Where outcrops are good enough for the distinction to be made seeps nearly always arise from cracks in the rock rather than from the rock matrix. Acoustic images of marine seepage pathways (see, for example, Hovland & Judd 1988) often show seep gas rising up fractures and dislocations in a similar way. The impression gained is that seepage is possible in a quite restricted set of geological conditions and that this set of conditions may recur only a few times over a large oilfield and not occur at all over a high proportion of smaller fields.

Future research directions

If the geological factors governing seepage can be established the scientific rewards will be considerable. Understanding seeps is primarily a geological problem, although review articles have tended to emphasize their geochemistry (e.g. Jones & Drozd 1983). The following topics deserve more attention:

(1) the geological and surface environments in which seeps occur;
(2) seep flow rates and their variations;
(3) factors which affect flow variations (weather, earthquake activity, etc.);
(4) seep composition and composition changes;
(5) seep biology;
(6) mineralogy of seep pathways.

Research results will define an important component of the global carbon cycle, with significance for both geochemical issues and for the current debate concerning environmental change. For example, to what extent are past and present atmospheric methane concentration changes affected by seepage?

A better understanding of seepage will help the design of petroleum exploration systems. Petroleum possesses remarkable physical properties which can form the basis of effective seep-finding survey methods. For example, BP has released information concerning an airborne laser fluorosensor which detects the intense fluorescence of light aromatics in fresh oil (Clarke et al. 1988). It is, however, not an easy task to devise an effective detection system for onshore work.

Finally, an understanding of seepage-related biological activity and mineralization will contribute to the understanding of certain types of matrix cements, vein fillings and commercial ore bodies.

We are grateful to many colleagues within BP for discussions and data, and to BP Exploration for permission to publish. N. L. Falcon very kindly drew our attention to Thomas' important *gach-i-turush* paper.

References

CLARKE, R. H., GRANT, A. I., MACPHERSON, M. T., STEVENS, D. G. & STEPHENSON, M. 1988. Petroleum exploration with BP's airborne laser fluorosensor. *Proceedings of the Indonesian Petroleum Association*, **IPA 88-12, 15**, 387–395.

HOVLAND, M. & JUDD, A. G. 1988. *Seabed pockmarks and seepages impact on geology, biology and the marine environment*. Graham & Trotman, London.

JONES, V. T. & DROZD, R. J. 1983. Predictions of oil or gas potential by near-surface geochemistry. *The American Association of Petroleum Geologists Bulletin*, **67**, 932–952.

KYLE, J. R. & PRICE, P. E. 1986. Metallic sulfide mineralisation in salt-dome cap rocks. *Transactions Institute of Mining and Metallurgy*, **95B**, 6–16.

LEES, G. M. & RICHARDSON, F. D. S. 1940. The geology of the oilfield belt of SW Iran and Iraq. *Geological Magazine*, **77**, 227–252.

LINK, W. K. 1952. Significance of oil and gas seeps in world oil exploration. *American Association of Petroleum Geologists Bulletin*, **36**. 1505–1539.

THOMAS, A. N. 1952. 'Gach-i-turush' and associated phenomena in southwest Persia. *VII Convegno Nazionale Del Metano E Del Petrolio Taormina, Section 1, preprint*.

Leakage from deep reservoirs: possible mechanisms and relationship to shallow gas in the Haltenbanken area, mid-Norwegian Shelf

E. VIK, O. R. HEUM, & K. G. AMALIKSEN

Statoil, Forushagen (UND-GE), Post books 300, 4001 Stavanger, Norway

Haltenbanken is a gas rich province. Investigations so far have shown that large amounts of thermogenic gas have been generated, and that the major source for this gas probably is the coal in the Lower Jurassic Åre Formation. Maturity of the Åre Formation is greatest in the western part of the area, where the formation is in the wet to dry gas window. The main part of the gas was probably generated there.

To understand migration and accumulation of hydrocarbons, it is essential to know the prevailing pressure relationships. The pressure distribution in the reservoir formations and in the overlying potential seal, is normally different and will be discussed separately. A rapid subsidence in the Pliocene caused the Cretaceous and also the lowermost Tertiary packages to be overpressured over the main part of the Haltenbanken area. Overpressured Cretaceous caprocks is the main sealing agent for several of the Jurassic reservoirs. The pressures in the Jurassic reservoir formations are more varied, with normal pressures in the eastern and central parts, and severe overpressures in the western part. Areas with severely overpressured reservoirs coincide with the areas where the Åre Formation is in the wet to dry gas window.

The overpressuring of the reservoirs in the western areas is partly due to this coincidence (volume expansion due to cracking) and partly due to structural features. The eastern and central part of the area is characterized by large, open structures, while the structures in the west are smaller and more heavily faulted. Assuming the faults are laterally sealing (fully or partly), the possibility for pressure release is more restricted in the western part than in the rest of the Haltenbanken area.

The overpressures in the Halten West reservoirs can be severe, reaching 1.82 EMW, and approaching the tensile strength of the cap rock, especially in zones of weakness related to faulting. Experience also shows that these reservoirs are dry, even when ideally located with respect to mature source rocks and migration. It is therefore believed that the gas has escaped upwards due to hydraulic fracturing of the cap rock. Modelling performed in the Smørbukk area by IFP, confirms this interpretation.

A regional mapping project on seismic chimneys on Haltenbanken concluded that well-defined seismic chimneys are numerous over structures and faults in the western parts of Haltenbanken where Jurassic reservoirs are overpressured. In normal pressured areas few well defined seismic chimneys are present. The seismic chimneys are interpreted to represent gas leakages, and they stop near top Cretaceous. This may be explained by the pressure build-up in the Palaeocene creating a barrier to cross-formational migration.

The Upper Cretaceous and Tertiary packages generally have westerly dips. The gas will obviously tend to migrate upwards along the pressure gradients, but will also tend to follow the more permeable dipping Cretaceous layers (where these occur). Hence gas leaking from deep structures in the Halten West area will migrate vertically until it reaches more permeable formations. Thereafter the migration will have a strong easterly component.

The geochemistry of the gas from the shallow traps indicates a biogenic origin. However, it is likely that more thermogenic gas has been generated than trapped in pre-Tertiary reservoirs. Thus one might expect to find thermogenic gas or biodegraded/differentiated thermogenic gas in shallow traps somewhere, especially on the Trøndelag platform. Geochemistry has, however, so far given no evidence for thermogenic gas in the shallow traps.

The main conclusions are:

(1) large amounts of gas are generated and migrated from the Åre Formation in the Halten West area;
(2) the reservoirs in the Halten West area are usually severely overpressured and dry;
(3) gas has leaked from these reservoirs as a result of hydraulic fracturing of the cap rock due to the overpressuring;
(4) migration in the covering rocks has a strong easterly component due to a regional westerly dip of all permeable strata;
(5) no direct link can be expected between deep reservoirs and occurrences of thermogenic gas in shallow traps.

From England, W. A. & Fleet, A. J. (eds), *Petroleum Migration*
Geological Society, Special Publication No. 59, p. 273.

Index